Energy Law in the United States of America

Energy Law in the United States of America

Lincoln L. Davies
Joseph P. Tomain

This book was originally published as a monograph in the International Encyclopaedia of Laws/Energy Law.

General Editors: Roger Blanpain, Frank Hendrickx
Volume Editor: Kurt Deketelaere
Associate Volume Editor: Bram Delvaux

Wolters Kluwer

Published by:
Kluwer Law International
PO Box 316
2400 AH Alphen aan den Rijn
The Netherlands
Website: www.wklawbusiness.com

Sold and distributed in North, Central and South America by:
Aspen Publishers, Inc.
7201 McKinney Circle
Frederick, MD 21704
United States of America
Email: customer.service@aspenpublishers.com

Sold and distributed in all other countries by:
Turpin Distribution Services Ltd.
Stratton Business Park
Pegasus Drive, Biggleswade
Bedfordshire SG18 8TQ
United Kingdom
Email: kluwerlaw@turpin-distribution.com

ISBN 978-90-411-6643-2

This title is available on www.kluwerlawonline.com

The Authors

Lincoln L. Davies is the James I. Farr Professor of Law, Presidential Scholar, and Associate Dean for Academic Affairs at the University of Utah S.J. Quinney College of Law. His research focuses on energy law and policy, as well as on administrative law, environmental law, water law, and law, religion and the environment.

Professor Davies is co-author of one of the two leading energy law textbooks in the United States, ENERGY LAW AND POLICY (West Academic, 2014), with Alexandra Klass, Hari Osofsky, Joseph Tomain, and Elizabeth Wilson. He has written extensively on energy law and policy, and in particular on renewables and alternative energy, nuclear power, and regulatory and technology innovation. His articles include *Feed-in Tariffs in Turmoil*, 116 WEST VIRGINIA LAW REVIEW 100 (2014) (with Kirsten Allen); *Power Forward: The Argument for a National RPS*, 42 CONNECTICUT LAW REVIEW 1339 (2010); *Alternative Energy and the Energy-Environment Disconnect*, 46 IDAHO LAW REVIEW 473 (2010); *Understanding Barriers to Commercial-Scale Carbon Capture and Sequestration in the United States: An Empirical Assessment*, 59 ENERGY POLICY 745 (2013) (with Kirsten Uchitel and John Ruple); *Beyond Fukushima: Disasters, Nuclear Energy, and Energy Law*, 2011 B.Y.U. LAW REVIEW 1937; and *Standing, On Appeal*, 2010 UNIVERSITY OF ILLINOIS LAW REVIEW 957 (with Amy J. Wildermuth).

In 2012, Professor Davies was awarded the McCloy Fellowship in Environmental Policy to conduct a comparative study of renewable energy law in the United States and Germany. He has written several reports on energy issues for the U.S. Department of Energy and the Korea Legislation Research Institute and is the former Chair of the Energy and Infrastructure Siting Committee for the American Bar Association.

Prior to joining the faculty at Utah, Professor Davies practiced law in the energy and natural resources group of Steptoe & Johnson LLP in Washington, D.C. and served as a clerk to Leonard H. Russon, Associate Chief Justice for the Utah Supreme Court. Davies was educated at Stanford Law School and the University of Michigan School of Natural Resources & Environment.

The Authors

Joseph P. Tomain is Dean Emeritus and the Wilbert and Helen Ziegler Professor of Law at the University of Cincinnati College of Law. His teaching and research interests focus in the areas of energy law, regulatory policy, contracts, and law and the humanities. Before returning to the faculty in 2004, he had previously served the College of Law as Dean for 15 years.

Dean Tomain has written extensively in the energy law field. His energy-related publications include *Clean Energy Politics: The Democratization of Energy* (forthcoming Cambridge University Press), *Ending Dirty Energy Policy: Prelude to Climate Change* (2011 Cambridge University Press), *Energy Law in a Nutshell, Regulatory Law and Policy, Energy Law and Policy for the 21st Century, and Nuclear Power Transformation, among others*. Dean Tomain also has published books entitled *Achieving Democracy: The Future of Progressive Regulation* (2014 Oxford University Press) and *Creon's Ghost: Law, Justice and the Humanities* (2009 Oxford University Press).

In addition to his teaching and scholarship, Dean Tomain serves on a number of professional and civic organizations. He is Chair Emeritus of the Board of the KnowledgeWorks Foundation and Chair of Knowledge Works Inc. He also serves on the Boards of the Ohio Justice and Policy Center, and the Center for Chemical Addictions Treatment. He has been actively involved with the ABA Section on Legal Education and Admissions to the Bar, as well as other state and local bar associations. Dean Tomain is a Life Member of the American Law Institute and a Fellow of the American Bar Foundation.

In his years in academia, Dean Tomain also has held positions as Visiting Environmental Scholar, Lewis & Clark Law School; Distinguished Visiting Energy Professor, Vermont Law School; Visiting Scholar in the Program of Liberal Studies, University of Notre Dame; Visiting Fellow, Harris Manchester College, Oxford University; Fulbright Senior Specialist in law in Cambodia; and National Endowment for the Humanities Summer Fellow, Stanford University.

Before beginning his academic career, Dean Tomain practiced general litigation in New Jersey. He received his J.D. from George Washington University and his B.A. from the University of Notre Dame.

Table of Contents

List of Abbreviations

AC	Alternating Current
ACES	American Clean Energy and Security Act of 2009
AEA	Atomic Energy Act
AEC	Atomic Energy Commission
AMI	Advanced Metering Infrastructure
ARRA	American Recovery and Reinvestment Act
BLM	Bureau of Land Management
BOEM	Bureau of Ocean Energy Management
BOEMRE	Bureau of Ocean Energy Management, Regulation and Enforcement
BSEE	Bureau of Safety and Environmental Enforcement
BSER	Best System of Emissions Reduction
CAA	Clean Air Act
CAFE	Corporate Average Fuel Economy
CAISO or Cal ISO	California ISO
CBM	Coalbed Methane
CCS	Carbon Capture and Sequestration
CEQ	Council on Environmental Quality
CERCLA	Comprehensive Environmental Response, Compensation, and Liability Act (Superfund)
CNG	Compressed Natural Gas
CPP	Clean Power Plan
CWA	Clean Water Act
CZMA	Coastal Zone Management Act
DC	Direct Current
DG	Distributed Generation
DOE	Department of Energy

List of Abbreviations

DOJ	Department of Justice
EA	Environmental Assessment
EGU	Existing Generating Unit
EIA	Energy Information Administration
EIS	Environmental Impact Statement
EISA	Energy Independence and Security Act of 2007
EOR	Enhanced Oil Recovery
EPA	Environmental Protection Agency
EPAct 1992	Energy Policy Act of 1992
EPAct 2005	Energy Policy Act of 2005
EPCRA	Emergency Planning and Community Right-To-Know Act
ERCOT	Electric Reliability Council of Texas
ERO	Electric Reliability Organization
ESA	Endangered Species Act
FCLAA	Federal Coal Leasing Act Amendments of 1976
FERC	Federal Energy Regulatory Commission
FIFRA	Federal Insecticide, Fungicide, and Rodenticide Act
FLPMA	Federal Land Policy and Management Act
FOOGLRA	Federal Onshore Oil and Gas Leasing Reform Act
FPA	Federal Power Act
FPC	Federal Power Commission
FRA	Federal Railroad Administration
FTC	Federal Trade Commission
FWS	Fish and Wildlife Service
GDP	Gross Domestic Product
HHI	Herfindal-Hirschman Index
HOV	High Occupancy Vehicle
ICA	Interstate Commerce Act of 1887
ICC	Interstate Commerce Commission
IOU	Investor Owned Utility
IPM	Integrated Planning Model
ISO	Independent System Operator
ISO-NE	New England ISO
ITC	Investment Tax Credit
KGS	Known Geologic Structure

LCOE	Levelized Cost of Electricity
LDC	Local Distribution Company
LNG	Liquefied Natural Gas
LUST	Leaking Underground Storage Tank
Mbbl	Thousand Barrels
MBTA	Migratory Bird Treaty Act
Mcf	Thousand Cubic Feet
MISO	Midcontinent ISO
MLA	Mineral Leasing Act
MWh	Megawatt Hour
MY	Model Year
NAAQS	National Ambient Air Quality Standard
NEPA	National Environmental Policy Act
NERC	North American Electric Reliability Corporation
NFMA	National Forest Management Act
NGA	Natural Gas Act
NGPA	Natural Gas Policy Act of 1978
NHPA	National Historic Preservation Act
NHTSA	National Highway Traffic Safety Administration
NRC	Nuclear Regulatory Commission
NWP 21	Nationwide Permit 21
NYISO	New York ISO
OATT	Open Access Transmission Tariff
OCS	Outer Continental Shelf
OCSLA	Outer Continental Shelf Lands Act
OPA	Oil Pollution Act of 1990
OPEC	Organization of Petroleum Exporting Countries
OSM	Office of Surface Mining
PHMSA	Pipeline Hazardous Materials Safety Administration
PJM	Pennsylvania-New Jersey-Maryland Interconnection
PMPA	Petroleum Marketing Practices Act
PSC	Public Service Commission
PTC	Production Tax Credit
PUC	Public Utility Commission
PUHCA	Public Utility Holding Company Act
PURPA	Public Utility Regulatory Policies Act

List of Abbreviations

QF	Qualifying Facility
RCRA	Resource Conservation and Recovery Act
RCT	Regional Coal Team
REC	Renewable Energy Certificate or Renewable Energy Credit
RFG	Reformulated Gasoline
RFS	Renewable Fuel Standard
ROD	Record of Decision
RPS	Renewable Portfolio Standard
RTG	Regional Transmission Group
RTO	Regional Transmission Organization
SAC	Standalone Cost
SEC	Securities and Exchange Commission
SEPA	Southeastern Power Administration
SIP	State Implementation Plan
SMP	Special Marketing Program
SNG	Synthetic Natural Gas
SPP	Southwest Power Pool
STB	Surface Transportation Board
Tcf	Trillion Cubic Feet
TSCA	Toxic Substances Control Act
USFA	United States Fuel Administration
WPT	Crude Oil Windfall Profit Tax Act

General Introduction

Chapter 1. General Background

1. Energy law in the United States must be understood in the broader context of policy. U.S. energy policy, in turn, can best be grasped with three ideas in mind. First, the policy long has had a general purpose of encouraging economic expansion as a part of domestic industrial policy. Second, the policy is influenced heavily both by the nation's available energy resources and the expansive geography of the United States. Third, the policy continues to evolve, especially as environmental, climate change, and clean energy concerns become more acute. This Chapter highlights each of these aspects of U.S. energy policy. It then notes three core tensions embedded within the policy today.

§1. Energy Use and Economic Policy

2. Historically, U.S. energy policy was primarily an industrial policy. Its aim was—and largely remains—to drive economic growth by ensuring an abundant supply of energy at a low cost.[1] Thus, an implicit assumption long embedded within U.S. energy policy is that gross domestic product (GDP) and energy consumption are intrinsically linked—that is, that more energy consumption by definition drives economic growth and industrial expansion.[2] Facilitating this economic pursuit has been the center of U.S. energy policy from its inception.

3. Although longstanding, U.S. energy policy's entrenched focus on economic growth also has been the source of much criticism.[3] A common political refrain in the United States is that the nation has no energy policy at all.[4] Under the banner of that refrain, numerous criticisms—often conflicting—have been made about U.S. energy policy. These include: U.S. energy policy insufficiently promotes energy

1. Joseph P. Tomain, *The Dominant Model of United States Energy Policy*, 61 U. Colo. L. Rev. 355 (1990).
2. Ugur Soytas & Ramazan Sari, *Energy Consumption and GDP: Causality Relationship in G-7 Countries and Emerging Markets*, 25 Energy Econ. 33, 33–37 (2003).
3. *See, e.g.*, Amory B. Lovins, *Soft Energy Technologies*, 3 Ann. Rev. Energy 477 (1978).
4. *See, e.g.*, E. Donald Elliott, *Why the United States Does Not Have a Renewable Energy Policy*, 43 Envtl. L. Rep. News & Analysis 10095 (Feb. 2013); *see also* Todd Ganos, *What Energy Policy?*,

extraction, particularly from domestic fossil fuel resources. U.S. energy policy fails to adequately account for environmental costs, specifically, by falling short in promoting clean energy and combating climate change. Or most malleably and perhaps most frequently, U.S. energy policy fails to achieve energy independence, making the United States' economic health reliant on foreign supplies, producers, and energy markets, all of which can be manipulated and disrupted.

4. The mere fact that U.S. energy policy elicits so many critiques should make clear that it is not devoid of content. Otherwise, there would be no target at which the critiques could aim. Indeed, careful analysis has shown that modern U.S. energy policy in fact follows a clear, "dominant model," one that has long persisted in domestic energy governance. That model rests on six pillars, which together seek to ensure: (1) abundant energy supplies that are delivered in a reliable, non-disruptable manner (2) at low costs to consumers (3) by large, archetype firms using (4) primarily traditional, fossil fuel resources which engage in (5) intra- and inter-fuel competition (6) under a regulatory structure that is federalist in nature.[5] We describe how this model came to be in more detail in § 2 below, but for our purposes here, note its general theme and arc. The model is not as much an *energy policy* model per se as it is a methodology for promoting economic growth, based on a clear presumption that greater energy consumption equates with additional economic development.

§2. Energy Resources, Consumption, and Geography

5. Beyond its economic-industrial focus, U.S. energy policy also is a function of the nation's physical resources. The United States, particularly compared with many other nations, benefits from an extensive and diverse energy wealth. The nation is blessed with large renewable and non-renewable resources. For much of the early twentieth century, the United States was a net oil exporter, and recent projections suggest that the nation may soon achieve that status again.[6] The nation also has been referred to as the "Saudi Arabia of coal,"[7] and the recent revolution of natural gas production using horizontal drilling and hydraulic fracturing has made the United States one of the largest producers of methane in the world.[8] Likewise, studies have consistently shown that the United States can produce at least 20% of its electricity

---

Forbes (Oct. 19, 2012), http://www.forbes.com/sites/toddganos/2012/10/19/what-energy-policy/. *But see* Lincoln L. Davies, *Tracing U.S. Renewable Energy Policy*, 43 Envtl. L. Rep. News & Analysis 10320 (2013).

5. Tomain, *supra* note 1.
6. Int'l Energy Agency, World Energy Outlook 1–2 (2012), *available at* http://www.iea.org/publications/freepublications/publication/english.pdf.pdf.
7. Warren Fiske & Jacob Geiger, *George Allen Says the U.S. Leads the World in Fossil Fuel Resources*, PolitiFact (Dec. 8, 2011), http://www.politifact.com/virginia/statements/2011/dec/08/george-allen/george-allen-says-us-leads-world-fossil-fuel-resou/.
8. U.S. Envtl. Prot. Agency, *Global Methane Initiative: Frequent Questions*, http://www.epa.gov/globalmethane/faq.htm (last visited Oct. 5, 2014).

using wind power, and another 27% from solar.[9] Hydroelectricity production is a mainstay in the United States, even if the era of big dam building now appears to be a relic of history.[10] Further, the United States produces more electricity from nuclear power than any other nation in the world (769 billion kWh, or 19% of the nation's electricity), including France (which produces over 75% of its electricity using nuclear), in part because of relatively abundant uranium deposits, particularly in the western United States.[11]

6. Thus, the United States relies on a combination of fossil fuels, nuclear power, and renewables to produce energy. Fossil fuels make up the great majority of the nation's energy production, currently representing 82% of primary energy consumption. Specifically, petroleum accounts for 36% of this consumption, natural gas 26%, coal 20%, and nuclear 8%.[12] Nevertheless, in recent years, the United States has begun to produce more energy using renewable resources, even though renewables still account for only 9% of all energy consumption. Of that share, hydropower comprises 35% of renewable energy production—a larger portion of energy production than wind and solar combined. Wood and biofuels also comprise a significant portion of this resource use, at a combined total of 48% of all renewables consumption. Wind accounts for 13% and solar for 2%.[13]

7. Just as the United States produces much energy, it also uses a large amount of energy. Roughly, this consumption divides the country into two broad energy sectors. The transport sector relies heavily on petroleum, and the electricity sector relies heavily on coal and nuclear. Natural gas bridges both sectors and is used for electricity generation as well as (and increasingly) for transportation. Natural gas also is used for heating in the United States.

8. Energy consumption in the United States can further be broken down into residential, commercial, industrial, and transportation sectors. The industrial sector uses the largest amount of energy (approximately 31%), and this proportion is expected to rise. By sector, transportation is the second largest consumer of energy

---

9. U.S. Dep't of Energy, 20% Wind Energy by 2030: Increasing Wind Energy's Contribution to U.S. Electricity Supply (July 2008); Int'l Energy Agency, *How Solar Energy Could Be the Largest Source of Electricity by Mid-Century*, http://www.iea.org/newsroomandevents/pressreleases/2014/september/how-solar-energy-could-be-the-largest-source-of-electricity-by-mid-century.html (last visited Oct. 19, 2014).
10. On the history of dam building in the United States, and modern resistance to it, see generally, for example, Robert W. Adler, Restoring Colorado River Ecosystems: A Troubled Sense of Immensity (2007); Daniel McCool, River Republic: The Fall and Rise of American's Rivers (2012); Richard White, The Organic Machine: The Remaking of the Columbia River (1996).
11. U.S. Energy Info. Admin., *How Much Electricity Does a Typical Nuclear Power Plant Generate?*, http://www.eia.gov/tools/faqs/faq.cfm?id=104&t=3 (last visited Oct. 6, 2014); World Nuclear Ass'n, *Nuclear Power in France* (July 2014), http://www.world-nuclear.org/info/Country-Profiles/Countries-A-F/France/.
12. Energy Info. Admin., *Monthly Energy Review* (Sept. 2014), http://www.eia.gov/totalenergy/data/monthly/pdf/sec1_7.pdf.
13. Univ. of Mich. Center for Sustainable Sys., *U.S. Renewable Energy: Patterns of Use*, (Oct. 2013), http://css.snre.umich.edu/css_doc/CSS03-12.pdf.

(28%), while residential is third (22%) and commercial fourth (19%).[14] As much energy as the United States consumes, its exports of oil exceeded the nation's imports in 2011 for the first time in many years.[15] The nation is also expected to become a net exporter of natural gas by 2018.[16] Nevertheless, on a per capita basis, the United States is currently the fifth highest consumer of energy in the world, using an estimated 312 million BTUs per person.[17]

9. In addition to its wealth of available energy resources, U.S. geography heavily influences the United States' energy profile and policy. The nation's vast expanse is part of what allows it to be so energy rich, but these resources are unevenly distributed. Coal deposits are primarily in the Appalachian region and parts of the West, particularly the Powder River Basin in Wyoming. Oil production occurs in many states, but Texas, North Dakota, California, Alaska, Oklahoma, and New Mexico lead the way, in that order.[18] Wind is richest in the Midwest and Great Plains, and solar is most readily available in the Southwest.

10. The geographic unevenness of energy resource availability in the United States has at least three key implications for U.S. energy governance. First, energy is often moved long distances in the United States, so the negotiation of how that occurs is one of the core dilemmas energy law addresses. Second, the need to traverse many—and large—areas of land to maintain energy stability means that energy governance often is carried out with a variety of jurisdictions imposing multiple, overlapping layers of regulatory authority on energy systems.[19] Third, different jurisdictions inherently have different interests and objectives, so energy governance in the United States often is not coordinated, and can be disjointed, disparate, and even in conflict. The national electricity grid is but one example. There is no single national U.S. grid. Instead, there are three: the Western Interconnection, roughly west of the Continental Divide; the Eastern Interconnection, roughly east of the Divide; and the Electric Reliability Council of Texas (ERCOT), occupying much—but not all—of that state. Theoretically, these grids all could operate and be regulated as a unified whole, and there might be benefits to such an approach.

14. U.S. Energy Information Admin., *supra* note 12.
15. U.S. Energy Info. Admin., *U.S. Petroleum Product Exports Exceeded Imports in 2011 for First Time in Over Three Decades*, http://www.eia.gov/todayinenergy/detail.cfm?id=5290 (last visited Oct. 19, 2014).
16. U.S. Energy Info. Admin., *AEO2014 Early Release Overview*, http://www.eia.gov/naturalgas/importsexports/annual/ (last visited Oct. 19, 2014).
17. U.S. Energy Info. Admin., *International Energy Statistics*, http://www.eia.gov/cfapps/ipdbproject/iedindex3.cfm?tid=44&pid=45&aid=2&cid=all,&syid=2007&eyid=2011&unit=MBTUPP (last visited Oct. 19, 2014).
18. U.S. Energy Info. Admin., *Rankings: Crude Oil Production, June 2014 (Thousand Barrels)*, http://www.eia.gov/state/rankings/?sid=US&CFID=13259961&CFTOKEN=4927be69fa02b9e6-2AAD71C6-25B3-1C83-5406E2FB4529D2CD&jsessionid=8430506e1ef71f55934343373e5f4a434327#/series/46 (last visited Oct. 7, 2014).
19. Hari M. Osofsky & Hannah J. Wiseman, *Hybrid Energy Governance*, 2014 U. Ill. L. Rev. 1 (2014).

Instead, the three grids operate independently, and within each of them, a true mélange of regulators influence their shape, function, and future.[20]

§3. U.S. ENERGY LAW AND POLICY IN TRANSITION

11. While U.S. energy policy long has been shaped by the nation's desire for economic and industrial expansion, several factors today are reshaping domestic energy policy and, in turn, U.S. energy law. One is a renewed emphasis on energy production. Technological innovation, such as hydraulic fracturing and both horizontal and offshore drilling, has allowed the United States to expand the amount of energy it produces, which by definition changes the outlook for energy policy. Indeed, increased energy production may have quite disparate impacts on the nation's energy profile. For instance, a greater abundance of domestic oil may slow down a transition to cleaner technologies in the transport sector, whereas access to previously unrecoverable supplies of natural gas may serve as a kind of bridge fuel to a clean energy transition both in electricity and transportation.[21]

12. Another emerging trend in U.S. energy policy is the increasing convergence of energy and environmental legal objectives. Historically, U.S. energy law and environmental law have been divorced, as discussed in more depth in Part V on the *Interaction between Energy Law and Environmental Law*.[22] Increasingly, however, energy and environmental law are becoming more coordinated. While the fields are far from merged, a number of recent innovations beginning in the late 1970s and continuing to today have realigned the fields, sometimes as to their objectives but also in terms of coordination between regulators and in execution.[23] Thus, laws such as the Public Utility Regulatory Policies Act (PURPA), corporate average fuel economy (CAFE) standards, and state renewable portfolio standards (RPSs) have begun to create common areas where energy law and environmental law now overlap. The U.S. Environmental Protection Agency's (EPA) Clean Power Plan (CPP) holds the potential to push this overlap even more.

13. A further manifestation of this trend is the movement toward a clean energy system in the United States. Climate change is arguably the strongest driver of this push. At the same time that U.S. energy policy is doubling down on pursuing expanded domestic fossil fuel production, there also is a growing consensus that the

20. These include state public service commissions or public utility commissions; the Federal Energy Regulatory Commission; independent system operators and regional transmission organizations, such as the California ISO and PJM; the North American Electric Reliability Corporation and its sub-entities; and large investor-owned utilities and holding companies, which themselves often span multiple states, such as PacifiCorp or American Electric Power (AEP).
21. JOSEPH P. TOMAIN, ENDING DIRTY ENERGY POLICY: PRELUDE TO CLIMATE CHANGE (2011).
22. Lincoln L. Davies, *Alternative Energy and the Energy-Environment Disconnect*, 46 IDAHO L. REV. 473 (2010).
23. Amy J. Wildermuth, *Is Environmental Law A Barrier to Emerging Alternative Energy Sources?*, 46 IDAHO L. REV. 509 (2010); Amy J. Wildermuth, *The Next Step: The Integration of Energy Law and Environmental Law*, 31 UTAH ENVTL. L. REV. 369 (2011) [hereinafter *The Next Step*].

nation's energy systems need to become more sustainable and resilient.[24] Thus, the amount of renewable resources being utilized is growing, vehicle fleets are switching to fuels like compressed natural gas (CNG), and the government continues to pursue climate mitigation measures, such as the Department of Energy's (DOE) backing of technologies like carbon capture and sequestration (CCS) and aggressive state measures like California's AB 20.[25] Many observers continue to note that the United States lags behind other jurisdictions, particularly the European Union (EU), that have been more forceful in trying to transform their energy systems to address climate concerns, but the fact remains that this is one area of U.S. energy policy where change clearly is afoot.

14. Finally, U.S. energy policy is changing because of its increased emphasis on efficiency. While, from a global perspective, U.S. energy consumption remains profligate, and while efficiency long has been a focus of U.S. energy policy, greater efforts on this front have begun to shift the overall direction of national energy policy. California is a key example. Although it is the nation's most populous state—and one with an economy larger than all but seven nations in the world—aggressive efficiency measures kept per capita energy consumption in the state effectively flat from 1979 until 2013.[26] National trends are beginning to follow suit, with energy sales in decline every year since 2007.[27] This may be partly due to the economic downturn, but as the economy has recovered, the use of energy has continued to decline.[28] Thus, legal and policy measures that promote efficiency are also growing. Tax credits and other incentives, such as free high occupancy vehicle (HOV) lane access, that promote fuel-efficient cars have emerged. State regulatory commissions have approved utility expenditures to subsidize efficient lighting purchases in lieu of constructing power plants. Building codes are tightening, federal tax incentives are available for weatherization, heating and cooling efficiency improvements, and net metering programs have been implemented to foster distributed generation (DG). Efficiency, in short, is one of the clear trends in how U.S. energy policy is being reshaped today.

---

24. NICHOLAS STERN, A BLUEPRINT OF A SAFER PLANET: HOW TO MANAGE CLIMATE CHANGE AND CREATE A NEW ERA OF PROGRESS AND PROSPERITY (2009); JEFFREY D. SACHS, COMMON WEALTH: ECONOMICS FOR A CROWDED PLANET (2008); ENERGY FUTURE COALITION, CHALLENGE AND OPPORTUNITY: CHARTING A NEW ENERGY FUTURE (2003).
25. For a discussion of some of these state and regional measures, see Hari M. Osofsky, *Diagonal Federalism and Climate Change Implications for the Obama Administration*, 62 ALA. L. REV. 237, 239 (2011).
26. Center For Continuing Study of the Cal. Econ., *California Posed to Move Up in World Economy Rankings in 2013* (July 2013), http://www.ccsce.com/PDF/Numbers-July-2013-CA-Economy-Rankings-2012.pdf, last visited Oct. 8, 2014; Nat'l Res. Def. Council, *California's Energy Efficiency Success Story: Saving Billions of Dollars and Curbing Tons of Pollution* (July 2013), http://www.nrdc.org/energy/files/ca-success-story-FS.pdf.
27. Am. Council for an Energy-Efficient Econ., *U.S. Electricity Use Is Declining and Energy Efficiency May Be a Significant Factor*, (Feb. 25, 2014), http://www.aceee.org/blog/2014/02/us-electricity-use-declining-and-ener.
28. Some possible factors include more energy efficient buildings and warmer winters. *Ibid.*

§4. TENSIONS IN U.S. ENERGY LAW

15. Given the countervailing forces at work in changing U.S. energy law and policy today, it should come as little surprise that there are a number of broad tensions that pervade the field. Three are perhaps most notable.

16. First, a core tension of U.S. energy law is whether to rely on markets or regulation. The default assumption in the United States often is that law should step back and allow markets to function. Of course, the very aim of energy law is to correct market failures that prevent the achievement of an economically efficient equilibrium. Thus, the mere presence of energy law presumes that regulation is needed to correct a market failure.[29] This has been true since nineteenth century property law attempted to deal with oil and gas extraction through the rule of capture, and it remains true today in the most avant-garde energy law enactments, from renewable fuel standards to energy efficiency requirements.[30] The question whether to regulate or to defer to markets pervades U.S. energy policy, and thus, U.S. energy law.

17. Second, the tension between environmental (primarily, clean energy) objectives and traditional energy (predominantly, fossil fuel interests) also persists throughout U.S. energy law and policy.[31] Perhaps most critically, this tension arises in how U.S. energy policy is set, and how our choices about consumption and production are made. These include debates over: whether to promote hydraulic fracturing for natural gas' air quality benefits, or to limit it because of risks to water quality;[32] whether to augment nuclear power's role in domestic electricity generation for its climate change advantages, or to pull back from its use because of the risk of disaster and the intransigence in finding long-term storage solutions;[33] and

---

29. LINCOLN L. DAVIES, ALEXANDRA B. KLASS, HARI OSOFSKY, JOSEPH P. TOMAIN & ELIZABETH J. WILSON, ENERGY LAW AND POLICY 165–97 (2014).

30. *See, e.g.*, Timothy A. Slating & Jay P. Kesan, *The Renewable Fuel Standard 3.0?: Moving Forward with the Federal Biofuel Mandate*, 20 N.Y.U. ENVTL. L.J. 374 (2014); Inara Scott, *"Dancing Backward in High Heels": Examining and Addressing the Disparate Regulatory Treatment of Energy Efficiency and Renewable Resources*, 43 ENVTL. L. 255 (2013).

31. *See generally* TOMAIN, *supra* note 21. As but one example, this tension arises in the context of how "clean energy" should be defined. Many policy advocates take a narrow view, confining it to renewables, such as solar and wind. Others take a broader view and include all low-carbon sources, such as nuclear. Still others push the definitional boundaries and add clean coal and similar technologies. Indeed, this divergence in definitions is not just philosophical; it manifests in codified law, including in state RPSs that now dictate, at least in part, the shape of electricity generation in thirty-seven states plus the District of Columbia. *See generally* Lincoln L. Davies, *Power Forward: The Argument for A National RPS*, 42 CONN. L. REV. 1339 (2010); *see also* Database of State Incentives for Renewables & Efficiency, *Renewable Portfolio Standard Policies* (Sept. 2014), http://www.dsireusa.org/documents/summarymaps/RPS_map.pdf.

32. Hannah Wiseman, *Fracturing Regulation Applied*, 22 DUKE ENVTL. L. & POL'Y F. 361, 363 (2012).

33. Fred Bosselman, *The Ecological Advantages of Nuclear Power*, 15 N.Y.U. ENVTL. L.J. 1 (2007); Lincoln L. Davies, *Beyond Fukushima: Disasters, Nuclear Energy, and Energy Law*, 2011 B.Y.U. L. REV. 1937, 1937 (2011); WILLIAM M. ALLEY & ROSEMARY ALLEY, TOO HOT TO TOUCH: THE PROBLEM OF HIGH-LEVEL NUCLEAR WASTE (2013).

challenges to renewable energy facilities due to aesthetic concerns or risks to endangered and avian species.[34] In U.S. energy law, in short, the energy-environment tension is pervasive.

18. Third, a central conflict in U.S. energy law is the governance question: Who regulates? Because the United States employs a federalist system of government, with some rights and powers granted to the federal government but others reserved to the individual states, energy law is carried out by numerous different actors at multiple levels of government. This multifaceted, multilayered governance system presents both challenges and opportunities. For instance, while some regulatory questions have been assigned to the federal government—such as natural gas pipeline eminent domain authority—others that could benefit from centralized oversight and planning remain regulated by the several states—including electric transmission line siting.[35] Likewise, many areas of energy regulation utilize divided governance, which can make coordination difficult or, at times, create conflict. The Federal Power Act (FPA) and Natural Gas Act (NGA), for example, grant some regulatory authority to the federal government while leaving much to the states. By contrast, allowing different levels of government to regulate the same energy questions can lead to experimentation and innovation, thus improving energy governance in the aggregate and over time. Some commentators suggest that this has occurred, at least to a degree, with respect to climate change governance in the United States in the absence of coordinated, federal congressional action.[36] Looking beyond the question of climate change, however, the problem of energy federalism is almost certain to remain one with important consequences for decades to come.

---

34. Kenneth Kimmell & Dawn Stolfi Stalenhoef, *The Cape Wind Offshore Wind Energy Project: A Case Study of the Difficult Transition to Renewable Energy*, 5 Golden Gate U. Envtl. L.J. 197 (2011); *see also* Animal Welfare Inst. v. Beech Ridge Energy LLC, 675 F. Supp. 2d 540 (D. Md. 2009).
35. Alexandra B. Klass & Danielle Meinhardt, *Transporting Oil and Gas: U.S. Infrastructure Challenges*, 100 Iowa L. Rev. 947 (2015).
36. *See, e.g.*, Ann E. Carlson, *Iterative Federalism and Climate Change*, 103 Nw. U. L. Rev. 1097 (2009); Hari M. Osofsky, *The Future of Environmental Law and Complexities of Scale: Federalism Experiments with Climate Change Under the Clean Air Act*, 32 Wash. U. J. L. & Pol'y 79 (2010).

Chapter 2. Basic Principles of Energy Law

19. Depending on the perspective, energy law is either very old or a rather new field of law. One concept of energy law is that it is basically an amalgamation of preexisting areas of law: the combination of oil and gas law, public utility law, with some modern environmental and safety regulation thrown in.[37] Another view might suggest that energy law is essentially a modern invention, an outgrowth of the oil crises of the 1970s and a concerted effort to address how energy production, consumption, and transport is governed in an industrial, urban, global economy.[38] Still others paint energy law quite broadly, describing it as both public and private law that governs the full life cycle of energy production to consumption, including the negotiation of public-public, public-private, and private-private exchanges. Such a wide-reaching definition of energy law thus would extend to all fields of law that touch on energy processes, including, for instance, utility regulation, resource law, environmental law, torts, contracts, property, and antitrust.[39]

20. Whatever the precise contours of the field, energy law fundamentally addresses a suite of problems that arise in the life cycle of energy extraction, production, distribution, and consumption. The problems are economic. They are also crosscutting. They appear at each stage of the energy life cycle, as well as across different types of energy resources. Specifically, there are four core economic problems with which energy law grapples.[40] The different principles of U.S. energy law thus might be thought of as seeking to solve these four economic dilemmas.

§1. Market Power

21. The first problem is market power. Market power is the ability of parties to influence competition in a way that leads to suboptimal economic outcomes, that is, outcomes that are not economically efficient.[41] Market power can be exercised by a single firm or a group of firms that act in concert. They do so by leveraging their purchasing or selling power, or by intentional market manipulation. Thus, several potential varieties of market power are possible in the energy context. These include: monopoly power (Standard Oil in the late 1800s and early 1900s is perhaps the best example),[42] natural monopoly power (public utilities), horizontal market power (controlling too much of a product market), vertical market power

37. Davies, *supra* note 22, at 479.
38. Kenneth A. Manaster, *An Introductory Analysis of Energy Law and Policy*, 22 Santa Clara L. Rev. 1151, 1158 (1982).
39. Rosemary Lyster & Adrian Bradbrook, Energy Law and the Environment (2006).
40. Davies et al., *supra* note 29, at 165 (2015).
41. U.S. Dept. of Justice & Fed. Trade Comm'n, *Horizontal Merger Guidelines* (Aug. 19, 2010), *available at* http://www.justice.gov/atr/public/guidelines/hmg-2010.html [hereinafter *Horizontal Merger Guidelines*].
42. Standard Oil Co. of New Jersey v. U.S., 22 U.S. 1 (1911); Daniel Yergin, The Prize: The Epic Quest for Oil, Money and Power (2008).

(controlling inputs into a product market), barriers to entry (*e.g.*, bottleneck facilities), and market manipulation (*e.g.*, Enron's manipulation of the western electricity markets at the turn of the twenty-first century).[43] Energy law attends to each of these problems in different ways,[44] and guarding against market manipulation is a long-standing emphasis by regulators and lawmakers in the United States.

22. There are numerous examples of how U.S. law seeks to regulate market power. The most prominent example may be natural monopoly regulation. A natural monopoly is a firm whose capital costs comprise such a high proportion of its economic structure that this effectively precludes the entry of a competing firm.[45] These often occur in energy, such as public utilities or oil and gas pipeline companies. The risk is that parties with market power will exercise that influence to exclude competitors, reduce production, or drive up prices.

23. U.S. law deals with market power in energy in a number of different ways. These include competition (antitrust) regulation, direct price regulation, facility access regulation, and regulation of market structures. For example, under both the Federal Power Act[46] and the Natural Gas Act,[47] utilities making wholesale sales of electricity and natural gas are required to do so only at "just and reasonable" rates[48]—that is, at prices approved by the Federal Energy Regulatory Commission (FERC). Likewise, those laws forbid utilities from charging rates that unduly discriminate against a particular customer or customer class.[49] Further, Section 203 of the FPA directly regulates market structure by mandating FERC preapproval when a utility proposes to merge or sell any jurisdictional facility of a sufficiently large value (currently, USD 10 million).[50] And FERC's landmark Order No. 636 in the natural gas industry and Order No. 888 in the electricity industry ordered natural gas pipeline owners and electricity transmission line owners, respectively, to provide service on their facilities on an "open access," non-discriminatory, first-come, first-served basis. Other rules in energy law similarly have compelled companies to open up their facilities to avoid market power problems, including when the Hepburn Act of 1906 amended the Interstate Commerce Act of 1887 (ICA) to make interstate oil pipelines in the United States common carriers. We return to each of these modes of energy regulation in detail in the ensuing Parts. For now, however, it is important to note that energy law in the United States uses a variety of tools to address the problem of market power.

---

43. David B. Spence & Robert Prentice, *The Transformation of American Energy Markets and the Problem of Market Power*, 53 B.C. L. Rev. 131, 154-175 (2012); Jacqueline Lang Weaver, *Can Energy Markets be Trusted? The Effect of the Rise and Fall of Enron on Energy Markets*, 4 Hous. Bus. & Tax L. J. 1, 6-25 (2004); Bethany McLean & Peter Elkind, The Smartest Guys in the Room: The Amazing Rise and Scandalous Fall of Enron (2013).
44. Davies et al., *supra* note 29, at 166–67 (2015).
45. Richard A. Posner, *Natural Monopoly and Its Regulation*, 21 Stan. L. Rev. 548 (1969).
46. 16 U.S.C. §§ 791a et seq.
47. 15 U.S.C. §§ 717 et seq.
48. 16 U.S.C § 824d; 15 U.S.C. § 717c.
49. *E.g.*, 16 U.S.C. § 824d(b).
50. 16 U.S.C. § 824b.

§2. WASTE

24. The second economic problem that energy law often addresses is waste. Waste occurs in two primary forms: physical and economic. Physical waste is perhaps what most people think of most often. Leftover food tossed in the bin following a football viewing party is an example of physical waste. The food is a resource that is being discarded rather than utilized in a beneficial way. Economic waste, on the other hand, manifests in a perhaps less obvious way. Economic waste is the "overproduction of a resource so that the resource is priced too low."[51] That is, economic waste occurs when financial resources are deployed "in excess of the usual, ordinary, and necessary sums for such projects."[52] For example, a consumer who pays USD 125 for a pair of running shoes from one website but who could have gotten the same shoes from another website for USD 65 has incurred a form of economic waste. The problem with waste is that it creates economically inefficient, or Pareto suboptimal, outcomes.[53] That is, waste occurs whenever a different allocation of resources would better maximize benefits to society.

25. Energy law seeks to limit waste in a variety of ways. Most fundamentally, these might be divided into two categories: laws that seek to promote *production* of resources, and laws that aim to ensure full *utilization* of resources.

26. Perhaps the best example of the first category is the rule of capture in oil and gas. This rule, which analogizes oil and gas to wild animals that can be "captured" and thus owned by anyone whose property they cross,[54] incentivizes the extraction of oil and gas because the failure of one property owner to do so opens up the opportunity for a neighboring property owner to take more than their fair share of the resource.[55] Likewise, other laws encourage resource extraction. For instance, the granting of mineral leases on federal land to private parties willing to extract oil, gas, or coal encourages resource extraction that might not otherwise occur.[56] Many states have similar laws, which oftentimes direct revenues from the leases to public goods or services, such as public education.[57]

27. While various aspects of U.S. energy law seek to avoid waste by encouraging production, the field also endeavors to restrain waste that occurs precisely because of such production. For instance, in response to the rule of capture, many

---

51. DAVIES ET AL., *supra* note 29, at 175.
52. Phillip Wm. Lear, *Utah Oil and Gas Conservation Law and Practice*, 1998 UTAH L. REV. 89, 106 (1998).
53. The idea of Pareto optimality is that "resources cannot be reallocated so as to make one person better off without making someone else worse off." NICHOLAS MERCURO & STEVEN G. MEDEMA, ECONOMICS AND THE LAW: FROM POSNER AND POST-MODERNISM AND BEYOND 21 (2d ed. 2006).
54. *See* Westmoreland & Cambria Nat. Gas. Co. v. De Witt, 18 A. 724 (Pa. 1889); *cf.* Pierson v. Post, 3 Cai. R. 175, 2 Am. Dec. 264 (N.Y. 1805).
55. *See* Barnard v. Monangahela Natural Gas Co., 65 A. 801 (Pa. 1907).
56. Gregory D. Russell, Robert J. Krummen, *Ohio's Experience with Preempting Local Regulation of Oil and Gas Development,* 19 TEX. WESLEYAN L. REV. 37 (2012).
57. Kadish v. Arizona State Land Dept., 747 P.2d 1183, 1184–85 (Ariz. 1987).

states have adopted oil and gas conservation statutes that attempt to limit the underutilization of oil fields (many times because of the rule of capture itself).[58] Likewise, corporate average fuel economy standards compel vehicle manufactures to meet heightened levels of fuel efficiency for automobiles.[59] Many efforts to modernize the electricity grid, or to encourage regional planning, like FERC's Order No. 1000,[60] aim to reduce the inefficiencies in the delivery of electricity to consumers. So too for governmental incentives, such as tax credits, that allow consumers to recoup funds for home energy efficiency improvements.[61] The underlying policy is pro-efficiency, or anti-waste.

§3. Negative Externalities (Pollution)

28. The third problem with which energy law grapples is negative externalities. "Externality" is an economic term. It is an action for which the actor does not bear the full benefits or costs of their choice. Thus, externalities can be either positive or negative. A neighbor who keeps up their yard creates a positive externality. They themselves benefit from the enjoyment value of the yard, but those around them also benefit from being able to see a tidy yard and, potentially, from increased property values. By contrast, a teenager who tosses his fast food wrappers out of the window of his speeding vehicle imposes a negative externality on society. He may receive a small benefit from avoiding the effort of finding a trashcan, but the litter defiles the landscape for others.

29. In the United States, much of the law that addresses negative externalities from energy is probably more aptly characterized as part of the field of environmental law than energy law per se.[62] Thus, much of the regulation that is relevant falls within the ambit of the United States Environmental Protection Agency or other agencies charged with protecting resources or the environment, such as the Fish and Wildlife Service (FWS). Likewise, the key relevant statutes are environmental and include the Clean Air Act (CAA),[63] the Clean Water Act (CWA),[64] the Resource Conservation and Recovery Act (RCRA),[65] the Surface Mining Control

58. Elliff v. Texon Drilling Co., 210 S.W.2d 558 (Tex. 1948); Westmoreland & Cambria Nat. Gas Co. v. De Witt et al, 18 A. 724 (Pa. 1889).
59. *See, e.g.*, Pub. Citizen v. Nat'l Highway Traffic Safety Admin., 848 F.2d 258 (D.C. Cir. 1988).
60. Order No. 1000, *Transmission Planning and Cost Allocation by Transmission Owning and Operating Public Utilities*, 18 C.F.R. Part 35 (July 21, 2011).
61. Internal Revenue Serv., http://www.irs.gov/uac/Energy-Incentives-for-Individuals:-Questions-and-Answers (last visited Mar. 3, 2015).
62. Todd S. Aagaard, *Environmental Law As A Legal Field: An Inquiry in Legal Taxonomy*, 95 Cornell L. Rev. 221 (2010).
63. Clean Air Act, Pub. L. 84-159, 69 Stat. 322 (codified as amended at 42 U.S.C. §§ 7401–7431).
64. Federal Water Pollution Control Act (Clean Water Act), Pub. L. 80-845, 62 Stat. 1155 (codified as amended at 33 U.S.C. §§ 1251–1274).
65. Resource Conservation and Recovery Act, Pub. L. 94-580, 90 Stat. 2795 (codified as amended at 42 U.S.C. §§ 6901–6908a).

and Reclamation Act,[66] the Endangered Species Act (ESA),[67] and the Migratory Bird Treaty Act (MBTA),[68] among others. Where these laws play an important role in regulating energy in the United States, we address them in the relevant chapters that follow. We also address the critical role of environmental regulation on energy in Part V below on the *Interaction between Energy Law and Environmental Law.*

30. Nevertheless, there are aspects of energy law proper that also address questions related to externalities. Part I of the Federal Power Act, for instance, which governs the regulation of hydroelectric dams in the United States, expressly includes consideration of environmental and recreational factors for dam licensing and relicensing.[69] The Atomic Energy Act (AEA), which gives the Nuclear Regulatory Commission (NRC) exclusive licensing authority for nuclear power plants and control over other radioactive materials, clearly aims at limiting the risk of radioactive contamination or other nuclear events, at the same time it expressly promotes the utilization of atomic energy. A driving motivation of CAFE standards is not just conserving fuel but also reducing pollution, including greenhouse gases. And renewable energy laws, like the production tax credit (PTC)[70] or state renewable portfolio standards,[71] promote a number of objectives, including job growth and fuel diversity and security, but also environmental goals in terms of reducing air and climate pollution from traditional fossil fuel electric power plants.[72]

§4. Positive Externalities (Free Riding)

31. Finally, energy law addresses the problem of "free riding." Free riding is another version of positive externalities. It gets its name from the analogy of someone catching a ride on a public bus or subway without paying—they get a free ride, for which the rest of the society is forced to absorb the cost.[73] The problem is that it is unfair to those who pay if some ride for free. Moreover, the fact that someone gets a free ride also may dilute the market signal of the need for that service, thus under-incentivizing its provision.

32. There are three key aspects of free riding that U.S. energy law addresses. First, the ability of one player to over-consume a resource in relationship to others is a kind of free rider problem. This occurs in oil and gas, for example, where

66. Surface Mining Control and Reclamation Act, Pub. L. 95-87, 91 Stat. 445 (codified as amended in scattered sections of 30 U.S.C.).
67. Endangered Species Act of 1973, Pub. L. 93-205, 87 Stat. 884 (codified as amended at 16 U.S.C. §§ 1531–1544).
68. Migratory Bird Treaty Act, 40 Stat. 755 (codified as amended at 16 U.S.C. § 703-12).
69. *See, e.g.*, Confederated Tribes and Bands of the Yakima Indian Nation v. FERC, 746 F.2d 466 (9th Cir. 1984).
70. 26 U.S.C. § 45.
71. Davies, *supra* note 31, at 1339; Felix Mormann, *Requirements for A Renewables Revolution*, 38 Ecology L.Q. 903 (2011).
72. Lincoln L. Davies, *State Renewable Portfolio Standards: Is There A "Race" and Is It "To the Top"?*, 3 San Diego J. Climate & Energy L. 3 (2012).
73. N. Gregory Mankiw, Principles of Economics 218 (7th ed. 2014).

multiple parties all own land with access to an oil field. The United States has adopted laws to deal with that problem, namely, state unitization and pooling laws. At one level, those laws seek to reduce physical and economic waste that can result from such a situation. At another level, however, these laws also seek to limit free riding: When the laws allocate revenues from the minerals' extraction, the very objective is to make sure that each landowner gets their "fair share" of the proceeds—that is, to avoid free riding.

33. Second, utility law long has dealt with the problem of cross-subsidization. In this version of free riding, one class of customers asserts that the way rates are designed is unfair, because the design forces that class to pay for costs that rightfully should be allocated to a different set of customers. This is a classic free riding problem—the idea that someone else benefits from your payment for a service. Utility law has developed a number of doctrines to deal with this issue, including the idea of non-discrimination, which is enshrined in both the FPA and the NGA, and cost-causation, which recently has made a strong resurgence in the case law, particularly in decisions involving regional transmission organizations (RTOs) and independent system operators (ISOs).[74]

34. Third, a very common aspect of free riding in energy is the under-incentivization of technology innovation and diffusion. That is, because someone who creates a new or better energy technology will not necessarily capture all of its benefits—or may face barriers in getting the technology adopted—there may be less incentive to explore changes in our energy systems than is socially optimal. Thus, U.S. energy law addresses this problem in many ways. A prominent one is by funding energy research and development.[75] Most recently, DOE-sponsored projects to explore the use of carbon capture and sequestration technology is a good example. Energy law may also promote the adoption of technology, such as by harnessing federally developed atomic military technology for civilian energy production, or by providing tax incentives for use of a particular kind of energy technology.[76] Or, law often mandates the use of technologies, such as through RPSs or CAFE standards.[77] All of these mechanisms aim at eliminating a free rider problem: under-innovation and under-utilization of certain energy technologies.

---

74. Midwest ISO Transmission Owners v. FERC, 373 F.3d 1361 (D.C. Cir. 2004); Illinois Commerce Comm'n v. FERC, 576 F.3d 470 (7th Cir. 2009); Illinois Commerce Comm'n v. FERC, 721 F.3d 764, 769 (7th Cir. 2013); Illinois Commerce Comm'n v. FERC, 756 F.3d 556, 556 (7th Cir. 2014).
75. Lincoln L. Davies, *Energy Policy Today and Tomorrow-Toward Sustainability*?, 29 J. LAND RESOURCES & ENVTL. L. 71 (2009); MARILYN A. BROWN ET AL., S.E. EFFICIENCY ALLIANCE, RENEWABLE ENERGY IN THE SOUTH 103 (Dec. 2010), *available at* http://nicholasinstitute.duke.edu/sites/default/files/publications/renewable-energy-in-the-south-paper.pdf; Sanya Carleyolsen, *Tangled in the Wires: An Assessment of the Existing U.S. Renewable Energy Legal Framework,* 46 NAT. RESOURCES J. 759, 771–76 (2006).
76. Pac. Gas & Elec. Co. v. State Energy Res. Conservation & Dev. Comm'n, 461 U.S. 190 (1983).
77. Felix Mormann, *Enhancing the Investor Appeal of Renewable Energy*, 42 ENVTL. L. 681 (2012); Felix Mormann, *Beyond Tax Credits: Smarter Tax Policy for A Cleaner, More Democratic Energy Future*, 31 YALE J. ON REG. 303 (2014) [hereinafter *Beyond Tax Credits*]; Mary Jean Bürer & Rolf Wüstenhagen, *Which Renewable Energy Policy Is a Venture Capitalist's Best Friend?: Empirical Evidence from a Survey of International Cleantech Investors*, 37 ENERGY POL'Y 4997, 4998 (2009).

Chapter 3. Historical Background of United States Energy Law and Policy

35. The energy policy of the United States is not contained in any single document. This is true even though, pursuant to the Department of Energy Organization Act of 1977,[78] the U.S. Department of Energy is required to propose a National Energy Plan biannually to the President who, in turn, is required to transmit the plan to Congress. The plan considers the fuel and energy needs of the country, paying particular attention to energy production, utilization, and conservation objectives. The goals of the plan include full employment, price stability, energy security, economic growth, environmental protection, nuclear non-proliferation, and the efficient utilization of public and private resources. The Act also requires the plan to identify strategies for reaching those objectives, as well as recommend legislative and administrative action.

36. Despite the comprehensiveness of these plans—and the requirement that they include coordination among various federal government agencies as well as the states—they do not constitute U.S. energy policy. Instead, U.S. energy policy is contained in a welter of legislation adopted over the last century. Prior to federal involvement in the energy sector, state and local governments began to regulate energy resources, their utilization, pricing, and distribution through a variety of laws and initiatives. Today, the body of statutes, regulations, and decisions that comprises U.S. energy law is found both in state and federal law, though federal law arguably predominates.

§1. Early Energy Law and the Emerging U.S. Energy Policy

37. The first major federal energy legislation was the Federal Water Power Act (now known as Part I of the Federal Power Act), passed in 1920, for the purpose of governing the nation's waterpower resources.[79] Prior to that time, though, the federal government had intervened in energy markets in two notable ways. First, the Hepburn Act of 1906 vested the now-defunct Interstate Commerce Commission (ICC) with the authority to govern common carriers engaged in the transportation of oil by pipeline.[80] The Hepburn Act granted the ICC the authority to set tariffs and rates for the transportation of oil in interstate commerce. The rates were required to be just and reasonable, and the Act prohibited special rates, rebates, and undue preferences.[81] Second, during World War I, the government created the United States Fuel Administration (USFA), which regulated prices, transportation, and

78. Department of Energy Organization Act, 42 U.S.C. §§ 7321–7322.
79. Federal Power Act, 16 U.S.C. §§ 791-823d.
80. 49 U.S.C. §§ 1 et seq. Subsequently, the Department of Energy Organization Act transferred oil pipeline authority to the Federal Regulatory Commission in 1977. Pub. L. No. 104-88. The ICC was abolished in 1995, and its remaining powers were transferred to the Surface Transportation Board.
81. *See generally*, Steven H. Brose et al., *Oil Pipelines, in* 4 Energy Law and Transactions ch. 85 (David J. Muchow & William A. Mogel eds., 2009).

distribution of energy resources. The USFA used those powers sparingly. Instead, the agency relied on decentralized administration and industry volunteerism to mobilize natural resources—mainly oil, natural gas, and coal—for the war.[82]

38. The 1920 Federal Water Power Act established the Federal Power Commission (FPC) (now the Federal Energy Regulatory Commission, or FERC). The Act authorized the FPC to issue licenses for the construction and operation of hydroelectric dams and associated facilities such as transmission lines. Licenses were for maximum periods (now, 50 years), could be renewed, and could only be transferred with the written approval of the Commission. Part I of the FPA has undergone numerous revisions since its adoption, including to require each licensee to develop a comprehensive plan to ensure that hydroelectricity facilities benefit interstate or foreign commerce; produce hydroelectric power; provide adequate protection of fish and wildlife; and deliver other beneficial public uses, such as irrigation, flood control, water supply, and recreation.[83] Nevertheless, this law stands as an early example of what many subsequent U.S. energy laws seek to do—use regulation to ensure that low-cost, reliable energy supplies are available to the nation.

39. Congress passed Part II of the Federal Power Act in 1935 in response to the national expansion of the electric industry.[84] As the delivery of electric power moved across state lines and as larger central power stations were constructed, the federal government intervened in the industry for the express purpose of regulating wholesale sales of electricity and the transmission of that electricity in interstate commerce. Thus, this portion of the FPA addressed the interconnection and coordination of electric transmission facilities, and the imposition of rates and charges for selling power in interstate commerce or moving it long distances over those facilities. The Act attempted to draw a bright line between federal and state jurisdiction, leaving retail ratemaking to state public utility commissions (PUCs) and public service commissions (PSCs) and giving the federal government authority over wholesale sales of electricity and rates for electricity transmission. Part II of the FPA, then, along with the Hepburn Act, embedded in U.S. law the idea that energy resources were provided in part as public goods, and that the price of those goods thus should be regulated. That idea persists today in the law, and was adopted across the states.[85]

40. In 1935, Congress extended federal electricity regulation to the industry's corporate structure. In that year, Congress passed the Public Utility Holding Company Act (PUHCA)[86] in response to a series of stock manipulations undertaken by

---

82. *See, e.g.*, John Clark, Energy and The Federal Government: Fossil Fuel Policies, 1940–46 (1987).

83. *Ibid* at § 803. Major amendments to the Act are included in the Electric Consumers Protection Act, Pub. L. No. 99-495; the Solar, Wind, Waste, And Geothermal Power Production Incentives Act of 1990, Pub. L. No. 101-575; and the Energy Policy Act of 2005, Pub. L. No. 109-190.

84. Federal Power Act, 16 U.S.C. §§ 824-791a.

85. *Ibid* § 824d.

86. 15 U.S.C. §§ 79 et seq.

Samuel Insull, the head of Commonwealth Edison of Chicago. The Act required holding companies whose subsidiaries were engaged in either the electric utility business or the retail distribution of natural gas or electricity to register with the Securities and Exchange Commission (SEC). The law authorized the SEC to examine the corporate and operational structure of these energy holding companies for the purpose of eliminating shareholder abuses. The SEC also had the authority to order divestiture if utility or non-utility operations posed a potential for financial abuse, or where electric utility operations were not integrated. PUHCA was largely repealed by the Energy Policy Act of 2005 (EPAct 2005),[87] though the SEC retained some residual record-keeping obligations.[88]

41. The federal government further regulated oil markets in 1935 with the passage of the Connolly Hot Oil Act.[89] This law sought to stabilize falling oil prices due to the discovery of flush fields, mainly in east Texas. State regulators played a significant role in protecting resources through unitization and proration. However, the federal government also intervened to impose production restrictions to conserve the resource and not flood the market with cheap oil.

42. The mid-1930s were also important for the natural gas industry. At that time, Congress became concerned with the exercise of monopoly power by the handful of interstate pipelines that carried natural gas. Congress thus directed the Federal Trade Commission (FTC) to investigate the extent of the industry's market power. The FTC, in turn, reported that natural gas pipelines had been exercising market power to the economic disadvantage of consumers. This report led to the passage of the Natural Gas Act of 1938.[90] The NGA gave the FPC the authority to regulate several areas of this industry, including: (1) the transportation of natural gas in interstate commerce; (2) the sale of natural gas for resale in interstate commerce; and (3) natural gas companies engaged in such transportation or sale. The NGA was modeled heavily on the FPA, and like the FPA before it, the NGA gave the FPC the authority to set rates on interstate sales of natural gas to ensure that they were just, reasonable, and nondiscriminatory.

43. Thus, federal energy legislation in the 1920s, and particularly during the New Deal in the 1930s, began to shape domestic energy policy. From 1938 until 1978, with only some exceptions, there was little federal energy legislation. Nevertheless, federal energy policy began to take shape. More particularly, U.S. energy policy sought to promote a general type of energy market in the nation—one epitomized by the "dominant" energy model of linking energy consumption with economic growth.[91] That is, U.S. energy policy largely sought to promote energy production and consumption, typically by fostering markets. Quite quickly, these

87. Pub. L. No. 109-58.
88. 42 U.S.C. §§ 16451-16471.
89. 15 U.S.C. § 715c.
90. 15. U.S.C. §§ 717 et seq.
91. *See* Tomain, *supra* note 1, at 355.

markets became shaped by large-scale, high-technology, capital-intensive, integrated companies that centralized energy production where resources and demand were greatest, relying primarily on fossil fuels.

44. The federal government, in turn, mirrored the structure of energy markets, adopting laws designed broadly to help sustain and facilitate those markets. Federal and state energy laws typically treated different resources separately, and often were adopted in a reactionary fashion to deal with problems as they emerged rather than planning in advance. Thus, while the 1920s and 1930s saw the early rise of these energy laws, others that later added to them created a wide swath of divergent statutes that together make up U.S. energy policy as a whole.

§2. Post-War Policy and the Changing Energy Markets

45. During World War II, the federal government continued to follow its earlier war mobilization policies. Congress created two basic regulatory agencies—the Petroleum Administration for War and the Solid Fuels Administration for War—which were given responsibility for promoting oil and coal resources for war purposes. Both agencies were largely guided by the industries themselves, even as industrial concentration continued.[92]

46. Between World War II and the energy crises of the 1970s, energy markets experienced four significant developments. First, coal use expanded to satisfy the demands of the electric industry. Second, natural gas markets became destabilized and then remained dysfunctional until the late 1970s and 1980s. Third, domestic production of oil failed to satisfy demand, and the nation became a net oil importer, making access to external supplies of that resource increasingly important. Fourth, the commercial nuclear power market was born out of the ashes of World War II.[93] Thus, over the next few decades, federal legislation would begin to address each of these issues.

47. Perhaps the most transformative of these issues was the development of nuclear power. During World War II, the Manhattan Project successfully developed the atomic bomb. The bombing of Hiroshima and Nagasaki to end World War II, however, remained troublesome from a national cultural perspective, particularly when coupled with the legacy of testing nuclear weapons in the arid American West. Specifically, in the immediate postwar years, many raised concerns about exclusive military control of this potentially vast energy resource. Accordingly, Congress adopted the Atomic Energy Act of 1946[94] for the purpose of apportioning power between military and civilian hands. The Act created a civilian Atomic Energy Commission (AEC) to regulate and control military use of atomic materials, as well

---

92. *See, e.g.*, Clark, *supra* note 82, at 1900–46.
93. *See* Tomain, *supra* note 1, at 355.
94. Pub. L. No. 585.

as to promote commercial development of this technology as an energy resource. Further steps would be needed, however, to push nuclear power into the mainstream of energy production and consumption in the United States.

48. Given the United States' general policy of preferring private ownership of corporate enterprises, rather than state-run businesses, it quickly became clear that the commercial nuclear power industry would not develop without private involvement. To that end, the Atomic Energy Act of 1954 authorized private ownership of nuclear power, and the resource was increasingly touted as the energy source of the future.[95] Pursuant to this Act, the AEC would issue construction and operation licenses for privately owned nuclear power reactors. The express purpose of the Act, then, was to provide a "program to encourage widespread participation in the development and utilization of atomic energy for peaceful purposes to the maximum extent consistent with the common defense and security and with the health and safety of the public."[96]

49. The 1954 Act, however, did not immediately generate great enthusiasm for commercial nuclear power. Following that law's passage, the AEC developed the Power Reactor Demonstration Program of 1955, which was intended to be a joint effort of government and industry to develop five separate reactor technologies through a competitive program. Once the reactors were developed, then private industry would take over ownership, and the government would step out of the process. Private industry, though, was reluctant to participate in the venture, given the risk of potentially immense liability from a nuclear accident. As one official from General Electric, one of the main reactor builders, noted, the company could not proceed with developing commercial nuclear power "with a cloud of bankruptcy hanging over its head." In response, Congress passed the Price-Anderson Act of 1957, which limited industrial liability while assuring compensation for the public in the event of an accident.[97]

50. While the nuclear industry was beginning to find its footing in the mid-1950s, the natural gas industry began to experience a dislocation. A series of events led to this result. First, prior to that time, the FPC had regulated only the rates of pipelines operating in interstate commerce; it had not regulated natural gas producers. The problem with this was that natural gas producers could set prices higher than a competitive market would have yielded, and interstate pipelines would then simply pass those prices through to consumers. The FPC decided that such pass-throughs were disadvantageous and sought to regulate producers who were selling gas in interstate commerce.[98] The FPC began to regulate producer prices, but rather than only benefitting the public, this quickly gave rise to a second dislocation in the gas market. Specifically, the FPC's choice to regulate producer rates bifurcated the

95. 42 U.S.C. §§ 2001 et seq.
96. 42 U.S.C. § 2013.
97. 42 U.S.C. §§ 2210 et seq.
98. That jurisdiction was upheld by the United States Supreme Court in 1954. Phillips Petroleum Co. v. Wisconsin, 347 U.S. 672 (1954).

market. Now, instead of there being one market, albeit with too-high prices, there were two separate markets with two different price points: a national market and a world market. This, then, became problematic for national gas supplies. The federally regulated interstate market set prices below the world market price, which meant that producers wanted to sell only into the world market. Intrastate gas prices, which were not regulated by the FPC, thus increased to the world market level, and a natural gas shortage ensued. The problem became entrenched, and it took both time and further regulatory efforts before the federal government could solve the problem.

51. The early 1950s also brought change to the nation's oil markets. At that time, demand increased rapidly as soldiers returned home, suburbs sprouted, and the personal motor vehicle began to dominate. Thus, the domestic supply of petrol became inadequate to meet all of the country's needs, and the United States increased its oil imports. At first, this was not problematic. The Arabian Peninsula had long been known for its substantial oil reserves, and the United States was importing relatively inexpensive foreign oil. This cheap oil, however, threatened domestic producers and, as a protective measure, President Eisenhower initiated the Mandatory Oil Import Quota Program in 1959, limiting the amount of foreign oil that could be imported into the country.[99] The program's purpose was to maintain domestic oil prices, on the theory that low prices would negatively affect the petroleum industry and the economy. The program had two direct effects. First, as intended, it succeeded in limiting imports. Second, in response to the first effect, the policy precipitated formation of the Organization of Petroleum Exporting Countries (OPEC). That organization, obviously, remains one of the key players in world oil markets today, and it should not be forgotten that its formation came in direct retaliation to U.S.-imposed import limits.

52. During the 1960s, no major federal energy legislation was passed. However, five significant energy-related events occurred during that decade. First, natural gas markets continued to be distorted, leading to a national shortage. Second, the cost of constructing nuclear power plants escalated. Third, the electric industry began experiencing overcapacity in generation. Fourth, domestic oil production peaked, as the nation began importing more foreign oil than it produced. And fifth, the environmental movement arrived in a flourish.

53. This final event had significant implications for energy law, because the growing amount of environmental regulation, particularly at the federal level, immediately began to impact how energy was produced and used in the United States. Indeed, on January 1, 1970, President Richard Nixon signed the National Environmental Policy Act (NEPA) into law.[100] NEPA provided a powerful tool for environmentalists. The Act requires that every federal action must go through an environmental review process. Additional environmental legislation passed around

99. Proclamation No. 3279, 24 Fed. Reg. 1,781 (1959).
100. 42 U.S.C. §§ 4321, 4331–4335, 4341–4347, 4361–4370h.

the same time included the Clean Air Act (1970)[101] and the Clean Water Act (1972)[102] (via major amendments to existing statutes), plus the Endangered Species Act (1973).[103] While NEPA provided environmental review procedures, the other legislation imposed substantive mandates and goals.

54. Together, this environmental legislation was significant. In the case of *TVA v. Hill*,[104] for example, environmentalists were able to halt construction of a nearly completed dam under the Endangered Species Act when it was discovered that continued construction would endanger the critical habitat of small fish—snail darters—in the area. Over USD 100 million had been spent in the dam's construction, but the Court ruled that the Act foreclosed its completion.[105] Ultimately, Congress reversed the Supreme Court's decision, but a line had been drawn. Environmental law was on the scene, and it would touch energy in many ways.

§3. The Energy Crises of the 1970s and Modern Energy Law

55. In addition to the early environmental laws of the 1970s, the United States experienced a significant energy crisis in 1973. In retaliation for the United States' decision to supply the Israeli military during the 1973 Arab-Israeli war, the Arab members of OPEC, together with Egypt, Syria, and Tunisia, imposed an embargo on oil shipped to the United States. Although the U.S. economy was already strained as a result of Vietnam War expenditures, the embargo put even more stress on the nation as the world price of oil quickly quadrupled from USD 10 per barrel to USD 40 per barrel. The direct consequence was that the United States experienced double-digit inflation and other economic dislocations.[106] As a result, the government began pushing very hard on the idea of energy independence as a national policy.

56. In an effort to wean the country from foreign oil, Congress passed the Emergency Petroleum Allocation Act of 1973[107] (and later the Energy Policy and Conservation Act of 1975[108]), which gave the president authority to regulate crude oil prices. Additionally, President Nixon inaugurated Project Independence. The project had three primary aims: (1) independence from foreign oil; (2) increasing use of the market to set energy prices; and (3) developing alternatives to traditional fossil fuels, including relying on increased conservation. In November 1974, the Project Independence report was published; that report then became the basis for an

101. 42 U.S.C. §§ 7401–7431.
102. 33 U.S.C. §§ 1251–1274.
103. 7 U.S.C. § 136; 16 U.S.C. §§ 1531–1544.
104. TVA v. Hill, 437 U.S. 153 (1978).
105. *See generally* Zygmunt Plater, Classic Lessons From a Little Fish in a Pork Barrel: Featuring the Notorious Story of the Endangered Snail Darter and the TVA's Final Dam (2013).
106. Davies et al., *supra* note 29, at 1–2.
107. Pub. L. No. 93-159.
108. Pub. L. No. 94-163.

omnibus energy bill known as the Energy Independence Authority Act of 1975—which was introduced in Congress but never enacted.[109]

57. The 1973 oil crisis put a bright light on energy questions in the United States. This led to efforts on multiple fronts, including legislation and administrative restructuring. Thus, in addition to the various substantive legislation passed in the 1970s, the Energy Reorganization Act of 1974[110] and later the Department of Energy Organization Act of 1977[111] shifted which agencies were regulating which portions of the energy industry at the federal level. Specifically, among other things, the Energy Reorganization Act transferred the authority of the Atomic Energy Commission to the Nuclear Regulatory Commission and the Energy Research and Development Administration, thus dividing matters of safety and industry promotion between the two new agencies. The Department of Energy Organization Act attempted to place all federal energy agencies and programs under the control of a single department. Under this Act, the Federal Energy Regulatory Commission replaced the Federal Power Commission. Additionally, the newly created DOE assumed the responsibilities of the former Federal Energy Administration and the Energy Research and Development Administration.

58. The economic consequences of the Arab Oil Embargo of 1973 continued to reverberate throughout the decade. President Jimmy Carter aggressively addressed energy through two major pieces of legislation: the National Energy Act of 1978 and the Energy Security Act of 1980. Both were omnibus bills that attempted to address every sector of the energy economy. In effect, then, this legislation set two precedents. First, it established the pattern of much of the energy legislation to come—big, omnibus or multi-part bills that address myriad parts of the industry and make large-to-incremental changes in regulation each time they are passed. Second, many would say, this legislation created the discipline now referred to as energy law, because prior to that time, energy resources, their distribution, and consumption were regulated through public utility law, natural resources law, and oil and gas law—but now single bills looked at the energy system as a whole.[112]

59. The National Energy Act of 1978 was comprised of five pieces of major legislation. Of these, two stand out. First, the Natural Gas Policy Act of 1978 (NGPA)[113] directly addressed the dislocation in natural gas markets by beginning to deregulate natural gas price controls on a scheduled basis. The NGPA's intent was to begin price decontrols, stimulate production to reverse the supply shortage, and unify domestic natural gas markets. It did so in stages. First, it imposed federal price controls on the interstate market. Second, it created a formula for monthly price increases for "new" (post-1978 discovered) natural gas. Third, it pegged the ceiling

109. *See generally* James E. Hickey, Jr. et al., Energy Law and Policy for the 21st Century ch. 6 (2000).
110. 42 U.S.C. §§ 5801, 5811–5821, 5841–5854, 5871–5879, 5891.
111. 42 U.S.C. § 7101 (codified in scattered sections in 42 U.S.C.).
112. Davies et al., *supra* note 29, at 3–4.
113. Pub. L. No. 95-621.

price of "new" natural gas to the price of refined oil, thus bringing it closer to the world price of gas. Finally, it provided for elimination of all price controls starting January 1, 1985.

60. The second particularly significant statute included in the National Energy Act was the Public Utility Regulatory Policies Act of 1978.[114] PURPA directed state public utility commissions to review their ratemaking procedures. For most of the century, while the electric industry experienced economies of scale as it continued to grow, state PUCs generally applied a declining block rate formula where prices got lower as users consumed more energy, which had the intended effect of increasing demand.[115] At the end of the century, however, electricity prices were rising, so conservation, not increased consumption, became more important.

61. PURPA had another significant impact on electricity markets. The law introduced a new program to encourage alternative energy providers to enter the generation market. Known as "qualifying facilities" (QFs) under the Act, these facilities, which included small renewables and cogenerators (facilities that use waste heat to produce electricity), were assured that if they entered the market, their electricity would be purchased. This was because PURPA required utilities to buy all of their excess electricity at the utility's "avoided cost," that is, at its marginal cost of producing the next unit of electricity. This worked. More QFs entered the market and, slowly, incumbent, vertically integrated utilities' grip on all aspects of electricity production and delivery began to show cracks.[116]

62. The National Energy Act also contained the Energy Tax Act,[117] the National Energy Conservation Act,[118] and the Power Plant and Industrial Fuel Use Act.[119] Each of these laws sought in different ways to move the nation's energy economy away from oil and toward other fuels, most notably natural gas and coal. More specifically, the Energy Tax Act used various taxes and tax credits to encourage conservation as well as to promote energy efficiency and renewable energy. The Power Plant and Industrial Fuel Use Act prohibited the construction of new baseload electric power plants that did not have the capacity for coal or other alternatives to oil to generate electricity. And the National Energy Conservation Policy Act directed the Department of Energy to set minimum energy performance standards, and to conduct energy surveys to help reduce consumption of non-renewable energy resources.

114. Pub. L. No. 95-617.
115. *See* FERC v. Mississippi, 456 U.S. 742 (1982).
116. *See, e.g.*, Joseph P. Tomain & Richard D. Cudahy, Energy Law in Nutshell 379–84 (2d ed. 2011).
117. Pub. L. No. 95-618.
118. Pub. L. No. 95-619.
119. Pub. L. No. 95-620.

§4. Energy Policy in the 1980s

63. As the 1980s arrived, the nation was still grappling in some ways with the oil crises of the 1970s, though regulators would soon turn their attention to other energy problems as well. As he was leaving office, President Carter again obtained passage of a major energy bill: the Energy Security Act of 1980,[120] which was comprised of six pieces of legislation, each aimed to promote alternatives to petroleum. The U.S. Synthetic Fuels Corporation Act, now defunct, created a corporation to help fund and promote the development of synthetic fossil fuels such as coal gas, tar sands, and oil shale. A variety of renewable resources were dressed in the Biomass Energy and Alcohol Fuels Act; the Renewable Energy Resources Act; the Solar Energy and Energy Conservation Act; the Geothermal Energy Act; and the Ocean Thermal Conversion Act. Also in 1980, in response to a request from President Carter, Congress passed the Crude Oil Windfall Profit Tax Act.[121] Because the world price of oil was being set at prices higher than the cost of domestic production, the thought was that those "windfall profits" could be captured and deposited in the U.S. Treasury. The Act, however, was not successful and was repealed in 1988.

64. Although the nation long had promoted nuclear power, one of the lingering problems with this technology was (and remains) the long-term disposal of nuclear waste. To begin addressing this question, Congress in 1982 passed the Nuclear Waste Policy Act.[122] The Act charged the Secretary of Energy with nominating several potential sites for the underground burial of nuclear waste. The DOE examined a number of sites, and in 1987 Congress amended the Act with a directive to study only a site at Yucca Mountain in Nevada.[123] DOE completed this study, and in fact began construction at Yucca, though political wrangling and court disputes slowed and, now, have stalled use of this site. In 2011, delivering on his campaign promise, President Obama defunded the Yucca site, returning the nation to its previous stalemate over where to permanently store nuclear waste.[124]

65. In 1986, the Electric Consumers Protection Act[125] amended the Federal Power Act to increase FERC's regulatory purview over hydroelectric plants. This included: (1) eliminating municipal preferences on relicensing; (2) stressing the importance of environmental considerations, with particular attention to fish and wildlife; (3) imposing stringent environmental conditions on hydroelectric projects at new dams; and (4) increasing FERC's enforcement powers. Generally, the law directed FERC to give consideration to energy conservation, recreational values, and the preservation of environmental quality.

---

120. Pub. L. No. 96-294.
121. Pub. L. No. 96-223.
122. 42 U.S.C. §§ 10101 (codified in scattered sections in 42 U.S.C.).
123. Pub. L. No. 100-203.
124. Blue Ribbon Comm'n on America's Nuclear Future, Report to the Secretary (2012).
125. 16 U.S.C. §§ 791a–823d.

66. In June 1989, President George H.W. Bush proposed—and, in 1990, achieved passage of—significant amendments the Clean Air Act.[126] These amendments addressed acid rain, urban air pollution, and toxic air emissions. More specifically, the law: (1) adopted market-based principles such as performance-based standards and emissions trading; (2) provided a framework for alternative clean fuels, including fleet standards for vehicle emissions; (3) promoted use of low sulfur coal and natural gas; (4) promoted a market for clean fuels derived from grain and natural gas; and (5) promoted energy conservation.

§5. Modern Energy Policy: The 1990s to the Present

67. Also during this first Bush administration, Congress passed the Energy Policy Act of 1992 (EPAct 1992). The Act comprised over 27 titles addressing clean energy, energy efficiency, energy independence, and the reform of prior legislation, including the Public Utility Holding Company Act of 1935 and the Public Utility Regulatory Policies Act of 1978. The Act was wide-ranging. Among other things, it addressed energy efficiency; conservation and management; natural gas imports and exports; alternative fuels; clean coal; and renewable energy. The Act promoted new entrants and electricity markets; sought to help open access to the electric grid by non-utility generators; encouraged state regulatory authorities to engage in integrated resource planning; and provided subsidies and tax relief for electric vehicles, synthetic fuels, renewable resources, and independent oil and gas producers. The Act also streamlined nuclear licensing procedures and created the United States Enrichment Corporation to provide uranium enrichment services in the international market.

68. The Clinton administration was not particularly active on the energy front. During his tenure, the DOE issued its *National Energy Policy Plan* (July 1995); a *Strategic Plan* (1997); and a *Comprehensive National Energy Strategy* (1998). Each of those documents promoted energy production and efficiency, national security, and environmental sensitivity. But none became enshrined in legislation. Also around this time, energy policy documents such as the *National Energy Strategy* (1991); *Sustainable America: A New Consensus for the Future* (1996); and *National Energy Plan* (2001) began to adopt the concept of "sustainability" as a part of overall U.S. energy policy.[127] Again though, such planning documents were not incorporated into law.

69. The next major energy legislation was the Energy Policy Act of 2005.[128] The most significant portions of that Act address the electric industry. More particularly, the Act created a federal role in electricity transmission siting decisions, granting FERC backstop authority to preempt states where state authorities withheld decisions on proposed lines in "national interest" electricity transmission corridors.

126. Pub. L. No. 101-549.
127. *See* World Comm'n On Env't and Dev., Our Common Future (1987).
128. Pub. L. No. 109-58.

The Act also charged FERC with appointing an "electric reliability organization" (ERO) to establish nationally-applicable reliability standards for the electric grid. Further, the Act directed states to consider requiring net metering options for customers, encouraged states to adopt demand response options, largely repealed the Public Utility Holding Company Act, lifted application of PURPA to many portions of the country, and extended Price-Anderson Act protection for the commercial nuclear power industry.

70. The last major energy legislation was the Energy Independence and Security Act of 2007 (EISA).[129] Like many of its predecessors, the EISA aimed to achieve greater energy independence and security through the production of renewable fuels, energy efficiency, and greenhouse gas capture and storage. Notably, the Act raised the corporate average fuel economy standards to 35 miles per gallon for cars. The legislation supported accelerated research and development for solar energy, biofuels, geothermal energy, and energy storage, as well as promoting carbon capture and sequestration. Additionally, the Act promoted efficiency standards for appliances and lighting, as well as for buildings and industry.

71. Looking back on U.S. energy policy from a modern perspective, federal legislation in this arena can be seen as coming in three major waves. First, during the New Deal, energy legislation sought to expand the national energy infrastructure, particularly in electricity and natural gas, and to control market power and abuses in those sectors. Second, after World War II, Congress facilitated the development and support of commercial nuclear power and made some changes to how other resources were regulated. Third, in response to the energy crises of the 1970s, a battery of legislation was passed in an attempt to coordinate the country's energy portfolio, with a particular emphasis on gaining independence from foreign oil.

72. Throughout all of these phases, U.S. energy policy generally aimed to sustain incumbent energy providers, which specifically included fossil fuel energy firms and the markets they utilized. From the 1970s onward, however, the policy also began to support conservation, energy efficiency, and renewable resources. Thus, one feature of U.S. energy law and policy is that even though the disciplines of environmental law and energy law both developed in the 1970s, they did so independently of each other and today remain largely separate. That separation, however, may have started to narrow, as the Obama administration has introduced the Clean Power Plan, which holds the potential to dramatically transform electricity production in the United States. We address that topic in depth in Part V on the *Interaction between Energy Law and Environmental Law.*

---

129. Pub. L. No. 110-140.

Chapter 4. Role of Government Institutions in the Shaping and Administration of Energy Law and Policy

73. Understanding how government institutions shape the implementation and administration of energy laws in the United States requires a grasp of three key insights about how government in the United States functions generally. These relate to the allocation of responsibilities among agencies, the division of powers among sovereigns, and baseline presumptions about how regulation should work—and how it does work—in this nation.

§1. Allocation of Agency Responsibilities

74. Agency responsibilities in the United States tend to be divided and disaggregated, rather than centralized. This ties directly to how the nation passes laws. Legislation in the United States tends to be somewhat reactionary. That is, rather than planning how to address a broad set of issues holistically from the beginning, much of U.S. legislation is adopted incrementally in response to specific issues or problems that arise. As a result, and particularly at the federal level, administrative power often is allocated amongst multiple agencies, depending on how any given Congress chooses to have legislation implemented. For instance, business regulation in the United States is not centralized in a single agency, but rather, is dispersed across multiple agencies, including the Federal Trade Commission, the Securities and Exchange Commission, the Consumer Financial Protection Bureau, the Commodity Futures Trading Commission, the Equal Employment Opportunity Commission, and the Occupational Safety and Health Administration. Each of these agencies has its own mission and charge, sometimes its own statutes (though sometimes the agencies overlap in terms of which parts of statutes they implement), and, depending on the agency, its own politics and political agenda. Likewise, environmental protection in the United States is centralized most heavily in the EPA, but the Fish and Wildlife Service, the National Marine Fisheries Service (NMFS), the Forest Service (USFS), the National Parks Service, the Bureau of Land Management (BLM), the Army Corps of Engineers (Army Corps), the Department of Energy, and to some extent FERC, among other agencies, also shoulder some of the burden for environmental protection in the United States.

75. The picture is no different for energy. Heavy amounts of federal authority over energy is vested in FERC and the Department of Energy. However, the EPA and National Highway Traffic Safety Administration (NHTSA) also deal with energy efficiency. Parts of the Department of the Interior address resource extraction, including oil and gas, most importantly perhaps the Bureau of Land Management and the Bureau of Ocean Energy Management, Regulation and Enforcement (BOEMRE). There are several resource-specific agencies, such as the Nuclear Regulatory Commission for atomic energy, FERC and the Bureau of Reclamation for hydroelectricity, and the Office of Surface Mining. Moreover, how environmental regulations are implemented heavily influences energy production, transport, and consumption. And at the state level, many governments have their own agencies that also regulate portions of the energy industry. Perhaps foremost, these include state public utilities commissions or public service commissions, but there are others as well, including state departments of natural resources, public lands administrations, and oil and gas divisions.

76. The implications of this dispersal of administrative responsibilities across a wide number of agencies are far-reaching for the energy industry. For one, it means that energy companies have a large number of governmental regulators with whom they must interface, so their duties of regulatory compliance can be extensive. More broadly, different agencies may pursue different objectives, which means that implementation of energy law may not be coordinated, or even less coordinated than the laws themselves are. Further, the division of responsibility among agencies means that different segments of the energy industry may be regulated in different ways, and with different objectives in mind. This last feature also creates the risk that some aspects of the energy cycle that might merit regulation do not receive it, precisely because coordinated, holistic regulation is lacking.

§2. Division of Sovereign Authority

77. Another fundamental facet of energy governance in the United States is that the nation divides the authority to govern among multiple players. This is by design. The federal Constitution, ratified in 1789 and amended only 27 times since, grants specific powers to the federal government. Perhaps most critically for energy, these include the authority to regulate interstate commerce.[130] All powers not given to the federal government by the Constitution, however, are reserved to the states as sovereign entities themselves. Within the states, localities—principally cities and counties—also maintain extensive governance authority within their bounds. And, the 566 recognized Native American tribes in the country also retain varying levels of sovereign authority and self-rule, and some of them have extensive tracks of lands and quite sophisticated governance systems.

78. In addition to this division of authority among sovereigns, there are key constitutional doctrines that limit which level of government can regulate what. Most critically, the Supremacy Clause of the U.S. Constitution declares that federal law trumps state law.[131] Thus, the courts have held that when state law directly conflicts with federal law, when federal law says that it controls over state law, or when federal law in an area is so extensive that there is no room left for state regulation, any state law in that area is unconstitutional. Likewise, the U.S. Supreme Court has interpreted the Commerce Clause of the United States Constitution to have a "dormant" or "negative" aspect.[132] This so-called dormant Commerce Clause renders invalid state laws that discriminate against interstate commerce, that try to regulate

130. U.S. Const. art. 1, § 8, cl. 3.
131. U.S. Const. art. 6, cl. 2.
132. *See, e.g.*, Wyoming v. Oklahoma, 502 U.S. 437 (1992).

outside the state's boundaries, and even neutral state laws that have a greater negative impact on interstate commerce than they deliver benefits locally to the state.

79. The result is that, in addition to the dispersal of authority across multiple agencies in a given area, governance in the United States also is divided vertically at various levels of government. This includes governance of single subject areas, such as energy, which then can have federal regulations, state law, and city, county, and potentially tribal authority come to bear on any given question or decision. The implications are the same as they are for the dispersal of authority across agencies but arguably more pronounced and profound. Decision-making may not be coordinated; it often is not centralized; and both conflicts and gaps in energy governance can and do develop. Moreover, because of the Supremacy Clause and dormant Commerce Clause, states and the federal governments sometimes get into turf wars over who governs in an area, and court challenges can become necessary when states try to invade an area where the federal government has sought to regulate, or when states or private actors believe the federal government has overstepped the bounds of its authority under the Constitution.

80. Many different examples in the energy context illustrate the implications that this division of sovereign authority can have, but the recent example of electricity transmission line siting may epitomize most facets of the dilemma. For decades now, siting of electricity lines has been the province of states, usually local governments at the city and county level. The result is that a utility that wants to build a new transmission line in the United States has to obtain numerous approvals, and a single jurisdiction that does not want a project to go forward can effectively halt the process with a unilateral veto. As a consequence, the electrical grid in the United States has become old, undercapitalized, and arguably insufficient. This, in turn, also has limited the way the nation's energy system can change, including its ability to integrate renewable and other new generation resources, and has made the cost of building new transmission lines significantly higher than it might otherwise be. To address this problem, Congress adopted a provision in the Energy Policy Act of 2005 that gave FERC "backstop" siting authority for key transmission lines.[133] The notion of centralizing planning and siting authority for key infrastructure—or, in this case, key infrastructure that is mired in localized interests and roadblocks—might make eminent sense. Nonetheless, the U.S. Court of Appeals for the Fourth Circuit construed FERC's authority in a quite constrained way,[134] which in turn effectively reverted transmission line siting in the United States to its original position. Thus, today as before, transmission lines remain subject to unilateral roadblocks, local "not-in-my-backyard" objections, and heightened legal costs, making transforming the grid a much slower, and more difficult, process than it might otherwise be. Perhaps most importantly, however, this kind of divided sovereign authority has similar impacts in many areas of energy law.

133. 16 U.S.C. § 824(p).
134. Piedmont Envtl. Council v. FERC, 558 F.3d 304 (4th Cir. 2009).

§3. Market Preferences and the Regulatory Lifecycle

81. A third key facet of the implementation and administration of energy law in the United States is two-pronged. First, the nation heavily favors markets rather than regulation as the default presumption. Second, once regulation is adopted, it tends to follow a particular lifecycle. Both are present in energy. Both help describe why law does and does not exist in some areas of the energy industry, and why and how energy law tends to change over time in the United States.

82. The notion that the United States favors markets above most else might be familiar to even the casual observer of American society. This nation is famous for its emphasis on capitalism and the "American dream" that hard work and innovation can catapult anyone to the pinnacle of society. The myth carries large elements of truth. The fact is that, in the United States, markets are the preferred form of social ordering. Thus, in many respects, the law favors markets over governmental regulation, planning, or control as a general proposition. And, as a consequence, U.S. law simply does not regulate broad swaths of the energy market. For instance, while electricity and natural gas delivery are heavily regulated, including as to price, neither natural gas nor oil prices are regulated today, though they once were.

83. The fact that regulations sometimes exist and then go away can be explained, in part, by the regulatory lifecycle. One might conceptualize this lifecycle in six steps: (1) market formation; (2) market failure; (3) government regulation; (4) regulatory failure; (5) regulatory reform; and (6) deregulation. Of course, not every area of regulation undergoes all six stages, and the lifecycle does not always proceed linearly. Nevertheless, energy markets provide examples of the lifecycle, so comprehending the lifecycle's course is useful for understanding how government institutions shape energy law and policy.

84. Energy markets are critical. A market might be defined as a product market (the thing or its substitutes that a given set of consumers buys) and a geographic market (the area where those products are available, or can compete with each other). The notion is that a market reaches economically efficient outcomes by minimizing costs and maximizing societal good through competition. For instance, if one gasoline supplier can provide the same quality product as another producer at a lower cost, it will do so because there are profits to be made. Likewise, if different consumers want different kinds of gasoline—with higher octane levels or different amounts of renewable resources in them—markets will detect that demand, differentiate the available products, and deliver those goods to society.

85. Of course, not all markets are the same. Neoclassical microeconomic theory starts with the idea of a perfectly competitive market. But the real world does not function like a hypothetical in a textbook, so it is more often the case than not that actual markets depart from a perfectly competitive market in meaningful ways. When economists say that a market is "perfectly competitive," or "free," what they mean is the market is either competitive or "contestable."[135] Competitive or contestable markets, in turn, exhibit several virtues, including: (1) distributive efficiency (putting resources to their most valued use); (2) maximizing producer and consumer surplus (allowing consumers to obtain products at their lowest cost); (3) stimulating innovation (via competition); and (4) generating profits and economic growth.

86. To achieve these virtues, however, markets have to bear certain attributes. These include: (1) numerous buyers and sellers; (2) product homogeneity; (3) a large enough quantity of goods so that no single buyer or seller has market power; (4) perfect product information; and (5) freedom of entry and exit from the marketplace. Needless to say, most markets lack one or more of these attributes, and that results in what is often called a "market failure"—a market defect that makes it less than perfect. We address how the substance of energy law deals with market failures in more detail in § 2 above, but the broader point is that when markets fail, regulation is needed. In turn, once regulation is introduced, it tends to follow a particular path, or lifecycle.

87. Government regulation comes in many forms, but key types in energy include: (1) *price-setting*; (2) *standard-setting*; (3) *registration*; (4) *disclosure*; (5) *licensing*; (6) *allocation*; (7) *codes*; (8) *financial supports*; and (9) *insurance and regulatory requirements*. Each of these will arise throughout the various Parts addressing specific energy resources, but it is useful to define them at the outset. In sum:

- Price-setting is the government process of determining the price an industry can charge for a good or service. In energy, it is most common in public utility regulation, where regulators seek to approximate the price that a competitive market would charge.
- Standard-setting includes regulations that prescribe a level of performance or achievement that must be met. For instance, under the Clean Air Act, EPA sets emissions standards that power plants must meet.
- Registration is the government operating as a gatekeeper to markets, by requiring that new entrants register with the regulator first. Public utilities register with FERC, for instance, and provide annual and quarterly financial information, just as do nuclear operators with the NRC.
- Disclosure is government-mandated provision of information to the market. For instance, labels on cars give mileage ranges. And as the smart grid develops, utilities provide more information about the cost of electricity to consumers.
- Licensing is a heightened version of registration. Such a process might impose particular qualifications to conduct an activity, such as financial capability, or compliance with engineering or safety standards. Nuclear power plant licensing may be the best example of this in the United States.

135. *See, e.g.*, WILLIAM BAUMOL, JOHN C. PANZAR & ROBERT D. WILLIG, CONTESTABLE MARKETS AND THE THEORY OF INDUSTRY STRUCTURE (1982).

- Allocation is the government's determination of how resources should be distributed. Because the United States so heavily favors markets, this is relatively uncommon. It does, however, occur. For instance, in the face of a natural gas shortage in the 1980s, the government established allocations of this resource to certain sectors of the economy, such as schools and hospitals.
- Codes standardize the way that systems function. For instance, the government might set a certain code or standard to assure uniformity in appliances, buildings, and the like. Such codes make resources compatible as well as safe.
- Financial supports include government spending to incentivize certain behaviors or technologies. These include research and development funding, tax incentives and expenditures, direct and indirect subsidies, loan guarantees and forgiveness, and the like. A good example is the production tax credit for renewable resources.
- Finally, insurance regulations seek to protect the public from damage from industrial and other operations. For instance, the Price-Anderson Act requires nuclear power operators to obtain insurance, and to pay into a national liability pool. Coal mine operators must also do so under the Surface Mining Control and Reclamation Act.

88. While regulation seeks to correct market failures, regulations also can fail, or fall out of line with industry as the world evolves around them. As a general matter, however, one might say that regulatory failure occurs either as the result of (1) a *mismatch* between the identified market failure and the regulatory response, or from (2) *inadequate regulation*. A good example of *regulatory mismatch* is the natural gas shortage of the 1970s and 1980s.[136] Quite simply, the government responded with the tool of price-setting, but exercising that authority, federal regulators set prices below what the market required, and shortages resulted. A good example of *inadequate regulation* of an energy industry is the *BP Deepwater Horizon* oil spill that occurred on April 20, 2010 in the Gulf of Mexico. While the United States passed the Oil Pollution Act of 1990 (OPA) in the wake of the *Exxon Valdez* disaster years before, that tool was insufficient for—and not adapted to deal with—the complex problems that could arise from deepwater drilling, as opposed to merely transporting oil by ship.[137]

89. In light of regulatory failure, reforms often are made in an effort to match the market failure with the correct regulatory response. Indeed, since the 1970s, energy regulations have been going through a wide range of reforms. In the electric industry, for example, new forms of rate design regularly have been explored. As our experience with hydraulic fracturing, particularly for the development of shale gas, continues, we have begun and will continue to look at a range of regulations to control methane leakage, water pollution, and other associated environmental harms. Similarly, the development of alternative fuels and alternative vehicles, the

---

136. *See, e.g.*, Paul W. MacAvoy, The Natural Gas Market: Sixty Years of Regulation and Deregulation (2000).

137. National Comm'n on the BP Horizon Oil Spill and Offshore Drilling, Deep Water: The Gulf Oil Disaster and the Future of Offshore Drilling (2011).

expansion of renewable resources, a greater emphasis on energy efficiency, and the development of new sources of energy all provoke—and require—new regulatory tools, as well as changes to those tools as their implementation reveals imperfections and further room for improvement.

90. Following regulatory reform, one response often is deregulation. The concept of deregulation can be used in two senses. Both are seen in energy. If there is complete deregulation of an industry, then the industry simply returns to the market. In the late 1970s, the Carter administration actively engaged in deregulating airlines, trucking, financial institutions, and energy industries. More often, however, deregulation is partial rather than complete. President Carter began deregulation of natural gas prices and oil prices. That deregulation, however, took place in stages. Similarly, the electric industry has seen large segments of it deregulated, most notably in generation and wholesale sales. Still, the wholesale electricity market is not completely deregulated, as FERC carefully monitors who is allowed to sell at market-based rather than regulated rates, and those entities that do must supply FERC with detailed information about their sales each quarter and their overall market position every three years.[138]

91. The lifecycle of regulation appears in many contexts in the energy sphere, and different energy resources are at different points of the cycle at different times. The general arc of U.S. energy policy has been to move first to regulation and then to regulatory reform or deregulation. For the past three decades, the nation generally has been in the stage of deregulation, though, again, how that manifests in any given part of the industry does not necessarily reflect what may happen in another. As deregulatory efforts have persisted, moreover, the need for new and additional regulations has arisen, demonstrating, perhaps, that for a resource as critical as energy, complete deregulation may never be possible. Overall, then, if there is a single lesson to learn about the U.S. energy economy, it is that the utilization of this resource is very much an explicit product of government and market interactions.

138. Order No. 697, Market-Based Rates for Wholesale Sales Electric Energy, Capacity and Ancillary Services by Public Utilities, 72 Fed. Reg. 39,000 (July 20, 2007), FERC Stats. & Regs. ¶ 31,252 (2007), *clarified*, 121 F.E.R.C. ¶ 61,260 (2007), *order on reh'g*, Order No. 697-A, FERC Stats. & Regs. ¶ 31,268, *clarified*, 124 F.E.R.C. ¶ 61,055, *order on reh'g*, Order No. 697-B, FERC Stats. & Regs. ¶ 31,285 (2008), *order on reh'g*, Order No. 697-C, FERC Stats. & Regs. ¶ 31,291 (2009), *order on reh'g*, Order No. 697-D, FERC Stats. & Regs. ¶ 31,305 (2010), *aff'd sub nom.*, Mont. Consumer Counsel v. FERC, 659 F.3d 910 (9th Cir. 2011), *cert. denied*, 133 S. Ct. 26 (2012).

Chapter 5. Sources of Energy Law

92. The body of energy law in the United States draws on many sources of law for its content. Substantively, as noted in § 2 above, modern energy law has several historical precursors. Public utility law, natural resources law, and oil and gas law all feed into the broad corpus that is energy law. Other areas of substantive law, including competition law (antitrust), tax law, and environmental law, also lay a heavy hand on how the modern energy industry functions. This has led at least one commentator to refer to U.S. energy law as a "synthetic" area of law,[139] and others to describe the scope of the field quite broadly.

93. Structurally, four sources of law feed into modern U.S. energy law. First in time if not importance is common law. Common law in the United States is judge-made law. It is law built on tradition, established by judicial precedent, and modified slowly over time on a case-by-case basis. The common law is sedimentary law—foundational bedrock created through historical accretion. In U.S. energy law, the common law continues to serve important functions. Core principles governing resource ownership and extraction, such as the rule of capture, are common law doctrines. Likewise, decisions about what authority lessees of mineral rights have to access the land, or to affect the surface estate, are often dictated by or originally derive from common law.[140] Indeed, many of the foundational principles of U.S. common law are critical in U.S. energy law, including rules affording private (rather than state) ownership of subsurface minerals, protection of private property rights, contracting rules, and the sanctity of contracts. These baseline principles form the foundational architecture on which modern U.S. energy law is built.

94. Second, much of energy law, like much of most modern U.S. law, is legislative in nature. Congress at the federal level and state legislatures at the subnational level adopt statutes—codified law—that govern energy actions and relationships. At the federal level, key statutes forming the content of modern U.S. energy law include the Federal Power Act,[141] the Natural Gas Act,[142] and the Public Utility Regulatory Policies Act.[143] Federal statutory law also involves numerous omnibus bills that address a vast array of energy topics, such as the Energy Policy Act of 1992[144] and the Energy Policy Act of 2005.[145] Many states have analogs to federal legislation, such as statutes that govern the local and state regulation of electric and natural gas utilities at the retail level, or the extraction of resources like coal and oil, or the siting of energy infrastructure, such as electricity transmission lines

139. Manaster, *supra* note 38, at 1158.
140. Troy A. Rule, *Property Rights and Modern Energy*, 20 Geo. Mason L. Rev. 803 (2013).
141. Federal Power Act, 41 Stat. 1063 (codified as amended at 16 U.S.C. §§ 792–823d.
142. Natural Gas Act, 52 Stat. 821 (codified as amended at 15 U.S.C. §§ 717–717d.
143. Public Utility Regulatory Policies Act, Pub. L. 95-617, 92 Stat. 3117 (codified as amended in scattered sections of 16 U.S.C.).
144. Energy Policy Act of 1992, Pub. L. 102-486, 106 Stat. 2776 (codified as amended in scattered sections of 42 U.S.C.).
145. Energy Policy Act of 2005, Pub. L. 109-58, 119 Stat. 594 (codified as amended in scattered sections of 42 U.S.C. § 15801).

or oil pipelines. State legislation also has been tremendously important in recent decades in transitioning the modern energy system. In the 1990s, for instance, state restructuring legislation forced many incumbent, vertically integrated electric utilities to sell off their generating assets, and in this century, state renewable portfolio standards have compelled utilities to acquire generation using different resources than they traditionally have utilized.[146]

95. The third structural source of modern U.S. energy law is executive law. In the modern administrative state, which has dominated U.S. regulation since at least the New Deal, much of the positive law that controls everyday activities is made not by the judiciary or legislatures, but rather by the executive branch. This law can include unilateral directives, such as executive orders, that compel agencies within the executive branch to take certain actions or act in a given way. But more common—and almost invariably more important—are rules, regulations, and policies and guidelines that agencies themselves put in place. Critically, agencies in the United States often have multiple functions, including those quasi-judicial and quasi-legislative. Thus, a U.S. agency charged with regulating an area of energy activity might promulgate regulations or decide disputes between parties under those regulations—or both. Administrative law consequently is another important overlay on modern U.S. energy law.[147] Arguably, the key federal agency in the United States for energy law is the Federal Energy Regulatory Commission, which is headed by five commissioners, including a chair. Also key at the federal level are the Nuclear Regulatory Commission, the Department of Energy, and a variety of other agencies that play more tangential but nonetheless important roles in making energy law, such as the Bureau of Land Management; the Environmental Protection Agency; the Bureau of Reclamation; the Bureau of Ocean Energy Management, Regulation and Enforcement; the Army Corps of Engineers; the Fish and Wildlife Service; and the Department of Justice (DOJ) and the Federal Trade Commission. Analogs at the state level similarly play an important role at that level of governance, such as state public utility or service commissions, oil and gas boards, and railroad and corporation commissions.

96. Finally, constitutional governance plays an important role in modern U.S. energy law, particularly through its negotiation of state-federal relations but in other ways as well, such as the Fourth Amendment's prohibition on the taking of private property by the government without just compensation.[148] A fundamental precept of the United States Constitution is that the federal government can exercise only the powers specifically given to it, and all other powers are reserved to the several states. However, the Supremacy Clause of the federal Constitution also mandates that where valid federal law conflicts with state law, federal law controls. This doctrine, known as federal preemption, plays a heavy role in U.S. energy regulation because the interface between state and federal laws on energy is both vast and

---

146. Davies, *supra* note 31, at 1339.

147. Administrative Procedure Act, Pub. L. 79-404, 60 Stat. 237 (codified as amended at 5 U.S.C. § 500–504).

148. U.S CONST. amend. IV.

extensive. For instance, many disputes about the Federal Power Act and the Natural Gas Act center on the scope of those laws, precisely because they create bright jurisdictional lines between local and federal regulation. Likewise, one of the U.S. Supreme Court's landmark energy decisions zeroed in on the jurisdictional dividing line between federal promotion of atomic energy and the state power to determine the makeup of local utilities' generation fleets.[149] Acknowledging both states' right to exercise traditional police powers and federal law's supremacy over state regulation, the Court allowed the state law to stand, but held that had the state attempted to regulate an area occupied by the Atomic Energy Act, it would have been foreclosed. Likewise, as described in § 4 above, the dormant Commerce Clause limits states' ability to discriminate against interstate commerce, or in favor of their own residents in economic terms.[150]

97. Together, these constitutional principles—and the federal energy laws that Congress enacts—create a rather unique form of energy governance within the United States. While federal environmental statutes typically adopt a "cooperative federalism" form of governance, where top-down federal regulation establishes the pollution standards that must be met but states receive broad latitude in how they do so, the often sharp jurisdictional lines that the Constitution and substantive energy statutes draw with respect to who governs create a much different context in the energy realm. Rather than pushing the federal government and states to coordinate, instead in the energy sphere often there are multiple decision makers—federal, regional, state, and local—acting independently, and thus sometimes at odds and other times in concert, but quite often in piecemeal fashion to regulate different aspects of a single problem. This has led some commentators to observe that energy federalism in the United States is not vertical, but rather, "diagonal,"[151] and that energy regulation and energy governance is hardly static, but rather "dynamic."[152] In any case, the constitutional overlay of energy governance in the United States makes the field of energy law a much more complex, if exciting, one than it otherwise would be.

149. Pac. Gas & Elec. Co. v. State Energy Res. Conservation & Dev. Comm'n, 461 U.S. 190 (1983).
150. *See, e.g.*, New England Power Co. v. New Hampshire, 455 U.S. 331 (1982).
151. Osofsky, *supra* note 25, at 237.
152. Hari M. Osofsky & Hannah J. Wiseman, *Dynamic Energy Federalism*, 72 MD. L. REV. 773 (2013).

Part I. Regulatory Framework Concerning Electricity

98. The United States' energy portfolio can be roughly divided between petrol for transportation and electricity for lighting, cooling, heating, and manufacturing. The division is significant because the two sectors operate largely independently of each other. Currently, less than 1% of electricity is generated using petrol in the United States, and less than 1% of electricity is used in the transportation sector.

99. The U.S. electricity sector, however, is increasingly gaining regulatory attention. That attention is due in part to the fact that electricity in the United States is predominantly produced by fossil fuels. Concerns about climate change thus shine a bright light on the sector. At the same time, concern about climate change impacts from the electricity sector are in tension with other ways in which the U.S. energy system is changing. Domestic oil and natural gas production are on a sharp increase in the United States. Consequently, regulation of the electricity industry is complicated by the fact that, on the one hand, the nation seeks to reduce greenhouse gas emissions and, on the other hand, cheap natural gas is available for electricity generation. This tension is worth noting at the outset because the U.S. electricity sector is very much one in transition.

100. There are four key themes that run through the history of electricity regulation in the United States. Understanding each provides a clearer picture of how this regulation functions today and how and why it is changing.

101. *First*, the overall industrial structure of the electricity sector has remained largely unchanged for nearly a century.[153] Beginning on September 4, 1882, when Thomas Edison first threw a switch that delivered electricity to 85 local customers, the electricity industry adopted a specific corporate form. Most simply, electricity was generated, transported along wires, and consumed by customers at the end of those wires.

153. For more on the history of the U.S. electricity industry, see, for example, ROBERT L. BRADLEY JR., EDISON TO ENRON: ENERGY MARKETS AND POLITICAL STRATEGIES (2011); RICHARD F. HIRSH, TECHNOLOGY AND TRANSFORMATION IN THE AMERICAN ELECTRIC UTILITY INDUSTRY (1989); RICHARD MUNSON, FROM EDISON TO ENRON: THE BUSINESS OF POWER AND WHAT IT MEANS FOR THE FUTURE OF ELECTRICITY (2005).

102. The most efficient corporate form, then, was vertical integration. These privately owned, or investor-owned utilities (IOUs), performed three specific functions: (1) electricity generation; (2) transmission of that electricity over long distances; and (3) final distribution of that energy to end-users. The key difference between the transmission and distribution segments of the industry is that electricity is sent through the transmission system at higher voltages and over much longer distances than when it is sent through the distribution system. From a generator, electricity is "stepped-up" in voltage for transmission, and then, at the distribution stage, "stepped-down" so that it can be used by consumers.

103. Vertical integration enabled IOUs to achieve economies of scale and economies of scope. As firms realized these economies, and as technological advances occurred, the transportation segment of the industry expanded—namely, when utilities moved from direct current (DC) transmission to alternating current (AC) transmission. With that expansion, industry consolidation began at the end of the nineteenth century.

104. Second, the legal tools used to regulate the electricity industry were adopted both to financially support, and to quell any ill effects of, the industrial structure just described. Specifically, a regulatory compact was created between industry and the state, meaning that the industry, though privately owned, was treated as a public utility—that is, an enterprise providing a *good* rather than just a *commodity* needed by society as a whole. This regulatory evolution began at the local level, quickly expanded to the state level, and, finally, became national in scope as well. Regardless of the level of governance, however, the regulatory approach was largely the same.

105. The "regulatory compact" that IOUs entered into with their regulators was, and largely remains, central to electricity regulation in the United States.[154] Under the terms of the compact, government regulators acquired the power to set the prices and profits of IOUs. The IOUs, in turn, were given exclusive service territories. In other words, IOUs were free from competition from other electricity providers in specific geographic markets. In exchange for that state-granted monopoly, IOUs also took upon themselves an obligation to serve all customers within that service territory.[155]

106. The key tool used by regulators to set prices and profits for the IOUs was cost-of-service ratemaking. This tool sought to replicate the price that a hypothetical competitive market would achieve: a price that would ensure the utility received a reasonable profit but also that customers were not overcharged. Most simply, IOUs' revenues were thus based on two items: (1) their prudently incurred operating expenses, dollar-for-dollar; and (2) a reasonable rate of return on their prudently

---

154. On the changing nature of the regulatory compact today, see Emily Hammond & David B. Spence, *The Regulatory Contract in the Marketplace*, VAND. L. REV. (forthcoming 2015), *available at* http://ssrn.com/abstract=2584619.

155. Transmission Access Policy Study Grp. v. FERC, 225 F.3d 667 (D.C. Cir. 2000).

made capital investments. Through this process, then, IOUs had an incentive to continue to make capital investments, and they did so for most of the twentieth century.

107. Third, the electric industry has been in the process of restructuring for the last four decades. In the last third of the twentieth century, the electricity industry hit a technological and economic plateau. Technologically, the size of an electricity plant had reached its maximum efficiency point, if not overshot it. Economically, the efforts to electrify the nation had succeeded; consequently, additional construction of electricity infrastructure (including generation) by traditional IOUs became increasingly expensive.[156] This became most pointed in the construction of nuclear power plants, resulting in intense fights between utilities and regulators in the last two decades of the century over "stranded costs" that had been incurred by utilities that started building then later abandoned nuclear power plants.

108. In short, traditionally structured IOUs, and their accompanying regulations, had reached a point at which the electricity produced under that system started to become more costly. Simultaneously, in part through regulatory initiatives, the entire industry began to realize that electricity could be produced more cheaply by non-traditional IOUs. Thus, over the last four decades, regulators have been experimenting—and instituting—new legal schemes aimed at bringing cheaper electricity from more producers to market. That is, the new construct in the United States is to move away from top-down regulation of the electricity industry toward more competition.

109. Fourth, new and possibly disruptive technologies—as well as new forms of regulations—have begun to emerge in the electricity sphere. For example, the "smart grid" can dramatically change the structure of the electricity industry. The Edisonian electricity system of old is essentially a one-way system in which electricity was produced and transported to customers who, in turn, paid for what they consumed. This was and is a system in which electricity is treated effectively as an invisible public good and users of that good are merely passive consumers. The smart grid, however, works differently. The smart grid is a two-way system that transports not only electricity back and forth between IOUs and end-users, but also conveys information back and forth between producers and consumers, thus beginning to transform the very structure of the electricity market. Smart grids begin to make electricity more of a commodity than merely a public good, and end-users more active market participants than merely passive consumers.

110. Under the smart grid, for instance, electricity producers will not just deliver electricity to end-users. They will also send more accurate price signals to those consumers, either seasonally, weekly, daily, hourly, or even instantaneously. Consumers, then, will receive price information that will enable them to choose how

156. *See* Richard F. Hirsch, Power Loss: The Origins of Deregulation and Restructuring in the American Electric Utility System (1999).

much electricity to use and when to use it. Further, consumers will be able to control their consumption through the use of smart appliances, programmable thermostats, and other smart devices that increase energy efficiency while decreasing demand. This system, which is beginning to emerge in the United States, is fundamentally different from the traditional Edisonian model that has been relied on for over a century.

111. As we move through the following Chapters, each of these four themes appears again and again. This is because the U.S. electricity industry is in the midst of experiencing remarkable and rapid transformation: increased competition, particularly in generation; the emergence of competition in realms where it has not meaningfully been seen before, such as merchant transmission firms; and rapid and evolving regulatory reforms affecting energy efficiency, fuel sources, and climate change, as well as other areas. Electricity regulation in the United States continues to lean heavily on the past. However, it is also increasingly pushing toward a different future.

Chapter 1. Production

112. There are five physical characteristics that affect both the electricity industry and the laws that regulate it. First, consumers expect electricity to be readily available literally at the flip of a switch. Any service breakdown causes economic losses and social disruption. Second, electricity is not easily stored, or at least not inexpensively stored. The need to charge laptops or mobile phones daily makes this point obvious. Third, electricity is a fungible product. There is no difference in quality of electricity produced by one resource or another, although there may be a difference in the functionalities different generation resources provide to the grid. Fourth, because electricity current is generated by moving electrons moves literally near the speed of light, it is difficult to clearly identify the actual physical path of the electricity that we consume.[157] Thus, electricity engineers are fond of saying, "You cannot color electrons." Fifth, pursuant to the second law of thermodynamics, a significant amount of energy is lost throughout the electricity system, including at each stage of generation, transmission, and distribution. To the extent that some of these losses can be recaptured, energy efficiency can increase. However, it is impossible to recapture all of them. Together, the combination of these characteristics means that the electricity in the overall system must always be in balance with the demand for that electricity (the so-called "load"); otherwise, parts of the system will fail, service will be disrupted, and economic losses will be incurred. This is precisely what happens when there is a brownout or blackout on any given electricity system.

113. As distinguished from natural gas or oil, electricity is produced from a variety of natural resources, including coal, natural gas, uranium (in the form of nuclear power), water (in the form of hydroelectricity or hydropower), and other renewable resources (such as solar, biomass, and wind). To date, the vast bulk of electricity in the United States—86%—is generated by fossil fuels and nuclear power. In the United States, electricity is produced from these resources in the following amounts: coal (40%); natural gas (26%); nuclear power (20%); renewables (13%); and oil (less than 1%).[158]

§1. Trends in U.S. Electricity Production

114. There are four notable trends currently emerging in electricity production in the United States. First, coal-fired electricity is gradually being replaced by natural gas. There are two reasons for this switch. First, the United States' domestic natural gas production is increasing dramatically and, therefore, natural gas has become a cheaper resource for electricity production than coal. And, second, increasing concerns about greenhouse gas emissions, especially carbon dioxide, are being translated into regulations to curb those emissions. The regulations have

157. New England Power Co. v. New Hampshire, 455 U.S. 331 (1982).
158. U.S. Energy Info. Admin., Monthly Energy Rev. 104 (Mar. 2015).

begun to impact, and threaten to heavily impact, coal-fired power plants, which are the least advantageous from a climate-change perspective.

115. Second, nuclear power has found a way to remain roughly steady in terms of the share of electricity production in the United States, which means that its total amount of production has increased significantly over the last decades. Nuclear power continues to account for approximately 20% of electricity generation. At its peak, the United States had 104 operating nuclear reactors. Over the lifetime of those plants, operating costs have declined significantly, making them desirable sources of baseload generation. This fleet of facilities has found a way to increase production over time by increasing efficiency and capacity factors (the amount of time per year a facility produces electricity rather than being off-line). However, the age of some nuclear plants has raised concerns about cost and safety that have led to several recent closures. Thus, today, there are currently 99 nuclear reactors in operation in the United States. That number, too, is shifting. Until recently, no new nuclear plants had been ordered since 1978, and all plants ordered after 1974 either were canceled or converted to coal-burning facilities. Currently, though, five new units have been licensed for construction and operation: two in Georgia, two in South Carolina, and one in Tennessee.

116. Third, renewable resources are increasing their share of electricity generation. As the cost of producing electricity from both wind and solar decline sharply, particularly for solar, and as regulations favor the development of these resources, renewables have begun to increase their share of electricity generation in the United States. In recent years, solar power costs have decreased 80%, and wind costs have declined over 30%.[159] Both are significant for the renewable sector. Indeed, in 2013, 37% of all new electricity generating capacity came from renewable energy—three times more new generation capacity than from oil, coal, and nuclear combined.[160]

117. Fourth, demand for electricity in the United States is largely flat. The Department of Energy estimates that between 2012 and 2040, the United States will experience an increase in demand of 29%, which translates into an increased demand of 0.9% per year, with electricity sales growing 0.8% per year.[161] These relatively flat demand projections, together with environmental concerns, put significant pressure on traditionally structured IOUs. In the past, IOUs had one business mission: sell electricity. And, because of the regulatory compact, IOUs were largely insulated from competition. Today, IOUs must consider how to respond to environmental laws and to restructuring regulations, which, combined, have pushed the traditional utility to rethink its business model. Thus, as utilities look to the future, they continue to reevaluate their business model, driven in part by efficiency gains in the United States that have kept electricity relatively flat.

---

159. Sonia Aggarwal & Hal Harvey, *Rethinking Policy to Deliver a Clean Energy Future*, 26 ELECTRICITY J. 7, 8 (Oct. 2013).

160. Am. Council on Renewable Energy, *Evolving Business Models for Renewable Energy: 2014 Industry Review* 15 (June 2014).

161. U.S. ENERGY INFO. ADMIN., ANNUAL ENERGY OUTLOOK 2014 WITH PROJECTIONS TO 2040 MT-16 (Apr. 2014).

§2. INDUSTRY STRUCTURE

118. The U.S. electricity industry is valued at over USD 700 billion, and is comprised both public and private firms. There are approximately 200 IOUs in the United States that together provide about 70% of the power consumed in the country. In addition, power is provided by public firms, mainly rural electric cooperatives (13%), municipal systems (11%), energy service providers (2.7%), and state projects and political subdivisions (3.5%).

119. Today, IOUs are comprised of traditionally structured utilities that vertically integrate all three electricity functions (production, transmission, and distribution), as well as "wires-only" companies that primarily provide transmission and distribution service and buy electricity on the open market or through power purchase agreements. In addition, there are now many non-utility generators that produce electricity for sale into wholesale markets. Some of these are merchant firms that entered the market by building their own facilities, or acquired units from others, including IOUs divesting generation in response to state restructuring laws in the 1990s and early 2000s.

120. In terms of electricity sales, electric companies serve three different customer classes: residential (37.9%), commercial (35.4%), and industrial (26.4%). It should be noted that each class has different demands for electricity and associated services that must ultimately be reflected in the electricity rates assigned to them.

121. The typical electric utility has specific characteristics. First, each utility has a different resource mix, often depending on where in the country the utility is located. For example, in the northwestern United States and in the northeastern United States, hydropower is more readily available than in other sections of the country. In the Southwest, wind power is an increasingly important source of electricity. And in the Midwest, coal and nuclear plants are of central importance.

122. A second characteristic is that because the electric system must stay in balance at all times, different types of electric plants comprise the nation's generation fleet. Electricity generation in the United States is utilized in what is known as a "dispatch stack," or "least cost merit order." That is, plants generally run first based on their lowest cost, and then are "stacked" as they are dispatched in turn. The idea is to minimize production costs by running the least expensive units the most often.

123. There are three chief categories of generation in the stack: "Baseload" plants, which remain in operation as much as possible, are almost always nuclear or coal plants and have low fuel costs. However, they cannot be turned on or off easily and are expected to run virtually all day and much of the year. This is how they earn their name. They are called "baseload" because they operate relatively constantly—at the base of the stack. "Intermediate load" plants, often older coal or natural gas plants, are more expensive to operate and are brought online depending upon demand. Some of these plants may only be used when demand is quite high, such as the original natural gas units at the Gadsby Power Plant in Salt Lake City, which historically have been used primarily to provide electricity for air conditioning on very hot days. At the top of the stack are "peaking" plants, which are used to meet system needs when there is highest demand, such as during the hottest day of the year or at the busiest hour of the day. Most frequently, peaking plants are natural gas plants and, increasingly, combined cycle natural gas plants that can be brought online or taken offline relatively quickly. Sometimes, plants in both the intermediate and peaking categories are referred to as "load-following" because, in contrast to baseload facilities, they ramp up and down as demand (or "load") decreases or increases throughout the day.[162]

124. The cost of electricity to end-users will vary according to (1) region; (2) the resource mix used to generate electricity; (3) a utility's customer costs and charges; (4) the types of electric plants in use; and (5) the amount of electricity consumed. Those costs, however, can be compared by a measurement known as the levelized cost of electricity (LCOE), which reflects the complete, or fully "rolled in," cost of construction, maintenance, and operation of a generating plant averaged over time.

125. Different resources have different LCOEs, and these, of course, vary depending on whether the facility is new or is older and has already recouped its initial capital costs. As of July 2012, the Department of Energy estimated the per-megawatt hour (MWh) LCOE for generation units in the United States as follows: traditional coal plants (USD 97.7/MWh); "clean coal" (USD 138.8/MWh); natural gas (USD 66.1/MWh); advanced nuclear power (USD 111.4/MWh); wind (USD 96.0/MWh); solar photovoltaic (USD 152.7/MWh); and hydropower (USD 88.9/MWh).[163] These figures, naturally, change as costs of resources shift over time.

162. DAVIES ET AL., *supra* note 29, at 302.
163. U.S. Energy Info. Admin., *Levelized Cost of New Generation Resources in the Annual Energy Outlook* 2012 (2012).

Chapter 2. Exploitation

126. The central issue in how electricity is exploited in the United States is cost. That issue, in turn, depends heavily on the regulatory structure used to set prices. Historically, cost-of-service ratemaking predominated, but in recent years, market-based rates have taken hold. Understanding both is thus critical to grasping how electricity prices are determined in the United States. While there are a growing number of non-regulated electric power firms, such as merchant generators, in the United States, much of the electric industry remains regulated at the state or federal level, or both, precisely for the purpose of carefully controlling electricity prices. To understand how electricity prices are determined in the United States, then, it is also necessary to understand both the jurisdictional boundaries of state and federal electricity regulation, as well as the ratemaking process. We first address the ratemaking question and then turn to key principles of federalism that govern jurisdiction over electricity regulation in the United States.

§1. Electricity Ratemaking Principles

127. Electricity rates are set by both state and federal regulators. Regardless of this dual authority, ratemaking follows similar principles in both the federal and state contexts. In brief, FERC and state public utility commissions enjoy ratemaking authority that aims to balance the respective interests of utility shareholders and consumers. Traditional cost-of-service ratemaking does this by attempting to establish electricity prices that approximate the price levels that would occur if electricity were sold in a competitive market.

128. The legal principles that guide electricity ratemaking in the states tend to mimic the standard that is applicable at the federal level. At the federal level, the Federal Power Act of 1935,[164] requires electricity rates to be "just," "reasonable," and not "unduly discriminatory."[165] Those three words convey substantial meaning about how ratemaking proceeds. The idea that rates must be "just" implies that the government cannot set rates so low as to unconstitutionally "take" the utility's investment.[166] The notion that they must be "reasonable" suggests that prices cannot be so high as to gouge customers. And, the mandate that they not discriminate imposes a requirement that utilities do not play favorites, create inside deals, or give undue preference to one class over other. This concept, then, reminds that the provision of electricity is not just a commercial action but a service imbued with a public nature to it. The provision is necessary because discrimination long has been a problem in energy, beginning with the practices of the Standard Oil trust in the late 1800s and continuing until recent history, when FERC found enough discrimination

164. Federal Power Act, 16 U.S.C. §§ 791–823d.
165. Federal Power Act. 16 U.S.C. § 824(d); 16 U.S.C. § 824(e).
166. FPC v. Hope Natural Gas Co., 320 U.S. 591 (1944).

in the provision of electricity transmission service to entirely reorganize how that service is provided.[167]

129. Traditionally, the methodology used to arrive at "just and reasonable" rates was via cost-of-service ratemaking. Cost-of-service ratemaking rewards prudently operated utilities. It effectively ensures that a utility's costs will be recovered from customers, and allows them to earn a reasonable return on their capital investment. In this way, utilities can attract investors while continuing to operate and making sufficient capital investments.

130. The FPA establishes ratemaking authority for FERC, and many state statutes parallel that law. Under the FPA, all rates and charges by an electricity company in connection with either the transportation or sale of electricity subject to Commission jurisdiction "shall be just and reasonable," and any rate that is not just and reasonable is unlawful.[168] Further, the FPA forbids jurisdictional electricity companies from granting any undue preference or advantage or maintaining unreasonable differences in rates between localities or classes of service.[169] The Act also requires utilities to put all rates on file with FERC, which makes them available to the public for inspection. This measure of transparency aims to further ensure that utilities do not enter into under-the-table deals. Indeed, once a rate is on file with FERC, the utility can charge only that rate, and it cannot collect more or less than the rate that is on file. Doing so is unlawful. This, then, is known as the "filed rate doctrine."[170]

131. Rates are set at FERC through a multi-step process. First, a utility files its proposed rate or contract with FERC. Then, there is a period in which anyone who may want to protest the rate may intervene in the proceeding. Next, FERC determines on the basis of the filing whether the rate is just and reasonable. If the rate is just and reasonable, then FERC enters an order putting the rate into effect. If, however, FERC cannot immediately determine whether the rate is just and reasonable, it may set the rate for hearing, which means that the agency will conduct a trial-type proceeding, complete with discovery, to investigate whether the rate is just and reasonable or not. At the end of that proceeding, FERC may then either accept the rate, modify it, accept it subject to refund and further proceedings, or reject it.

132. This process, and the filing of rates at FERC, affords parties a chance to challenge electricity rates. They may do so in the initial proceeding itself, or they may come back after the fact to challenge existing rates. Under FPA § 206, an entity that suspects a rate has become unjust, unreasonable, or unduly discriminatory may file a complaint at FERC initiating an investigation into that rate.[171] The burden is

167. New York v. FERC, 535 U.S. 1 (2002).
168. 16 U.S.C. § 824d.
169. 16 U.S.C. § 824d.
170. *See, e.g.*, Montana-Dakota Utils. Co. v. Northwestern Pub. Serv. Co., 341 U.S. 246 (1950); United Gas Pipe Line Co. v. Mobile Gas Serv. Corp., 350 U.S. 332 (1956); FPC v. Sierra Pac. Power Co., 350 U.S. 348 (1956).
171. 16 U.S.C. § 824e.

on the challenging party to show that an existing rate is no longer lawful, or that a new rate it proposes should be adopted in place of the existing rates, which are presumptively just and reasonable. Parties often include clauses in power purchase agreements abdicating this right under § 206 to challenge existing rates. When they do so, they may still challenge the rate later as "against the public interest," but the courts have repeatedly held that this is almost an insurmountable burden to meet.[172]

133. Significantly, courts generally defer to FERC and PUC decisions regarding electricity rates.[173] Nevertheless, as electricity regulation became more complex, that deference was tested. As utilities began to abandon nuclear power plant construction in the latter part of the twentieth century, for instance, particularly after the meltdown at Three Mile Island, they found themselves in the position of having to choose among canceling a nuclear plant, continuing to invest in it, or converting a nuclear power plant to a coal-fired generating unit. The problem with each of these options was that capital investments typically far exceeded what initially was planned. Consequently, ratemaking authorities had to decide whether to allow utilities to recoup those excess investments from ratepayers.

134. A variety of approaches were available, but none was satisfactory. There were three general choices ratemakers considered: They could deny excess investments under the "used and useful" test because ratepayers received no electricity for the investment—thus making the facilities neither used nor useful. They could allow recoupment of the excess investments if they were "prudently" made at the time, even if they later turned out to be imprudent. Or, they could split the costs between shareholders and ratepayers through some form of amortization. Regulators struggled with which approach to adopt, and courts in turn had difficulty determining whether the choices made by unelected administrative officials should be given leeway or quite carefully scrutinized.[174]

135. Perhaps unsurprisingly, courts reached different outcomes, pointing up the complexity of cost-of-service ratemaking and the difficulties often experienced by regulators operating under an older statutory scheme being stretched to address unexpected events, changes, and results. Of course, the very nature of cost-of-service ratemaking—which attempts to replicate how a *hypothetical* market would function—might have forewarned of such problems. And those problems, in turn, helped foment a move to a different form of ratemaking altogether.

136. Cost-of-service ratemaking dominated electricity price setting in the United States for much of the twentieth century. It still is used by FERC for transmission rates and some wholesale power sales, as well as in many states, particularly those

172. *See* United Gas Pipe Line Co. v. Mobile Gas Serv. Corp., 350 U.S. 332 (1956); FPC v. Sierra Pac. Power Co., 350 U.S. 348 (1956); Papago Tribal Util. Auth. v. FERC, 628 F.2d 235 (D.C. Cir. 1980).
173. FPC v. Hope Natural Gas Co., 320 U.S. 591 (1944).
174. Jersey Central Power & Light v. FERC, 810 F.2d 1168 (D.C. Cir. 1987).

that have not restructured and still receive electricity from traditional vertically integrated IOUs.

137. However, toward the end of the twentieth century, FERC began modifying its rate designs in an effort to create rate structures that were more competitive. Cost-of-service ratemaking served the country well as long as energy industries were expanding. But once the utility industry had built out its infrastructure and reached certain technological plateaus, cost-of-service ratemaking was less effective because of its built-in bias to encourage more capital investment. Consequently, FERC and some state PUCs began to use other ratemaking methods for electricity.

138. The first shock to the ratemaking system was the Public Utility Regulatory Policies Act, or PURPA, adopted by Congress in 1978.[175] That statute encouraged state regulators to consider different rate formulas to promote conservation and better manage consumption. Specifically, the statute required state regulators to consider rate designs other than declining block rates, which effectively encouraged consumption. PURPA also put in place a requirement that incumbent utilities purchase power from smaller renewable facilities and other "qualifying facilities," such as combined heat-and-power facilities. This portion of PURPA thus had a dual aim to promote diversity in generation and to open up the wholesale electricity market to new and different competitors. It took some time, but PURPA, combined with other regulatory innovations, began to have an effect.

139. Significantly, around the same time PURPA was adopted, FERC had been experimenting with market-based rates. The idea behind market-based rates was that if a single firm lacks market power, it is likely that competitive forces will establish a fair price—a price that satisfies the just and reasonable requirement of the FPA.[176] Thus, market-based rates could replace cumbersome cost-of-service ratemaking and, hopefully, reduce administrative costs, promote competition, and reach more efficient prices than cost-of-service ratemaking could.

140. Today, the vast majority of wholesale sales of power are made at market-based rates. FERC grants market-based rate authority to electricity companies on a case-by-case basis. In granting such authority, FERC examines a variety of factors, including: (1) whether the entity and its affiliates lack market power in the generation segment of the market or have adequately mitigated that market power; (2) whether they lack market power or have mitigated it in the transmission segment of the market; (3) whether they have the power to erect barriers to entry; and (4)

175. Public Utility Regulatory Policies Act, Pub. L. 95-617, 92 Stat. 3117 (codified as amended in scattered sections of 16 U.S.C.).

176. California ex rel. Lockyer v. FERC, 383 F.3d 1006 (9th Cir. 2004); Montana Consumer Counsel v. FERC, 659 F.3d 910 (9th Cir. 2011); Zephyr Wind, LLC, 141 F.E.R.C. ¶ 61,113 (2012); AEP Power Marketing, Inc., 124 F.E.R.C. ¶ 61, 274 (2008).

whether there is any evidence of affiliate abuse or sweetheart deals among affiliates.[177] Once a utility obtains market-based rate authority, it must take several actions. It must file a market-based rate tariff with FERC. It must submit quarterly electronic reports of all the transactions it enters into under that tariff. And, it must file updated analyses of its potential for exercising market power every three years. Market power is also discussed in Part VII, *Interaction between Energy Law and Competition Law.*

141. One prerequisite for receiving market-based rate authority is that transmission-owning public utilities must file an open access transmission tariff (OATT). This tariff, which FERC mandated for all jurisdictional utilities in its path-breaking Order No. 888, allows utilities to use transmission capacity on their systems first for their long-time retail customers (their so-called "native load"), but then requires them to sell off all excess capacity to the market on a first-come, first-served basis using an internet interface.[178] The idea was to eliminate electric utilities' ability to discriminate against non-favored parties, or to keep out competition from other generation-owning companies. Thus, Order No. 888 prescribes standard offer terms that utilities must give to all comers for transmission service, opening up the market to generators not owned by incumbent IOUs. Today, many of these markets are overseen by regional transmission organizations and independent system operators. Both Order No. 888 and regional coordination through RTOs and ISOs are discussed in more detail below in Chapter 3 on Transportation.

142. While market-based rate authority has virtually taken over rate-setting for wholesale power sales before FERC, the picture on the retail side of rate regulation is much more mixed. At the same time states began to restructure the electricity industry in the late 1990s and early 2000s, many started to move to retail competition as well. The idea was parallel to the wholesale world. Regulators believed that if consumers were given a choice of the provider from whom they received electricity, the market would drive down costs and foster the offering of different electricity products desired by consumers. Instead, many states halted their efforts to promote retail electricity competition when California faced an energy crisis at the turn of the century and Enron and others were found to have manipulated the market.[179] Many other states who had been contemplating moving to retail competition decided not to, and today only a few states have effective competition for electricity

---

177. *See supra* note 138; *see also, e.g.*, William T. Miller & Randolph Lee Elliott, *Rates and Tariffs*, *in* 4 ENERGY LAW AND TRANSACTIONS, *supra* note 81, at § 80.01[3][b].

178. Order No. 888, Promoting Wholesale Competition Through Open Access Non-discriminatory Transmission Services by Public Utilities; Recovery of Stranded Costs by Public Utilities and Transmitting Utilities, F.E.R.C. Stats. & Regs. ¶ 31,036 (1996), 61 Fed. Reg. 21,540 (1996) (codified at 18 C.F.R. pts. 35, 385) [hereinafter Order No. 888], *order on reh'g*, Order No. 888-A, F.E.R.C. Stats. & Regs. ¶ 31,048 (1997), *order on reh'g*, Order No. 888-B, 81 F.E.R.C. ¶ 61,248, 62 Fed. Reg. 64,688 (1997), *order on reh'g*, Order No. 888-C, 82 F.E.R.C. ¶ 61,046 (1998), *aff'd in relevant part sub nom.*, Transmission Access Policy Study Group v. FERC, 225 F.3d 667 (D.C. Cir. 2000), *aff'd sub nom.*, New York v. FERC, 535 U.S. 1 (2002).

179. SPECIAL COMM. ON RESTRUCTURING OF THE ELEC. INDUS. 2006 ANNUAL REPORT, ABA ENVIRONMENTAL ENERGY & RESOURCES LAW.: YEAR IN REVIEW 328 (2006).

at the retail level. Indeed, Texas may be the only state that stands out as a true success story in this regard.[180]

143. In addition to market-based rates, another important form of innovative rate design is known as marginal cost pricing. In brief, marginal cost ratemaking better reflects the true cost of a utility's product. Under the cost-of-service method, utility rates were set based on historical average costs—typically represented by a historic "test year." Thus, costs that already had been incurred were the basis for setting rates. This created perverse economic incentives for utilities, because the more they spent, the more they could recover. Indeed, the problem with using historical costs cuts both ways. When a utility is realizing economies of scale, it is overcompensated because its costs are declining. When, however, market costs are rising, a utility suffers unnecessary losses. The more sound economic approach is to set rates based on current costs—that is, on the marginal cost of producing the next unit.[181]

144. Thus, under the marginal cost approach, a utility's costs are more accurately reflected in the rates that are set, and price signals to consumers are more accurate. There are several variations of marginal cost pricing, including time-of-day rates that reflect the actual cost at the time of use.

145. Other innovations in ratemaking include incentive rates. These can be structured in various ways. If a particular utility, for example, is operating efficiently and saving costs, then it may be rewarded with a slightly higher rate of return. Similarly, to the extent that regulators want to encourage investments in technological innovations or capital improvements, rates can reflect such investments by including a portion or all of those investments in the rate base or, alternatively, by allowing a slightly higher rate of return.

146. Further, through rate design, regulators can encourage or discourage consumption. As energy markets were expanding and as infrastructure needed to be constructed, regulators often used declining block rates. Rates would be set based upon blocks of usage. In the initial blocks, the price per kWh was high, which allowed utilities to quickly recover their fixed costs. Then, as consumption increased, the rates would be lowered in subsequent blocks, thus encouraging consumption. Again, this type of rate design was seen favorably when regulators were trying to encourage expansion of the electricity system. However, as electricity prices began to rise, regulators became concerned, and many imposed requirements that utilities turn the blocks in their rates upside down—that is, impose "inverted block" rates that increase with consumption, and thus, discourage overuse of electricity. As an even more aggressive form of discouraging consumption, some regulators also experimented with demand "decoupling," in which an electric utility would recover its fixed costs independently of the amount of electricity that it sold.

180. Distrib. Energy Financial Group LLC, Abaccus: An Assessment of Restructured Electricity Markets (2011).

181. In re Madison Gas & Electric Co., 5 P.U.R. 4th 28 (Wisc. PSC 1974).

147. Another innovative rate design is known as the "straight fixed variable" (SFV) rate, which is a form of decoupling. Under this method, all of the utility's fixed costs are placed into one component of the consumer utility bill. Variable costs, however, such as fuel, are charged on the bill separately, so that customers receive more accurate price signals. This method is distinctive because, particularly for residential and commercial consumers, electricity bills traditionally were "bundled," which means that the electricity, customer, and demand charges were all contained in the single rate. The drawback with the SFV method, however, is that it provides no incentive for consumers to conserve. To correct that inefficiency, some regulators adopted an approach known as a "feebate," in which customers who conserve by using off-peak electricity receive a rebate, while customers who consume peak electricity pay a higher fee. With a feebate, there is no additional cost to ratepayers or to shareholders; instead, these cost are apportioned among different groups of ratepayers.

148. Another approach to innovative ratemaking seeks to promote demand response programs, which are addressed in more depth in § 5 below. Briefly, demand response is the process of a utility managing system load not by ramping up additional generation, but rather, by being able to reduce demand for energy during key time periods. This poses a problem for utilities because, to the extent their revenues are based on sales, use of demand response reduces their sales and thus their revenues. To promote demand response, then, some state regulators have used particular rate mechanisms, namely, "lost revenue recovery" or "shared savings" devices. Both methods use the same idea. They allow utilities, to the extent they are able to promote energy savings using demand response, to share those savings between customers and shareholders, rather than just letting customers recoup the benefits.[182]

149. Regulators' choice to innovate in ratemaking by adopting devices like these has transformed how electricity is exploited in the United States today. Indeed, the movement away from cost-of-service ratemaking, together with increased attention to energy efficiency and conservation, is forcing utilities to rethink their business models. For most of the twentieth century, utilities were virtually guaranteed a profit, and thus, were encouraged to keep increasing electricity sales. A focus on conservation, however, means the utility risks losing sales. Therefore, electric utilities must, in addition to selling electricity, also consider providing energy efficiency products and services. We return to this theme below in § 5 on Consumption.

182. *See, e.g.*, In re Portland General Elec. Co., 142 P.U.R.4th 240 (1993); *see also* Elec. Power Supply Ass'n v. FERC, 753 F.3d 216 (D.C. Cir. 2013), *cert. granted sub nom.*, EnerNOC, Inc. v. Electric Power Supply Ass'n, 135 S. Ct. 2049 (2105) (reviewing FERC Order No. 745, *Demand Response Compensation in Organized Wholesale Energy Markets*, 134 FERC ¶ 61,187 (2011), which instituted demand response incentives in RTOs and ISOs).

§2. Electricity Federalism

150. While traditional ratemaking principles tend to parallel each other at both the federal and state levels, jurisdiction over the U.S. electricity industry is drawn in sharp lines. The federal government has authority over wholesale power sales, transmission of electricity in interstate commerce, disposition of certain facilities, and hydropower licensing. State governments have authority over retail sales, distribution services, energy planning, and almost all siting questions for power plants and transmission and distribution lines. This division of authority extends to ratemaking as well. Thus, FERC has ratemaking authority in the wholesale and interstate transmission areas, and states have it at the distribution and retail levels of the industry.

151. Federal authority is delineated by the Federal Power Act,[183] which was adopted to fill the "*Attleboro* gap" that was created when the Supreme Court ruled that states could not regulate electric utilities beyond their boundaries.[184] Thus, the FPA sought to centralize some regulatory authority in the federal government, in turn promoting development of and coordination between large utilities, while retaining some powers for the states.[185] Courts have interpreted federal jurisdiction rather broadly. For example, the United States Supreme Court has held that FERC jurisdiction extends to the sale of electricity by one utility within a state to another utility within the same state, so long as some of that energy originated out of state at some point.[186] However, the restructuring of the electric industry is making the dividing line between federal and state authority less clear, raising significant governance issues regarding appropriate jurisdiction.[187] By way of example, the need for a smart grid is a matter of federal, as well as state, concern; yet federal siting authority for the smart grid is circumscribed.

152. Questions about the division between federal and state authority in electricity often arise in two ways: through the doctrine of preemption and through application of the "dormant" or "negative" Commerce Clause.

153. Preemption is application of the Supremacy Clause of the U.S. Constitution.[188] That clause dictates that federal law controls over state and local law in the United States. Courts have developed three tests to determine whether there is such a problem, that is, whether federal law preempts state law. Preemption occurs if: (1)

---

183. Federal Power Act, 16 U.S.C. §§ 791–823d.
184. Pub. Utilities Comm'n of R.I. v. Attleboro Steam & Elec. Co., 273 U.S. 83(1927).
185. 16 U.S.C. § 824a.
186. FPC v. Southern California Edison, Co., 376 U.S. 205 (1964); FPC v. Florida Power 7 Light Co., 404 U.S. 453 (1972).
187. New York v. FERC, 535 U.S. 1 (2002); Elec. Power Supply Ass'n v. FERC, 753 F.3d 216 (D.C. Cir. 2014) *cert. granted in part sub nom.* EnerNOC, Inc. v. Elec. Power Supply Ass'n, No. 14-841, 2015 WL 223557 (U.S. May 4, 2015) and *cert. granted in part*, No. 14-840, 2015 WL 223555 (U.S. May 4, 2015).
188. U.S. Const. art. VI.

federal law expressly says that it preempts state law; (2) state law conflicts with federal law such that it frustrates the accomplishment of specific federal objectives or makes compliance with both federal and state law impossible; or (3) federal law so occupies the field of regulation in an area that it leaves no room for state regulation.[189]

154. Preemption often arises with respect to electricity ratemaking.[190] For example, once FERC determines that a rate is just and reasonable, states are prohibited from reviewing those rates, ignoring them, or modifying them—even if the state PUC disagrees with FERC's decision, or itself deems the rate to be unjust or unreasonable.[191] Thus, courts have recognized that "FERC's jurisdiction over interstate wholesale rates is exclusive."[192] "Congress meant to draw a bright line easily ascertained, between state and federal jurisdiction," the U.S. Supreme Court has said. "This was done in the [FPA] by making [FERC] jurisdiction plenary and extending it to all wholesale sales in interstate commerce except those which Congress has made explicitly subject to regulation by the States."[193] This regulatory regime "leaves no room either for direct state regulation of the prices of interstate wholesales of [energy], or for state regulations which would indirectly achieve the same result."[194]

155. By contrast, courts have allowed states to make decisions that may be in tension with FERC's determinations but that do not actually override them or prevent their effectuation. For instance, state PUCs are allowed to make calls on whether expenses incurred by a utility for a given contract were prudent from an investor's perspective, thus impacting whether the utility's shareholders must bear those costs rather than their ratepayers, so long as such a determination does not impact a FERC ruling on the justness and reasonableness of a rate.[195] Likewise, courts have ruled that state PUCs can walk up to the line drawn in favor of FERC's ratemaking jurisdiction, so long as they do not cross over it. Thus, one court has allowed its state PUC to evaluate the terms and conditions of wholesale power contracts *before* they are entered into, so long as that does not occur after they become subject to FERC scrutiny.[196]

---

189. City of Morgan City v. South Louisiana Elec. Coop. Ass'n, 31 F.3d 319, 321–22 (5th Cir. 1994).
190. In *California* v. *Federal Energy Regulatory Commission*, 495 U.S. 490 (1990), for example, the Supreme Court held that state regulations imposing minimum flow rates on rivers used to generate hydroelectric power were preempted by the Federal Power Act.
191. *See, e.g.*, Nantahala Power & Light Co. v. Thornburg, 476 U.S. 953 (1986); Mississippi Power & Light Co. v. Mississippi ex rel. Moore, 487 U.S. 354 (1988); Entergy Louisiana, Inc. v. Louisiana Pub. Serv. Comm'n, 539 U.S. 39 (2003).
192. Appalachian Power Co. v. Pub. Serv. Comm'n, 812 F.2d 898, 902 (4th Cir. 1987).
193. Nantahala Power & Light Co. v. Thornburg, 476 U.S. 953, 966 (1986) (quoting FPC v. So. Cal. Edison Co., 376 U.S. 205, 215–16 (1964)).
194. No. Natural Gas Co. v. State Corp. Comm'n, 372 U.S. 84, 91 (1963).
195. *See, e.g.*, Public Serv. Co. of New Hampshire v. Patch, 167 F.3d 29 (1st Cir. 1998), *cert. denied*, 526 U.S. 1066 (1999); Public Serv. Co. of New Hampshire v. Patch, 221 F.3d 198 (1st Cir. 2000), *cert. denied*, 531 U.S. 1145 (2001); *cf.* Kentucky W. Va. Gas Co. v. Pennsylvania Pub. Util. Comm'n, 837 F.2d 600 (3d Cir. 1988), *cert. denied*, 488 U.S. 941 (1988).
196. State ex rel. Utilities Comm'n v. Carolina Power & Light Co., 614 S.E.2d 281 (N.C. 2005).

156. The Supreme Court's decision in *Mississippi Power & Light Co. v. Mississippi ex rel. Moore* provides a good example of the jurisdictional lines the Federal Power Act draws between FERC's and states' respective ratemaking authorities.[197] In that case, FERC had ordered a utility to purchase a specific percentage of a generator's electricity. The utility then asked state regulators to approve an increase in its retail rates in order to allow it to recover the costs imposed by FERC's order. The state declined. It declared it retained authority to review whether the power purchases required by FERC's order were prudently incurred, and thus, whether they could proceed at all. The Supreme Court disagreed. Observing that the question of preemption does not turn on "whether a particular matter was actually determined in the FERC proceedings," the Court held that the "reasonableness of rates and agreements regulated by FERC may not be collaterally attacked in state or federal courts. The only appropriate forum for such a challenge is before the Commission or a court reviewing the Commission's order."[198]

157. Jurisdictional tensions also have arisen as technology has changed and the industry has evolved. For instance, when many U.S. utilities decided to abandon planned nuclear power plants, there were strenuous fights before the agencies and in courts over who would bear the costs of those partially built plants. None of the methods available for resolving this problem were perfect, and regulators were concerned that ratepayers might be saddled with significant price increases without receiving benefits for those payments. Pennsylvania, for example, passed legislation that adopted a used-and-useful approach, under which ratepayers would only pay for investments from which they actually received electricity. In that case, the costs of canceled plants were borne entirely by shareholders, and the legislation was upheld by the United States Supreme Court.[199]

158. A second way that federalism principles influence who can regulate in the energy arena in the United States, including for electricity, is application of the "dormant" or "negative" Commerce Clause. The Commerce Clause of the United States Constitution gives Congress the authority to regulate activities in interstate commerce.[200] However, relying on the notion that the nation "must sink or swim together,"[201] the U.S. Supreme Court has interpreted this affirmative grant of power to Congress to also have a "dormant" or "negative" aspect to it. Thus, the Court has consistently ruled that states may not affirmatively discriminate against commerce from each other, may not control commerce extraterritorially outside their jurisdiction, and may not even regulate evenhandedly if the burden of that regulation on interstate commerce outweighs the local benefits accrued from that regulation.[202] Numerous court decisions have applied these doctrines to production and exploitation of electricity.

---

197. 487 U.S. 354 (1988).
198. *Ibid* at 374; *see also, e.g.*, PPL EnergyPlus, LLC v. Nazarian, 753 F.3d 467 (4th Cir. 2014).
199. Duquesne Light Co. v. Barasch, 488 U.S. 299 (1989).
200. U.S. Const. art. I, § 8, cl. 3.
201. Baldwin v. G.A.F. Seelig, Inc., 294 U.S. 511, 523 (1935).
202. C & A Carbone, Inc. v. Town of Clarkstown, N.Y., 511 U.S. 383 (1994); Pike v. Bruce Church, Inc., 397 U.S. 137 (1970).

159. New England Power Co. v. New Hampshire is perhaps the leading case.[203] There, a public utility, New England Power Co. (NEPCO), provided electricity primarily to residents of Massachusetts, though it also sold some power to Rhode Island and New Hampshire. NEPCO owned six hydroelectricity plants in New Hampshire, which together could produce about 420 MW of power at very low cost and provided about 10% of NEPCO's electricity needs. New Hampshire long had on the books a law that gave the state public utilities commission authority to decide whether hydroelectricity could be transmitted outside the state. For decades, the New Hampshire PUC let NEPCO do this, but in 1980, it ordered NEPCO to stop exporting the cheap hydroelectricity and to sell it instead to companies and residents of New Hampshire. As a reason for its decision, the PUC said that "the Public Service Company of New Hampshire, the State's largest electric utility, had generating costs about 25% higher than those of New England Power; and that if New England Power's hydroelectric energy were sold exclusively in New Hampshire, New Hampshire customers could save approximately USD 25 million a year."[204]

160. The U.S. Supreme Court had little difficulty striking down this order as forbidden by the dormant Commerce Clause. That doctrine forecloses balkanization of interstate markets by states, and that is exactly what New Hampshire's law sought to do. "The order of the New Hampshire Commission, prohibiting New England Power from selling its hydroelectric energy outside the State of New Hampshire, is precisely the sort of protectionist regulation that the Commerce Clause declares off-limits to the states," the Court wrote. "The Commission has made clear that its order is designed to gain an economic advantage for New Hampshire citizens at the expense of New England Power's customers in neighboring states."[205] That, the Court said, states cannot do.

161. A good example of the effects this division of authority has on electricity markets arises in the context of renewable energy. As noted above, one of the key trends in U.S. electricity production and exploitation is the movement toward greater reliance on renewable resources. However, another byproduct of divided jurisdiction over the U.S. electricity sector is that while FERC controls wholesale power markets, states retain authority over electricity generation fleets.[206] Thus, as renewable energy prices have come down, it has primarily been state regulators that have promoted these technologies as a way to transform electricity exploitation in the United States.

162. Governments that want to promote any given energy source have various options to do so. They might fund basic research and development to help innovate or get the technology off the ground. They might offer financial incentives for the

203. 455 U.S. 331 (1982).
204. *Ibid* at 336.
205. *Ibid* at 339.
206. Pac. Gas & Elec. Co. v. State Energy Res. Conservation & Dev. Comm'n, 461 U.S. 190 (1983).

technology's use, such as through subsidies or tax credits. They might compel purchase of the energy resource at incentive prices, such as PURPA does. Or they might mandate use of a specific technology.

163. Because of the way that the FPA carves up the electricity market, state regulators are constrained in which of these choices they may utilize. The federal government in the United States has, somewhat off and on, offered a production tax credit for the exploitation of renewable resources to produce electricity.[207] However, states are limited by preemption principles from offering incentive rates for renewable energy production, other than under the auspices of PURPA.[208] Likewise, because the FPA gives exclusive authority to FERC to regulate the price of wholesale power contracts, states are further limited in how they might promote renewables.[209] Thus, states primarily are limited to promoting renewables through their control over the composition of electricity generation fleets.

164. States have exercised this authority by adopting what are known as "renewable portfolio standards," or "RPSs." RPSs generally include four components: (1) a target percentage of renewables that the state intends to reach; (2) a target date for reaching the percentage goal; (3) the identification of qualifying resources such as wind, solar, biomass, and hydroelectricity; and (4) rules for enforcing the RPS and for imposing it on different kinds of electricity providers. Because these laws are state-based, they vary somewhat substantially from jurisdiction to jurisdiction. By way of example, however, a number of states adopted RPSs that mandate 20% of electricity production from renewable resources by the year 2020.

165. One way to facilitate RPS operation is to tie the law's renewable energy mandate to a market. Thus, most RPSs typically include multiple ways that utilities can satisfy the law's mandate. Utilities can build new renewable energy facilities, buy renewable electricity from other providers, or purchase the renewable energy *attribute* of that electricity on an open market. These attributes are referred to as renewable energy credits or renewable energy certificates (RECs), which typically represent 1 MWh of electricity production from qualifying renewable energy resources. The idea behind RECs is simple. They facilitate achievement of the RPS mandate at the lowest cost, because the providers who can produce the next MWh of electricity at the lowest cost will be incentivized to do so by the willingness of others to pay for that REC rather than produce the electricity themselves at a higher cost.

---

207. 26 U.S.C. § 45.

208. Am. Ref-Fuel Co., Covanta Energy Grp., Montenay Power Corp., & Wheelabrator Technologies Inc., 105 F.E.R.C. ¶ 61,004 (2003); ARIPPA v. Pennsylvania Pub. Util. Comm'n, 966 A.2d 1204 (Pa. Commw. Ct. 2009).

209. California Pub. Utilities Comm'n, 132 F.E.R.C. ¶ 61,047 (2010). The other policy device that many nations use to promote renewable energy is feed-in tariffs (FITs), which, like PURPA, guarantee access to the grid, obligate utilities to buy the relevant energy, and guarantee a premium price for the power, usually for a long (*e.g.*, 10- or 20-year) period. The problem for FITs in the United States is that the price that can be used for them is capped by PURPA's avoided cost rate and, moreover, FERC has exclusive jurisdiction over wholesale power sales.

166. Although RPS programs have been popular, certain challenges remain. First, each state program is distinct, with different goals, different qualifying resources, and different target dates. Moreover, some state programs are mandatory and others are voluntary, or close to voluntary. As a result, a uniform national REC market is unlikely to develop. Second, because each state is distinct, interstate sales of renewable electricity is likewise not uniform.[210] Third, to the extent that one state program favors in-state utilities and renewable electricity providers, such favoritism may violate the constitutional prohibition against discrimination in interstate commerce.[211]

167. Nonetheless, RPSs appear to be changing how resources are exploited to produce electricity in the United States. As noted, renewable energy now constitutes the greatest portion of new installations in the United States, by a fairly wide margin.[212] Likewise, since 2002, over sixty percent of non-hydroelectric renewables installations in the United States have been in RPS states, with that figure much higher in some years.[213]

168. RPSs, then, represent another important way that the jurisdictional divide in the United States affects electricity regulation. As of March 2015, 37 states plus the District of Columbia had some version of an RPS.[214] However, the federal government has not been able to pass a national RPS, despite widespread support for the RPS device amongst the states. Dozens of bills have been introduced to create a federal RPS, but to date none has passed both houses of Congress. Thus, as with much of electricity regulation, renewables regulation remains largely divided as it has been in the past, even as the nation moves forward into the future.

---

210. Davies, *supra* note 31, at 1339.

211. Illinois Commerce Comm'n v. Fed. Energy Regulatory Comm'n, 721 F.3d 764 (7th Cir. 2013).

212. Am. Council on Renewable Energy, *supra* note 160, at 15.

213. Ryan Wiser & Galen Barbose, Lawrence Berkeley Nat'l Lab., Renewable Portfolio Standards in the United States: A Status Report with Data Through 2007, at 12 (2008), *available at* http://eetd.lbl.gov/ea/ems/reports/lbnl-154e-revised.pdf.

214. *See* Database of State Incentives for Renewables and Efficiency, *Renewable Portfolio Standard Policies* (Mar. 2015), http://ncsolarcen-prod.s3.amazonaws.com/wp-content/uploads/2014/11/Renewable-Portfolio-Standards.pptx.

Chapter 3. Transportation

169. Over the past twenty years, the transportation, or transmission, segment of the electric utility industry has become the epicenter of industry and regulatory changes. This is for several reasons. First, regulators, most notably federal regulators, have been experimenting with various forms of deregulation and industry restructuring specifically for the purpose of encouraging market competition. Second, the move to a more competitive electricity industry could not occur without accompanying changes in the regulation of transmission. The provision of generation from new market entrants and alternative sources could not take place unless those entities can get their product to market. Third, in addition to an expansion of non-utility generated electricity, the increasing use of variable resources, most notably solar and wind power, also require integration into the traditional electric grid. In order to achieve that integration, changes to the grid must be made, including building out new transmission lines that can deliver the power to market. Fourth, the potential for the introduction of "disruptive" technologies, including the smart grid, have the potential to radically change the electric sector. More specifically, the wide-scale introduction of electric vehicles and long-term electric storage will substantially change the industry and its regulation. This ties into the role that transmission plays in the industry.

§1. PURPA and the Origins of Electricity Restructuring

170. The changes in electric industry regulation being experienced today began decades ago. As noted earlier, in many ways, the discipline of energy law was a response to energy crises of the 1970s. President Carter's major energy legislation consisted of a variety of statutory provisions, one of which was the Public Utility Regulatory Policies Act of 1978.[215] The primary purposes of that Act was to encourage state rate regulators to move away from declining block ratemaking, to encourage integrated resource planning, and to foster competition from new and different market participants. Thus, the law encouraged independent power production through cogeneration and small power generation (80 MW or less) as alternatives to large-scale IOUs.[216]

171. The statute was upheld as constitutional by the United States Supreme Court,[217] and these independent power producers, known in the statute as "qualifying facilities," proliferated to an unanticipated extent. Integral to the effectiveness of these new electricity generators, PURPA gave FERC authority to require electric utilities to provide transmission services to new generation competitors. Prior to that

215. Pub. L. No. 95-617 codified in various sections of 16 U.S.C.

216. *See, e.g.*, Floyd L. Norton et al., *Electricity: Open Access, Transmission & Comparability*, *in* 4 Energy Law and Transactions, *supra* note 81, at § 82.01.

217. FERC v. Mississippi, 456 U.S. 742 (1982).

time, FERC lacked jurisdiction to order involuntary "wheeling" (*i.e.*, transportation) of electricity.[218] Under PURPA, however, FERC gained limited interconnection authority in order to promote competition in electricity. Such an interconnection order would be issued when: (1) it was in the public interest; (2) competitive conditions were maintained; and (3) either energy would be conserved or energy efficiencies would be realized.

172. PURPA took a bold step. It required local utilities to purchase QF electricity at the purchasing utility's "full avoided cost." The local utility, in other words, was required to connect with co-generators and small power producers and purchase the excess electricity produced by them at the purchasing utility's marginal cost. This had the effect of promoting competition within the market. The local utility paid its higher cost to purchase QF power, which in turn encouraged QFs to enter the market. This pricing arrangement was also upheld by United States Supreme Court.[219]

173. QFs found this market very attractive, and the number of QFs proliferated. Utilities, in turn, complained that these new competitors were abusing the system—that some of them were only producing electricity to take advantage of the higher prices the statutory scheme afforded, and that it was not efficient or effective to buy power from these producers. Incumbent utilities eventually gave such producers an unseemly moniker: "PURPA machines." At the same time, though, the ability of QFs to operate in the market showed others that competition in generation could work. Thus, as improvements were made in the technology for gas-fired generation, these cleaner, more efficient, and easier-and-cheaper-to-build units also started entering the market.

174. These new market participants created new demands on the transmission network. Both more capacity and additional interconnections were required. Congress, however, embracing an overall move in the nation toward competition and deregulation, endorsed the emerging restructuring of the electricity industry. Thus, in the Energy Policy Act of 1992,[220] it gave these non-utility generators further legal support by exempting them from the Public Utility Holding Company Act of 1935.[221] Such entities became known as EWGs, or "exempt wholesale generators." By exempting these entities from PUHCA, Congress lowered their regulatory costs and encouraged their entry into the market.

175. As a result of these developments, energy policymakers began to rethink regulation of the transmission segment of the industry. Put simply, these new classes

218. *See* Otter Tail Power Co. v. United States, 410 U.S. 366 (1973).
219. American Paper Inst., Inc. v. American Elec. Power Serv. Corp., 461 U.S. 402 (1983).
220. Energy Policy Act of 1992, SEC. 711, Pub. L. No. 201-486.
221. Public Utility Holding Company Act of 1935, 15 U.S.C. §§ 79–79z-6 (largely repealed by the Energy Policy Act of 2005).

of electricity generators needed access the grid.[222] EPAct 1992, then, began addressing transmission access in significant ways.

176. The Act gave FERC additional authority, subject to a finding of "public interest," to order transmission entities to wheel power for wholesale transactions. Specifically, a generating unit, such as a QF, could apply for access to the transmission system under a FERC-mandated wheeling order. Under the Act, such wheeling was mandatory unless it could be shown that the utility's customers would be unduly disadvantaged, or that reliability would be unduly impaired. Also under the Act, FERC had the authority to review the rates of transmitting facilities, some of which were previously outside of its jurisdiction.

177. While transformative, EPAct 1992's provisions were also limited. The kind of wheeling FERC could order required a particular applicant to make a request for wheeling, meaning that FERC would have to act on a case-by-case basis, which could be both cumbersome and expensive for new entrants. Thus, if transmission was to be truly open, something more would be required.

§2. Order No. 888: Open Access Transmission

178. FERC brought that something more in 1996. In that year, the agency promulgated its groundbreaking rule, Order No. 888,[223] which specifically sought to open up the transmission segment of the market to all comers. The rule was the final linchpin. By saying that all entities could use transmission lines, the agency effectively converted these bottleneck facilities into common carriers. This, in turn, allowed a significant restructuring of the electricity industry precisely because, for the most part, transmission and distribution lines were privately owned by IOUs. Now, however, new competitors could use IOUs' lines too.

179. In issuing Order No. 888 and its companion rule, Order No. 889,[224] FERC relied on its authority under FPA §§ 205 and 206 to remedy undue discrimination.[225] Specifically, Order No. 888 required jurisdictional transmission firms to do two things.

---

222. *See, e.g.*, Reinier H.J.H. Lock & Marlene L. Stein, *Electricity Transmission*, *in* 4 Energy Law and Transactions, *supra* note 81, at § 81.01[1]–[4].

223. *See supra* note 178.

224. Order No. 889, Open Access Same-Time Information System (Formerly Real-Time Information Networks) and Standards of Conduct, FERC Stats. & Regs. ¶31,049 (1996) (and supplementary orders).

225. Prior to this order, FERC started to open transmission access by closely scrutinizing transmission tariffs, as well as merger requests, to ensure that transmission services were not unduly discriminatory. As a condition of a merger, FERC would require that transmission services of the merged entity be offered to third parties on a comparable basis as the merged entity itself could use the services. Norton et al., *supra* note 216, at § 82.03[1].

180. First, utilities were required to file an open access transmission tariff containing terms and conditions for nondiscriminatory transmission service; this was known as the pro forma tariff, or the pro forma open access transmission tariff. The OATT contained standardized terms that contained (1) a set of general provisions applicable to all services; (2) provisions for network integration service; and (3) provisions for point-to-point transmission service.[226]

181. Second, utilities were required to "functionally" unbundle their wholesale power services by charging uniform wholesale transmission rates to all third parties. Additionally, utilities were required to set separate rates for wholesale generation, transmission, and ancillary services. Thus, functional unbundling fell short of requiring divestiture of utilities' generation assets or other, more drastic forms of corporate restructuring, but it nonetheless was a significant change.

182. In order to facilitate the smooth functioning of transmission markets, FERC also promulgated Order No. 889, which created an electronic computer system called the Open Access Same-Time Information System (OASIS). The order set out the information that was necessary to provide on OASIS, including "available transmission capacity, total transmission capacity, points of delivery and receipt, posted paths, curtailment and interruption occurrences, information concerning constrained and unconstrained paths, transmission request denials and acceptances, and pricing."[227]

183. The intent of Orders No. 888 and 889 was to create a competitive wholesale market by requiring public utilities to provide nondiscriminatory transmission service to all eligible customers. The direct consequence of this scheme, then, was to further open wholesale markets to competition. Order No. 888 also allowed for the recovery of "stranded" costs incurred by utilities as they moved from the old regulatory regime to the new, more competitive one. FERC's overall effort in Order No. 888 ultimately was upheld by the courts.[228]

§3. Order No. 2000: ISOs and RTOs

184. Order No. 888 was important in itself, but it also led to further changes in the market. As electricity markets had grown, so too had the way that utilities coordinated with each other. Thus, many utilities in the same geographic regions had become interconnected and started to coordinate how they ran their operations, in what were known as "power pools" or "tight power pools."[229] Power pools entered into contractual arrangements with each other for the coordination of the operation

226. *Ibid* at § 82.03[2][a].

227. *Ibid* at § 82.03[3][b](2010); *see also* 18 C.F.R. § 37.6.

228. Transmission Access Policy Study Group v. FERC, 225 F.3d 667 (D.C. Cir. 2000) *aff'd sub nom.*, New York v. FERC, 535 U.S. 1 (2002).

229. Depending upon how complex the arrangements were, these pools were known as either "tight" power pools with a high degree of coordination, or "loose" power pools, with a lower degree of coordination.

and maintenance of their transmission systems. The idea was to allow participating utilities to most efficiently and economically dispatch their electricity loads, in part by relying on the generation and transmission resources of their neighbors.

185. Eventually, the idea of power pools grew. FERC began championing the idea that power pools themselves could combine, yielding efficiencies from greater coordination over even larger areas. FERC also thought that areas that lacked tight power pools should increase their coordination. Thus, starting with EPAct 1992, FERC began to consider regional solutions to the transmission issue, with the purpose of further opening access to markets and encouraging more competition. In 1993, for example, FERC issued a non-binding policy statement regarding Regional Transmission Groups (RTGs).[230] Then, with Order No. 888, FERC began to consider how to bring power pools into compliance with that order.

186. As part of Order No. 888, FERC provided guidance regarding the adoption of multi-party transmission arrangements through independent system operators. The idea of an ISO was simple but revolutionary. It was that transmission owners would turn over operation of their system to an independent third party. This, FERC said, would allow for greater coordination and also distance the incumbent utility from its transmission system, thus further reducing the risk of discriminatory treatment of competitors. FERC did not require mandatory participation in an ISO in Order No. 888, but sought to encourage voluntary membership as a way of complying with the requirements of nondiscriminatory transmission tariffs.

187. FERC then set out principles to guide how ISOs would function. In this guidance, FERC stated that a proper ISO would: (1) have independent governance; (2) ensure that employees had no financial interest in any market participant; (3) provide transmission service using nondiscriminatory rates; (4) take responsibility for short-term system reliability; (5) identify constraints on the system; and (6) provide information to market participants through an ISO-run OASIS.[231]

188. The industry thus began to experience significant changes. Just as FERC was pushing for restructuring, states were too. As a result, many IOUs were divesting some or all of their generating assets in response to state mandates to do so. There also was a noted increase in mergers between electric utilities and between electric and gas utilities. And, as noted, independent and affiliated power marketers and generators had begun to enter the industry. All of these changes put pressure on the transmission system by increasing loads, creating more generators that had to be integrated into system operation, and by generally making balancing operations more complex. In response, FERC's idea of a regional solution to transmission took on new life.

---

230. Michael A. Yuffee & J. Peter Ripley, *Regional Transmission Organizations/Independent System Operators*, *in* 4 Energy Law and Transactions, *supra* note 81, at § 89.02[2][b].
231. *Ibid* § 89.03[2].

189. In 1999, FERC issued Order No. 2000,[232] which aimed to promote the creation of regional transmission organizations across the nation. Doing so, FERC said, would better promote competition, which in turn would be "the best way to protect the public interest and ensure that electricity consumers pay the lowest price possible for reliable service."[233] This was largely because central to both ISOs and RTOs is the functional unbundling of generation and transmission from vertically integrated IOUs. The insertion of an independent party to run the transmission grid, FERC said, should remove longstanding concerns about incumbent IOUs exercising control over the transmission grid to keep out, or discriminate against, competitors. Thus, FERC required this functional unbundling as a condition of approval to form an RTO.

190. FERC also asserted that RTOs would enhance system operations and reliability and "eliminate any residual discrimination in transmission service that can occur when the operation of the transmission system remains in the control of a vertically integrated utility."[234] FERC thus summarized the benefits it believed RTOs could provide: "[RTOs can] (1) improve efficiencies in transmission grid management; (2) improve grid reliability; (3) remove remaining opportunities for discriminatory practices; (4) improve market performance; and (5) facilitate lighter handed regulation."[235]

191. To this end, Order No. 2000 established minimum characteristics and functions for RTOs. At a minimum, Order No. 2000 requires an RTO to: (1) be independent; (2) govern an appropriate geographic region; (3) have operational authority for all transmission facilities under their control; and (4) have short-term reliability responsibility and authority.

192. Additionally, RTOs must perform a variety of functions, including: (1) transmission tariff development and administration; (2) adoption of congestion management procedures; (3) serve as provider of last resort for all ancillary services; (4) operate the OASIS for all transmission under its control; (5) monitor markets and measure market power; (6) plan and coordinate transmission upgrades and additions; (7) develop coordination mechanisms with other regions; and (8) develop and implement loop flow parallel path procedures.[236] Initially, FERC sent signals that it would require all jurisdictional utilities to join RTOs. However, when utilities in the West and Southeast—and politicians from those states—objected, vehemently, FERC backed off that plan and used Order No. 2000 simply to encourage the formation of RTOs.

232. Order No. 2000, Regional Transmission Organizations, F.E.R.C. Stats. & Reg. ¶ 31,089 (1999), 65 Fed. Reg. 809 (1999), *order on reh'g*, Order No. 2000-A, F.E.R.C. Stats. & Regs. ¶ 31,092 (2000), 65 Fed. Reg. 12,088 (2000), *aff'd sub nom.*, Pub. Util. Dist. No. 1 v. FERC, 272 F.3d 607 (D.C. Cir. 2001).
233. *Ibid* at 3.
234. *Ibid.*
235. *Ibid.*
236. Yuffee & Ripley, supra note 230, at § 89.03[3].

193. While RTOs must perform many of the same functions, each organization differs according to its region and how it was formed. Some, for example, were formed voluntarily; others were formed pursuant to state legislation. The Pennsylvania-New Jersey-Maryland Interconnection (PJM) is a good example. It was approved by FERC in 1997 as an ISO.[237] Under the terms of that arrangement, PJM was responsible for: (1) directing the operation and maintenance of certain transmission facilities; (2) physically operating and maintaining each transmission owner's facilities; and (3) constructing and financing needed modifications to facilities based on studies conducted by the ISO.[238] In 2002, PJM applied to FERC under Order No. 2000 to act as an RTO. FERC found that PJM satisfied its stated criteria and granted it RTO status.[239] Pursuant to that order, PJM's role expanded. It is now responsible for coordinating the movement of wholesale electricity; managing wholesale and other electricity markets; and applying reliability rules developed by a regional council.[240]

194. By contrast, the California ISO (Cal ISO or CAISO) was formed by government action as that state began electricity restructuring in the mid-1990s. At that time, the California legislature adopted a restructuring bill that required creation of the California ISO to run the grid and a California Power Exchange (Cal PX), with the intent of developing market-based prices for electricity. The three major utilities in the state thus applied to FERC to create an ISO by transferring operational control of transmission facilities while retaining ownership of them. Pursuant to that proposal, the ISO became responsible for: (1) administering nondiscriminatory tariffs for access to transmission facilities within its control; (2) directing operation of all facilities in the grid that affect system reliability; and (3) approving requests for the removal of transmission equipment from service, as well as for returning facilities to service when necessary.[241]

195. Although each RTO and ISO is unique, they also share similar characteristics. They are nonprofit organizations with governance rules that oversee management and operation. RTOs typically count as their members traditional IOUs, merchant generators, electricity marketers, and other electricity providers. ISOs and RTOs generally perform three key functions. First, they physically run, operate, and help plan the transmission grid within their areas. IOUs continue to own the lines that they built, but the RTO or ISO handles day-to-day system operations. Second, these organizations create formalized markets for energy and ancillary services and coordinate power sales among the parties, including both day-ahead and spot markets. Third, they assume responsibility for ensuring grid reliability in the short-run.

---

237. Pa.-N.J.-Md. Interconnection, 81 F.E.R.C. ¶ 61,257 (1997), *order on reh'g & clarification*, 92 F.E.R.C. ¶ 61,282 (2000).
238. Yuffee & Ripley, *supra* note 230, at § 89.04[2][a].
239. Order Granting PJM RTO Status, PJM Interconnection, L.L.C., 101 F.E.R.C. ¶ 61,345 (2002).
240. For a detailed discussion of PJM operations *see* Yuffee & Ripley, *supra* note 230, at § 89.04[2][b]–[e].
241. Pacific Gas & Elec., Co., 77 F.E.R.C. ¶ 61,077 (1996); Pacific Gas & Elec., Co., 80 F.E.R.C. ¶ 61,128 (1999) and 81 F.E.R.C. ¶ 61,122 (1997). For a detailed discussion of CASIO operations, see Yuffee & Ripley, *supra* note 230, at § 89.04[3][a]-[e](2013).

"The RTO/ISO acts as the traffic controller, determining which generators have access to the transmission grid, on what terms, and which load is served over what lines."[242] Thus, these organizations are responsible for ensuring that the grid is operated in an open and reliable manner with fair, nondiscriminatory rates.

196. Today, RTOs manage about 60% of electric power supply in the United States. Including PJM and the California ISO already mentioned, there are seven ISOs and RTOs now operating in the United States. The additional entities are: (1) the New England ISO (ISO-NE), which grew out of a tight power pool known as the New England Power Pool; (2) the Midcontinent ISO (MISO, formerly Midwest ISO), which is actually a not-for-profit RTO responsible for maintaining transmission in fifteen states and Canada; (3) the New York ISO (NYISO); (4) the ERCOT ISO, which manages much of the power grid in Texas; and (5) the Southwest Power Pool (SPP), which operates as an RTO in a large geographic expanse.[243]

197. RTOs and ISOs file both their transmission tariffs and their market rules with FERC as rate schedules, giving FERC authority to oversee their operations and RTO participants a chance to object that the rules are not just and reasonable. Initially, some RTOs sought to use these filings to effectively make their existence permanent, by limiting participant utilities' ability to withdraw once they joined. The courts, however, ruled this impermissible under the FPA.[244] Accordingly, FERC does not have the authority to force a utility to join an RTO, nor does it have the authority to prohibit a utility from leaving one.

198. Although RTOs have formalized electricity markets and helped add many efficiencies to the industry, they are not without their problems. One particular problem that has repeatedly arisen is cost allocation: determining who among the RTO's members pays for new facilities and services. While each case differs, these disputes are driven by a general principle that costs should be paid by those who benefit—what courts refer to as "cost causation." The cost causation principle is a reasonable one, although sometimes difficult to apply to specific facts. In recent years, a number of cost allocation battles have arisen as RTOs seek to reduce transmission congestion, optimize operations, and integrate renewable resources into the grid by building large, high-capacity transmission lines. Indicative of the problems these cases raise—determining whether such lines really benefit particular customers when, in fact, they always benefit the system as a whole, albeit to different degrees—the courts grappling with them have reached differing results, and often have provoked spirited dissents in doing so.[245]

---

242. Yuffee & Ripley, *supra* note 230, at § 89.01.

243. For a more detailed discussion of these regional organizations, *see Ibid* at § 89.04[4] to [8][a]–[e].

244. Atlantic City Elec. Co. v. FERC, 329 F.3d 856 (D.C. Cir. 2003).

245. *See, e.g.*, Illinois Commerce Comm'n. v. FERC, 576 F.3d 470 (7th Cir. 2009) (ICC I); Illinois Commerce Comm'n. v. FERC, 721 F.3d 764 (7th Cir. 2013) (ICC II); Illinois Commerce Comm'n. v. FERC, 756 F.3d 556 (7th Cir. 2014) (ICC III); Midwest ISO Transmission Owners v. FERC, 373 F.3d 1361 (D.C. Cir. 2004).

§4. Additional Transmission Issues: Reliability, Modernization, and Planning

199. Without question, the primary change in transmission regulation over the past several decades in the United States was to open access to the grid and then foster greater coordination and competition through ISOs and RTOs. Nevertheless, several other issues have received substantial attention from regulators in recent decades, including Congress and FERC. These include efforts to address grid reliability; to expand transmission capacity and modernize the grid; and to facilitate better planning and to circumvent state and local barriers to grid expansion, including NIMBY (not in my backyard) resistance to transmission line siting.

200. Grid reliability was thrust back into the national spotlight on August 14, 2003, when the Northeast United States and portions of Canada experienced an extensive blackout due to a transmission line failure. This massive, cascading blackout affected an estimated 50 million people and imposed economic costs somewhere between USD 4 billion and USD 10 billion.[246] This event focused attention keenly on the need for facilities upgrades and on the need for greater transmission planning. Without adequate planning, the grid can lack reliability, perform inefficiently, and fail to fully integrate new generation capacity.

201. In response to that blackout, Congress, as part of EPAct 2005, ordered FERC to establish an electric reliability organization with oversight authority to ensure that utilities maintain their systems with adequate reliability. FERC then designated the North American Electric Reliability Corporation (NERC) as the ERO. Following that designation, NERC created regional EROs that implement NERC reliability standards.[247] Those standards govern a wide array of system operations and maintenance issues, including interconnections and operating reserves, system resource balancing, tree trimming and line clearances, sabotage reporting, and grid safety.

202. In addition to seeking greater grid reliability, EPAct 2005 also sought to cut through much of the entrenched local resistance that has hampered the nation's ability to upgrade and expand its now aging transmission system. To this end, the Act provided FERC with more, albeit limited, siting authority for transmission lines. Specifically, under the statute, DOE was charged with identifying national priority areas for transmission development known as "national interest electric transmission corridors." If a state failed to approve a proposed transmission line within a particular period of time in an NIETC, FERC then had authority under the law to step in and site the line itself.

---

246. U.S.-Canada Power System Outage Task Force, Final Report on the August 14, 2003 Blackout in the United States and Canada: Causes and Recommendations (Apr. 2004).
247. *See* 18 C.F.R. § 40.3.

203. This was what became known as FERC's transmission line siting "backstop authority," and it gave many in the country great hope that a new era of transmission expansion would soon emerge. Quickly, however, two U.S. Court of Appeals decisions sapped the force of these hopes. The U.S. Court of Appeals for the Ninth Circuit ruled that DOE could not designate NIETCs without engaging in extensive consultation with the states, thus constraining that process.[248] And the Fourth Circuit ruled that FERC's backstop authority extended only to states' failure to act at all on siting proposals.[249] Thus, even when states flatly deny critical transmission line siting proposals, FERC cannot step in with its "backstop" authority to ensure proper grid expansion.

204. Perhaps in part because the courts limited FERC's authority under EPAct 2005, the agency took other measures to help advance a more modern grid. To further encourage transmission construction, FERC, in Order No. 679, authorized higher rates of return on facilities that met certain qualifications.[250] Specifically, under that order, FERC created financial incentives to encourage new investment, including: (1) a return on equity at the "upper end" of zone reasonableness; (2) the ability to include 100% of construction work-in-progress in the rate base and expensing prudently incurred pre-commercial costs; (3) permission to base rates on hypothetical capital structures rather than actual data; (4) accelerated depreciation for new transmission investments; (5) full recovery of abandoned transmission projects, as long as investments were prudently incurred and the decision to abandon was not made by the developer; (6) the ability of some companies subject to rate moratoria to defer cost recovery; (7) the ability to file separate tariffs for new transmission without reopening existing tariff schedules; and (8) a higher return on equity for companies participating in either RTOs or ISOs.[251]

205. Most recently, FERC has attempted to further transmission planning on a regional basis. It began this effort in Order No. 890,[252] which required public utilities to participate in open and transparent transmission planning processes, and then expanded that initial effort in Order No. 1000.[253] Order No. 1000 mandated regional planning and coordination among utilities. In service of that goal, it established three specific requirements for utilities: (1) participation in a regional transmission planning process that satisfies the requirements of Order No 890; (2) implementation of a process for identifying transmission needs and proposed solutions to those needs; and (3) coordination with public utility transmission providers in neighboring transmission planning regions to determine cost-effectiveness. In 2014,

248. California Wilderness Coal. v. U.S. Dep't of Energy, 631 F.3d 1072 (9th Cir. 2011).

249. Piedmont Envtl. Council v. FERC, 558 F.3d 304 (4th Cir. 2009).

250. *See* Promoting Transmission Investment Through Pricing Reform, Order No. 679-A, 72 Fed. Reg. 1152 (Jan. 10, 2007), FERC Stats. & Regs. ¶ 31,236 (2007).

251. Mary Anne Sullivan, *The Many Challenges of the "Full Portfolio" Approach: Utilities Prepare for Climate Change Regulation*, 2008 No. 3 RMMLF-Inst. Paper No. 15.

252. *Final Rule: Preventing Undue Discrimination and Preference in Transmission Service*, F.E.R.C. Docket Nos. RM05-25-000, RM05-17-000 (2007).

253. Transmission Planning and Cost Allocation by Transmission Owning and Operating Public Utilities, 136 F.E.R.C. ¶ 61,051 (2011).

the U.S. Court of Appeals for the District of Columbia largely upheld Order No. 1000.[254]

206. FERC's foray into transmission planning and implementation may make important strides into helping the nation's grid modernize and become more nimble as generation assets change and more renewables need to be integrated into the system. However, state and local interests remain strong, as evidenced by the court decisions limiting FERC's direct siting authority under EPAct 2005. Thus, federal authority over electricity transmission planning and siting remains incomplete and continues to be shared with state regulators. How that authority is implemented—and how it may shift—in the years to come thus remains a critical question for the future of the U.S. transmission grid.

254. South Carolina Pub. Serv. Auth. v. FERC, 726 F.3d 41 (D.C. Cir. 2014).

Chapter 4. Distribution

207. The significant developments in the transmission sector over the last several decades were not unaccompanied by change in the distribution segment of the industry. As open access initiatives took force and as electricity markets appeared to be more competitive, state regulators also began to restructure their power markets.

208. Many state regulators began requiring utilities to unbundle generation, that is, either to divest it to unrelated entities or to separate their transmission-owning arms and generation-owning arms into different corporate entities. A number of states also sought to replicate what was occurring in wholesale markets at the retail level, implementing a wide variety of approaches to promote retail electricity competition. State efforts, however, were uneven, and they were largely brought to a close as a result of what is known as the California electricity crisis.

209. In 2000, state regulators in California attempted to deregulate wholesale and retail electricity markets within their jurisdiction. The regulatory design for doing so, however, proved to be disastrous. Regulators set a ceiling on the *retail* rates that utilities could charge consumers and, simultaneously, imposed no ceiling on prices that producers paid in *wholesale* electricity markets. As a result of this design, and the failure of regulators to adequately monitor markets, gross market manipulation ensued. Thus, electricity shortages broke out, California became subject to rolling brownouts and blackouts, and the financial stability of local utilities was threatened.[255] This sent chills down the spines of policymakers nationwide, and many states that were considering adopting retail competition abandoned their discussions. Others that had already started backed away or even reversed course. Indeed, "Texas was the only state to proceed with its retail market in 2002—no other state has bothered to introduce retail competition or restructure its electricity market over the past decade, and few are contemplating to do so."[256]

210. A further consequence of the California debacle was that regulation of the U.S. electricity industry became even more fragmented than it was before. Now, not only was regulatory authority split between states and the federal government, a mishmash of state regulatory systems emerged, with some utilities either functionally unbundled or having divested generation assets, sometimes operating in competitive markets such as those run by RTOs and sometimes not, and others remaining subject to traditional regulation, including cost-of-service ratemaking. Today, some states are known as "regulated," while others are known as "non-regulated," jurisdictions. Therefore, state regulators had to figure out how to regulate traditional, still-vertically integrated IOUs, plus unbundled firms, and firms that were either a part, or not, of regional transmission arrangements.

---

255. *See, e.g.*, Tomain & Cudahy, *supra* note 116, at 408–420.
256. *U.S. Retail Competition Is Alive, and Seemingly Managing Well*, 26 Electricity J. 2, 5 (2013).

211. One remaining division at the distribution level are states that have moved to retail competition and those that have not. Regulation of the distribution grid in the United States, as noted, is performed by the states, whereas the federal government regulates access to and rates on the interstate transmission grid. Nevertheless, principles governing distribution regulation and transmission regulation historically were quite aligned. State regulators used cost-of-service ratemaking to determine electricity prices, which included distribution service, just as FERC used cost-of-service ratemaking to determine the justness and reasonableness of power purchase agreements, which also often included transmission service. State regulators applied a prudence test to investments in distribution facilities, just as FERC allowed cost recovery only for used and useful facilities in its rates. And, state regulators allowed utilities to maintain exclusive control over their distribution facilities in their service territory; would-be competitors could not demand access simply to enter the market. FERC treated transmission lines the same way.

212. As FERC opened up transmission lines to promote wholesale competition, however, states that decided to move to retail competition also opened up distribution line service to promote competition, or guaranteed uniform distribution costs to level the playing field for new electricity providers. States that moved to retail competition also had to determine what to do if customers did not engage in the regulatory experiment. The standard response was to designate a "provider of last resort," or "POLR," that would supply electricity if a customer did not switch or switched to a new competitor that failed. Inevitably, POLRs were the incumbent utilities that had provided electricity locally all along.

213. States, of course, adopted different forms of retail competition, and these differences were wide-ranging. They spanned from reverse auctions of "tranches" of demand in New Jersey to full-throated competition between suppliers in Texas. Texas, in fact, pushed competition particularly hard by ensuring a legislatively-set "price-to-beat," which was instituted at 6% lower than the price in effect on January 1, 1999, adjusted to reflect fuel costs as approved by the Texas PUC. POLRs then had to provide that discounted cost-of-service rate until they lost 40% of their customers, at which time the price-to-beat was lifted and they could compete with other entrants in the market.[257]

214. The policy design choices that states made in deciding to move to retail competition mattered. Somewhere between 16 and 20 states moved to retail competition, but their success has been widely divergent. In general, retail competition has not reshaped the distribution of electricity in the United States anywhere near the way that the interlinking moves to wholesale competition and open access transmission have changed those segments of the industry. According to one recent study, the leading states for retail competition in the United States, as measured by the number of residential customers who have switched providers, are: Texas (100%);

257. Reliant Energy, Inc. v. Pub. Util. Comm'n of Texas, 62 S.W.3d 833 (Tex. Ct. App. 2001); *see also* Phillip G. Oldham & Joseph P. Younger, *Lighting the Lone Star: The Texas Experience with a Competitive Electricity Market*, 40 WAKE FOREST L. REV. 709 (2005).

Connecticut (44.1%); Ohio (42.2%); Pennsylvania (31.5%); New York (22.7%); Illinois (22.4%); Maryland (22.1%); Maine (21.6%); Massachusetts (14.3%); New Jersey (14.3%); and the District of Columbia (12.3%).[258]

215. Retail switching, however, is not the only measure of distribution-level competition's success. In many such states, "[c]ommercial and industrial (C & I) customers have more choices and options, [though] in several states large industrial customers have become increasingly dissatisfied with retail market prices."[259] Moreover, while additional states have not decided to move to distribution-level competition since the California energy crisis and the accompanying Enron debacle, those states that have maintained their "restructured retail electricity markets are advancing and making progress," and "[n]one of the states is moving back toward reregulation or a reversal of electricity restructuring."[260] Likewise, a number of states—including Connecticut, Delaware, Illinois, Maine, Maryland, Massachusetts, New Jersey, New York, Ohio, Pennsylvania, Texas, and the District of Columbia—have been deemed to have "workable" competition in the C & I sector, meaning that they have "relatively large percentages of C & I consumers who have shopped for power and switched to a competitive supplier."[261] And, Connecticut, Illinois, Maine, Maryland, Massachusetts, New York, Ohio, Pennsylvania, and Texas all might be deemed to "have achieved significant competitive activity in the residential sector."[262] Still, in many jurisdictions with distribution-level competition, "prices have not decreased as expected, and the range of new options and services is often limited."[263]

216. Thus, the status of competition at the distribution level in the United States remains somewhat paradoxical. It is a line that continues to divide states from each other, and that has made the operation and delivery of electricity in the nation more complex. It mirrors, in some ways, the experience of utilities on the wholesale sale and transmission segments of the market, but it diverges sharply from that story in others. Its success is quite mixed, working exceedingly well in some states, like Texas, and barely moving the needle in others. And, of course, it significantly affects

258. Distributed Energy Financial Group LLC, 2012 ABACCUS: An Assessment of Restructured Electricity Markets 7, *available at* http://defgllc.com/publication/2012–abaccus-electricity-restructuring-scorecard/.

259. The Electric Energy Market Competition Task Force, Report to Congress on Competition in Wholesale and Retail Markets for Electricity: Pursuant to Section 1815 of the Energy Policy Act of 2005, at 6–7 (2007), *available at* http://www.ferc.gov/legal/fed-sta/ene-pol-act/epact-final-rpt.pdf.

260. Distributed Energy Financial Group LLC, 2012 ABACCUS: An Assessment of Restructured Electricity Markets 7, *available at* http://defgllc.com/publication/2012–abaccus-electricity-restructuring-scorecard/.

261. *Ibid.*

262. *Ibid.*

263. The Electric Energy Market Competition Task Force, *supra* note 259, at 6–7. *But see* Adam Swadley & Mine Yücel, Federal Reserve Bank of Dallas, Did Residential Electricity Rates Fall After Retail Competition? A Dynamic Panel Analysis 1 (2011) ("The effects of a competitive retail electricity market are mixed across states, but generally appear to lower prices in states with high participation rates.").

not just how the distribution grid is operated but also how end-users ultimately consume power as well. These changes, then, combined with innovations that regulators have begun to make to ratemaking in the United States, say much about how electricity is used in the nation today.

Chapter 5. Consumption

217. The key phrase today in U.S. electricity consumption regulation is efficiency. For most of the twentieth century, U.S. energy law promoted electricity consumption as a way to foster economic expansion and growth. By the 1970s, however, three developments caused reevaluation of the extant regulatory system promoting consumption: the energy crises, rising electricity costs, and environmental concerns. These concerns soon manifested in legislation that promoted conservation and new methods of electric industry regulation at the consumptive level. Today, efforts to increase efficiency in domestic electricity consumption arise in several respects: first, through demand response; second, via development of the smart grid; third, through use of distributed generation; and, finally, through integration of electric vehicles, which promotes overall energy efficiency even if it does not drive down electricity consumption itself.

218. As discussed above, PURPA first promoted energy conservation and efficiency by encouraging state PUCs to move away from rate structures that increased demand. That statute also created a market for non-utility generators, with the intent of generally increasing energy efficiency. Subsequent laws and actions by FERC to promote competition also aimed to elevate efficiency, and by the end of the twentieth century, the shape of the industry had measurably changed. As that century turned, EPAct 2005 further encouraged the development of non-utility generation and also promoted energy efficiency and the development of renewables, as discussed in greater length in Part V on the *Interaction between Energy Law and Environmental Law*.

219. Energy conservation and efficiency are also seen as increasingly important elements to a comprehensive energy portfolio. Conservation can be considered simply as consuming less energy. Energy efficiency, by contrast, has two dimensions. First, physical resource efficiency can be measured as an increase in the amount of electricity produced for every unit of input from a natural resource, such as coal or uranium. Second, policy efficiency might be measured by the degree to which policy choices, such as rate design decisions, promote higher uses of resources or lower consumption of electricity, such as by providing information for customers to make more informed consumption choices.

§1. Demand Response

220. One key, if broad, approach to both energy conservation and efficiency is known as "demand response" or "DR." According to the U.S. Department of Energy, demand response programs provide opportunities for both consumers and producers. Consumers can reduce or shift their electricity usage during peak periods in response to time-based rates or other forms of financial incentives. Utilities, in turn, can use DR programming as a way to better balance supply and demand and, in some instances, reduce investments in new plants. For instance, utilities in some states have been allowed to invest in rebates for consumers to buy high efficiency lightbulbs in lieu of building new generation facilities.

221. From the consumers' end, DR programs are effectuated through rates designs as well as through the use of smart appliances. From the utilities' end, DR programs can be a way for power companies to cycle electricity on and off during periods of peak demand in exchange for financial incentives and lower electricity bills. Further, DR programs can be used to enhance grid modernization through advanced metering infrastructure, such as through a more extensive use of sensors that can identify peak load problems and either divert or reduce power in strategic locations, thus reducing the risk of power failure.[264] Some utilities in the United States have done this at the consumption level by, for instance, entering into agreements with large industrial customers to have them shut down during hours of high system demand, or by placing devices on residential customers' air conditioning units allowing the utility to shut them off for a portion of an hour during high demand periods.

§2. Smart Grid

222. Another key efficiency effort in the United States is the smart grid. In a real sense, the developments in transmission and distribution have coalesced into a need for grid modernization in the United States. In popular parlance, this is what has become known as the "smart grid." Many policymakers insist that development of the smart grid is essential. And, slowly, in part through innovation in both federal and state regulation, it is beginning to emerge.

223. The smart grid is a suite of technologies both in front of and behind the meter. From the utility side of the meter, technologies are introduced for the purpose of sending more accurate price signals on a seasonal, weekly, daily, hourly, or even more frequent basis. This in-front-of-the-meter technology is often referred to as an advanced metering infrastructure (AMI). At the consumers' end, the technology consists of smart meters, programmable appliances, and other energy-saving devices. Therefore, by using the smart grid, utilities can, in addition to selling electricity, enter the business of selling *energy efficiency* through demand response programs and the like.

224. Several benefits are attributed to the adoption of the smart grid. A smart grid should: (1) provide more efficient transmission of electricity; (2) improve grid reliability; (3) reduce the threat of cybersecurity; (4) reduce operation and management costs for utilities; (5) help reduce demand and, therefore, lower electricity costs; and (6) improve the integration of renewable resources into the system. All of

264. Energy.gov, *Demand Response*, *available at* http://energy.gov/oe/technology-development/smart-grid/demand-response (last visited Apr. 15, 2015).

these benefits, however, require changes in the business model for traditionally structured utilities, including the regulations used to oversee them.[265]

225. At the federal level, legislation already has been adopted to support development of the smart grid. The Energy Policy Act of 2005,[266] for example, directed the Department of Energy and FERC to conduct assessments for demand response potential, including on smart meters, and to submit reports to Congress with their findings. That Act also required state PUCs to determine whether utilities and other electricity providers should be required to offer time-based rates and for utilities to provide smart meters to customers requesting them. Additionally, the Act required that advanced meters, to the extent possible, be used in all federal buildings.

226. The Energy Independence and Security Act of 2007 also promoted development of the smart grid.[267] Section 1301 of that Act declared it national policy to modernize the country's electricity transmission and distribution systems. Among the goals set out in that statute are: (1) increasing use of digital information and control technology; (2) deploying and integrating distributed resources, including renewables; (3) reducing greenhouse gas emissions; (4) developing and incorporating demand response programs into the grid; (5) integrating smart appliances and consumer devices; (6) developing and integrating advanced electric storage systems, including plug-in and hybrid electric vehicles; and (7) developing interoperability standards for appliances and equipment connected to the grid. Additionally, the Act created a federal Smart Grid Task Force[268] and authorized an allocation of USD 100 million per year for five years for regional smart grid demonstration projects.

227. Development of interoperability standards is essential for the smooth functioning of the smart grid. In the EISA, the National Institute of Standards and Technologies,[269] together with DOE and others, was assigned the responsibility for developing a smart grid interoperability framework. Today, the standards are still under development. Nevertheless, work continues with support from FERC[270] and the White House.[271] From the perspective of FERC, in addition to the need for interoperability standards[272] and cybersecurity standards, the development of the

265. *See, e.g.*, TOMAIN, *supra* note 21, at ch. 6.
266. Pub. L. No. 101-575.
267. Pub. L. No. 110-140.
268. *See, e.g.*, Energy.gov, *Federal Smart Grid Task Force*, *available at* http://energy.gov/oe/technology-development/smart-grid/federal-smart-grid-task-force (last visited Apr. 14, 2015).
269. *See* NATIONAL INSTITUTE OF STANDARDS AND TECHNOLOGIES, NIST FRAMEWORK AND ROADMAP FOR SMART GRID INTEROPERABILITY STANDARDS, RELEASE 2.0 (Feb. 2012).
270. *See* Smart Grid Interoperability Standards, 136 F.E.R.C. ¶ 61,039 (2011).
271. See EXECUTIVE OFFICE OF THE PRESIDENT OF THE UNITED STATES, A POLICY FRAMEWORK FOR THE 21ST CENTURY GRID: ENABLING OUR SECURE ENERGY FUTURE (JUNE 2011); EXECUTIVE OFFICE OF THE PRESIDENT OF THE UNITED STATES: A POLICY FRAMEWORK FOR THE 21ST CENTURY GRID: A PROGRESS REPORT (FEB. 2013).
272. Smart Grid Policy, 74 Fed. Reg. 37,098 (July 27, 2009).

smart grid also requires attention be paid to demand response[273] and cost allocation.[274]

228. The American Recovery and Reinvestment Act of 2009 (ARRA) also provided substantial funding for smart grid development and deployment. Under that Act, Congress dedicated USD 4.5 billion in funding to smart grid development, including installation of smart meters, pilot projects, and projects intended to integrate different technologies into a larger system. These federal investments were leveraged by the private sector, which invested USD 1 billion for demonstration projects, matched by USD 620 million from the federal government.[275]

§3. Distributed Generation

229. A third change emerging at the consumption end of the electricity system to promote efficiency is growth in the use of distributed generation. Quite simply, DG consists of a range of small-scale technologies and devices that are designed to provide electricity at locations close to consumers. DG generates electricity from both fossil fuels as well as renewable resources. The array of technologies is broad and includes solar energy, combined heat and power, micro-turbines, industrial gas turbines, micro-grids, wind turbines, and fuel cells, among other technologies.

230. Proponents of DG note that it offers many benefits to the electricity system. Because it is closer to end-users, transmission costs—and losses—are significantly reduced. Because DG is produced by a variety of actors, competition in the electric market is increased, and therefore, price competition increases as well. This increase in electricity providers should also help relieve congestion on the grid and increase local grid reliability. Further, a broader array of electricity providers reduces the amount of possible damages as a result of a breach in cybersecurity. Additionally, the expansion of the electricity market should lower pressure on traditional IOUs to invest in more plant and transmission facilities.

231. Despite the promise of such benefits, DG also raises concerns. The predominant one is expressed by utilities, who are aware of the consequent loss of electricity sales on their books. In part, the expansion of DG puts pressure on utilities to reshape their business models in ways that differ starkly from practices of the past. Further, the expansion of DG puts pressure on regulators to monitor electricity markets to ensure that all customers are served fairly, and that some customer classes are not put in a position of subsidizing DG users.[276]

---

273. Policy Statement, Smart Grid Policy, 128 F.E.R.C. ¶ 61,060 (2009).

274. Transmission Planning and Cost Allocation by Transmission Owning and Operating Public Utilities, 136 F.E.R.C. ¶ 61,051 (2011).

275. *See generally* U.S. Dept. of Energy, Energy.Gov, Office of Electricity Delivery and Energy Reliability, http://energy.gov/oe/technology-development/smart-grid.

276. For an analysis of the relationship between DG and IOUs, *compare* Elisabeth Graffey & Steven Kihm, *Does Disruptive Competition Mean a Death Spiral for Electric Utilities?*, 35 Energy L.J. 1

232. Because DG works at the consumption level, FERC has disclaimed jurisdiction over it, leaving regulation to the states.[277] The extent to which any one state promotes DG, then, of course, differs widely from state to state. Typically, states do this through use of what is called a "net metering" policy—laws that allow consumers who locate DG on their properties to deduct from their monthly bills the amount of electricity consumption they reduce via distributed generation.

233. Currently, 44 states plus the District of Columbia have some kind of net-metering rules. The only exceptions are Alabama, Mississippi, South Dakota, and Tennessee, plus Idaho and Texas, which do not have statewide rules but include some utilities that allow net-metering. While net-metering can promote DG, it still presents critical issues for regulators. These include how to set rates that both promote DG and also ensure maintenance of the grid and utility solvency, since DG naturally reduces utility revenues. Regulators must also be attuned to the need not to overly subsidize DG users by letting them "free ride" on others for the cost of maintaining the grid, which is shifted more heavily to non-DG users when customers with DG start paying lower bills.[278] For this reason, some regulators and utilities have begun imposing standalone fees to be imposed on DG users for their use of the grid.[279]

§4. Integration of Electric Vehicles

234. Finally, electric vehicles promise to be the most disruptive technology in the energy sector. Today, very little electricity is used in the transportation sector and, correspondingly, very little oil is used to generate electricity. Electric vehicles have the potential to change that situation by breaching the divide between electricity and transportation. Additionally, electric vehicles offer the potential for further weaning the United States from its dependence on oil, reducing greenhouse gas emissions, reducing fuel costs of driving, and stabilizing the electric grid.

235. At this point, the penetration of electric vehicles into the market is relatively small but is increasing significantly. Several challenges remain to greater market penetration. First, consumers are concerned about "range anxiety," that is, how far car will travel on one electric charge. Additionally, "refueling" (recharging) presents another challenge to expanding that market. Finally, cost is a factor. This is, in

(2014), *with*, David Raskin, *Getting Distributed Generation Right: A Response to "Does Disruptive Competition Mean a Death Spiral for Electric Utilities?,"* 35 Energy L.J. 262 (2014). *See generally* Joseph P. Tomain, *Traditionally-Structured Electric Utilities in a Distributed Generation World*, 38 Nova L. Rev. 473 (2014).

277. *See* MidAmercian Energy, 94 FERC 61,340 (2001); Sun Edison LLC, 129 FERC 61,146 (2009). *But cf.* So. Cal Edison v. FERC, 603 F.3d 996 (D.C. Cir. 2010) (rejecting FERC's monthly billing theory used to disclaim jurisdiction); Calpine Corp. v. FERC, 702 F.3d 41 (D.C. Cir. 2012).

278. *See, e.g.*, David B. Raskin, *The Regulatory Challenge of Distributed Generation*, 4 Harv. Bus. L. Rev. Online 38 (2013).

279. *See, e.g.*, In re Arizona Public Service Co. Application of Net Metering Cost Shift Solution, 2013 WL 6384419 (Dec. 2, 2013).

part, because most electric vehicles cost more than combustion engine vehicles, and also because the cost and weight of the batteries used in electric vehicles is high. Despite these challenges, electric vehicles also have much potential, particularly with respect to the smart grid. To the extent that the grid develops its two-way capacity for information and energy sales, electric vehicles, particularly when they are recharging at off-peak times, might effectively serve as storage devices (or batteries) to the grid.

236. Two issues arise regarding government support for the electric vehicle market. First, under the American Recovery and Reinvestment Act of 2009,[280] tax credits were available for persons who purchase new electric vehicles. The size of the credit was based upon the capacity of the vehicle's battery measured in terms of kWh; the credit began at USD 2,500, with a maximum of USD 7,500. Thus, there is an issue of the extent to which the government will choose to promote development of this market. Second, as the electric vehicle market grows, it inevitably will present costs to the system. Chief among these is the need for an electric vehicle recharging infrastructure. Thus, state regulators will be faced with questions of whether their governments should invest in this infrastructure, as well as whether to allow for its cost recovery if local utilities want to participate in building it out. Currently, state PUCs have supported utility investments in pilot projects for recharging investments, and states have supported investments by consumers who have prepared their homes for charging electric vehicles with favorable rate treatment.[281]

280. American Recovery and Reinvestment Act of 2009, Pub. L. No.111-5.

281. *See, e.g., Central Maine Power Co. Proposed Elec. Vehicle Pilot Project*, 2013 WL 4433468 (Me. P.U.C., Aug. 5, 2013) (utility may recover investments in a pilot project involving promotional projects to purchase electric vehicles); *Re Vehicle Charging Stations and Related Svcs.*,306 P.U.R 4th 331 (Md. P.S.C., Aug. 12, 2013) (utility investment in pilot project for recharging will receive rate treatment); *Re Alternative –Fuels Vehicle Tariffs*, 2013 WL 34642351 (Cal. P.U.C. June 27, 2013) (rate treatment was afforded utilities for investments in residential service facility upgrades in preparation for electric vehicles).

Part II. Regulatory Framework Concerning Natural Gas

237. The chief ingredient of natural gas is methane, a compound composed of one carbon atom and four hydrogen atoms. Methane is contained in the remains of plant and animal life that have built up in layers over millions of years. Over time, natural gas migrates into surrounding geologic formations. Thus, natural gas is found in various onshore and offshore formations. Natural gas located in sandy rock layers is relatively easy to extract. However, gas deposited in shale formations, which have low permeability, is more difficult to extract. This gas, which is now known as shale gas, is the target of modern hydraulic fracturing operations.

Chapter 1. Production

238. The United States consumes approximately 25 trillion cubic feet (Tcf)[282] of natural gas annually, which constitutes 25% of U.S. energy consumption. The U.S. Energy Information Administration (EIA) projects that natural gas consumption will increase to 31.6 Tcf by 2040.[283] Most gas consumed in the United States is produced domestically, and U.S. exports are starting to increase.

239. Domestic consumption occurs in all sectors of the economy. In the residential and commercial sectors, it is used mostly for heating and cooking. In the industrial sector, natural gas is used to produce steel, glass, paper, chemicals, and other products. Natural gas also fuels a slight portion of the transportation sector in the United States. In recent years, due to the discovery and production of low-cost gas from shale formations, natural gas has begun to replace coal in electricity production, and its share of fuel for electricity production is increasing. Today, natural gas accounts for approximately 33% of the energy used to generate electricity in the United States.[284]

240. Natural gas in the United States is extracted from three primary sources. First, much natural gas historically was acquired from "associated" wells—that is, wells that produce oil and natural gas together. Second, and increasingly, natural gas is acquired from independent wells dedicated to natural gas production. Third, natural gas is acquired from coal beds; this type of gas is known as "coal bed methane." Perhaps, however, the most significant recent change in the production of U.S. natural gas is the combination of two technologies: directional drilling and hydraulic fracturing. Today, these production techniques are rapidly changing the United States energy landscape by making natural gas trapped in shale formations accessible.

241. The discovery of domestic shale gas has been labeled a "game changer" for the United States.[285] In no small part, shale gas, together with burgeoning domestic oil production from shale formations, have reduced U.S. gas and petroleum imports, thus strengthening U.S. energy independence—a longstanding pillar of U.S. energy policy. Additionally, increased production from shale gas extraction projects, or "plays," has driven down the price of natural gas substantially. Low-price shale gas,

---

282. A Tcf is enough natural gas to heat 15 million homes for one year, or to generate 100,000,000,000 kWh of electricity.

283. U.S. Energy Info. Admin., Annual Energy Outlook 2014 with Projections to 2040 MT-21 (Apr. 2014), *available at* http://www.eia.gov/forecasts/aeo/pdf/0383(2014).pdf [hereinafter 2014 Energy Outlook].

284. U.S. Energy Info. Admin., Monthly Energy Review 4 (Jan. 2015), *available at* http://www.eia.gov/totalenergy/data/monthly/archive/00351501.pdf; U.S. Energy Info. Admin., *Natural Gas Explained*, http://www.eia.gov/energyexplained/index.cfm?page=natural_gas_home (last updated Apr. 11, 2014) [hereinafter *Natural Gas Explained*].

285. *See* Int'l Energy Agency, *supra* note 6, at 1; Alan J. Krupnick et al., Resources for the Future, The Natural Gas Revolution: Critical Questions for a Sustainable Energy Future 1 (Mar. 2014), *available at* http://www.rff.org/RFF/Documents/RFF-Rpt-NaturalGasRevolution.pdf.

together with natural gas' lower emissions profile, in turn has helped gas increasingly replace coal as the fuel of choice for electricity generation in the United States.

242. As with other direct regulation of oil and gas extraction,[286] shale gas regulation is largely a matter of state rather than federal law. States, for example, regulate the location and spacing of wells, impose setback requirements, regulate drilling methods and casing requirements, oversee the plugging of wells, and control the disposal of oil and gas wastes and site restoration.[287] These numerous state laws regarding shale gas derive from late nineteenth and early twentieth century oil and gas laws and have been updated and modified to accommodate the changing technology and boom in gas development. Because the laws are state-based, however, they necessarily vary from jurisdiction to jurisdiction. This causes some complications, particularly where states adapt their old laws in different ways,[288] or rely on courts to do so. It also has presented some difficulties in states such as New York, Pennsylvania, and Ohio, which have rich shale gas plays but that have not needed to heavily regulate oil and gas for decades. In other words, some state regulators are not familiar with regulatory regimes over this new natural resource, even though best practices appear to be developing.

243. Shale gas is simply natural gas found in particular shale formations. Those formations formed 300-400 million years ago and are constituted by the organic remains. Natural gas that seeped into sandy rock layers is relatively easy to extract. However, natural gas that seeped into shale formations with low permeability is more challenging to extract because those formations have to be broken up, or "fractured," to release the trapped natural gas.

244. While shale gas plays have been found throughout the United States, some of the largest plays to date are in the Northeast (the Marcellus Shale), Texas (the Barnett Shale and the Haynesville Shale), and North Dakota (the Bakken Shale). Currently, shale gas production is taking place in sixteen states. According to the Energy Information Administration, the United States has an abundance of natural gas—an abundance that is expected to continue for the foreseeable future. In 2012, the EIA estimated that there are between 482 trillion cubic feet of technically recoverable shale gas resources and an estimated 542 Tcf of proved and unproved resources in the United States, constituting 25% of the U.S. total of 2,203 Tcf.[289]

245. Over the last decade, shale gas production has increased more than tenfold, a trend that is expected to continue. Production is expected to rise from 9.7 Tcf in

---

286. *See infra* Ch. 2.

287. *See* Nathan Richardson et al., Resources for the Future, The State of State Shale Gas Regulation 5 (June 2013).

288. *Compare, e.g.*, Coastal Oil & Gas Corp. v. Garza Energy Trust, 268 S.W.3d 1 (Tex. 2008), *with*, Stone v. Chesapeake Appalachia, LLC, 2013 WL 2097397 (N.D. W. Va. 2013), *vacated*, 2013 WL 7863861 (N.D. W. Va. 2013).

289. U.S. Energy Info. Admin., Annual Energy Outlook 2012 Early Release Overview 9 (Jan. 2012); U.S. Energy Info. Admin., Annual Energy Outlook 2012 with Projections to 2035 93 (June 2012), *available at* http://www.eia.gov/forecasts/aeo/pdf/0383(2012).pdf.

2012 (40% of U.S. dry gas production) to 19.8 Tcf in 2040 (53% of U.S. dry gas production).[290] According to EIA, shale gas contributes significantly to our natural gas portfolio—a portfolio that has numerous important benefits on a societal basis: (1) generating 25% of U.S. electricity; (2) providing heat to 56 million residences and businesses; (3) providing 35% of the energy and feed stocks needed by U.S. industry; (4) supplying over 2 million jobs, with shale gas alone accounting for over 600,000 jobs; and (5) generating over USD 250 billion in various revenues and taxes.[291] Gas is also often seen as the "bridge" fuel to the future for the United States, both because it helps with energy security and because its lower climate impact when burned may help provide a way to pivot away from more environmentally damaging resources, such as coal.[292]

246. Despite all these benefits from natural gas, there are several environmental issues associated with developing this resource, particularly through hydraulic fracturing. As noted, the proliferation of shale gas production is a result of two technological innovations: hydraulic fracturing and horizontal drilling.[293] Hydraulic fracturing, also sometimes known as "fracking" or "fraccing," involves pumping fluid under high pressure into a shale formation, which fractures the formation and thus allows the natural gas to be extracted. The fluid used in the fracturing process "consists primarily of water along with the proppant (usually sand) and around 2% of other chemical additives."[294] The chemical composition of these additives has been controversial and come under much scrutiny in the United States.

247. Indeed, numerous concerns have been raised about potential water pollution from hydraulic fracturing operations. First, hydraulic fracturing requires massive amounts of water, which have to be withdrawn from groundwater or surface supplies. A typical well can consume between two and four million gallons of water. Second, the hydraulic fracturing process uses various chemicals, which are mixed into the water during drilling, so any spills of these chemicals can cause environmental harms, including on the surface. Third, because of the use of chemicals in the process, concerns have been raised that hydraulic fracturing can impact drinking water resources. Fourth, wastewater must be transported, so spillage either during transport or storage can have health effects. Finally, the voluminous amounts of

---

290. 2014 ENERGY OUTLOOK, *supra* note 2, at MT-23.

291. NATURAL GAS FROM SHALE: QUESTIONS AND ANSWERS, ENERGY.GOV, *available at* http://energy.gov/sites/prod/files/2013/04/f0/why_is_shale_gas_important.pdf (last visited Mar. 25, 2015).

292. *See, e.g.*, Richard J. Pierce, Jr., *Natural Gas: A Long Bridge to a Promising Destination*, 32 UTAH ENVTL. L. REV. 245 (2012); John D. Podesta & Timothy E. Wirth, *Natural Gas: A Bridge Fuel for the 21st Century*, CENTER FOR AM. PROGRESS 3 (2009), *available at* https://cdn.americanprogress.org/wp-content/uploads/issues/2009/08/pdf/naturalgasmemo.pdf;Joel Kirkland, *Natural Gas Could Serve as 'Bridge' Fuel to a Low-Carbon Future*, SCI. AM. (June 25, 2010), *available at* http://www.scientificamerican.com/article/natural-gas-could-serve-as-bridge-fuel-to-low-carbon-future/.

293. *See generally* GROUND WATER PROTECTION COUNCIL, MODERN SHALE GAS DEVELOPMENT IN THE UNITED STATES: A PRIMER ES-1 (Apr. 2009), *available at* http://energy.gov/sites/prod/files/2013/03/f0/ShaleGasPrimer_Online_4-2009.pdf.

294. Andrea J. Chambers et al., *Shale Gas Development and Hydraulic Fracturing, in* 3 ENERGY LAW AND TRANSACTIONS, *supra* note 81, at § 76.01.

wastewater created from hydraulic fracturing must be treated and disposed of, and thus, can also impact drinking water.[295]

248. The potential for water pollution involved in the hydraulic fracturing process is significant. Much of the water that is injected into wells—anywhere from 10% to 50%—flows back to the surface, and is thus known as "flowback" waste. That waste contains both the chemicals used during the operation as well as other materials it absorbed beneath the surface. Wastewater chemicals are often toxic. They can include organic pollutants, heavy metals, and radioactive materials.[296] Moreover, there is some question about what is actually included in the water hydraulic fracturing operators inject in their operations. Industry has been reluctant to disclose the composition of these materials, arguing that their formulations need to be protected as proprietary trade secrets.

249. Nevertheless, these flowback wastewater must be properly managed and treated, which obviously is easier if the wastewater's contents are known. One option for regulation, then, is to require disclosure of the exact composition of the injection fluids. Many producers have decided to self-report both the chemical makeup of their fluids and the volumes they produce; the public website, FracFocus,[297] is an important resource in this regard. Another key question is how to dispose of the flowback water. In many regions, injection into deep geologic formations can be used to dispose of the flowback and produced water. However, in some regions, including the Marcellus Shale, deep geologic injection is not available or is limited, so the flowback water must be otherwise disposed of, which includes treatment and possible recycling.

250. The Department of the Interior has responsibility for managing operations on public and Indian lands. Under that authority, the Department has issued a proposed rule requiring companies to disclose the chemicals used in hydraulic fracturing. The disclosure would not be required until all operations have been completed.[298]

251. Currently, the federal government has limited authority over water pollution associated with oil and gas drilling. The Energy Policy Act of 2005 prohibited

---

295. *See, e.g.*, Thomas W. Merrill & David M. Schizer, *The Shale Oil and Gas Revolution, Hydraulic Fracturing, and Water Contamination: A Regulatory Strategy*, 98 MINN. L. REV. 145 (2013).

296. *See* OHIO EPA, DRILLING FOR NATURAL GAS IN THE MARCELLUS AND UTICA SHALES: ENVIRONMENTAL REGULATORY BASICS (Nov. 2012).

297. *FracFocus 3.0*, FRACFOCUS: CHEMICAL DISCLOSURE REGISTRY, https://fracfocus.org/ (last visited Mar. 26, 2015).

298. Oil And Gas; Well Stimulation, Including Hydraulic Fracturing, on Federal and Indian Lands, 77 Fed. Reg. 92 (proposed May 11, 2012) (codified at 43 C.F.R. pt. 3160); Press Release, U.S. Department of the Interior, Bureau of Land Management, Interior Releases Updated Draft Rule for Hydraulic Fracturing On Public and Indian Lands for Public Comment (May 16, 2013), *available at* http://www.blm.gov/wo/st/en/info/newsroom/2013/may/nr_05_16_2013.html; Press Release, U.S. Department of the Interior, Bureau of Land Management, BLM Extends Public Comment Period on Proposed Hydraulic Fracturing Rule (June 7, 2013), *available at* http://www.blm.gov/wo/st/en/info/newsroom/2013/june/nr_06_07_2013.html.

EPA from using the Safe Drinking Water Act[299] to regulate hydraulic fracturing operations, although this exclusion does not apply when diesel fuel is used in the fracturing process. Further, the Act amended the Clean Water Act to exclude water that is injected into wells to facilitate oil or gas production from that statute's definition of "pollutant"; it also excluded water associated with oil and gas production in states that approve the well and determine that injection or disposal will not degrade groundwater or surface waters.[300] Despite its limited authority, the EPA has initiated a rulemaking to establish standards for wastewater discharges from shale gas production.[301] Further, state regulators, environmental groups, and industry have collaborated to devise a voluntary program, using the acronym STRONGER, to develop guidelines for better management and disposal practices for oil and gas wastes.[302]

252. Notwithstanding EPAct 2005, in 2010, Congress directed EPA to study hydraulic fracturing's effects on drinking water. EPA issued a progress report on its study in December 2012.[303] The final report is expected in 2016.

253. Hydraulic fracturing can also be a significant source of air pollution. The operations themselves have air pollution effects, as they utilize diesel engines, rigs, trucks, and other equipment that produce conventional as well as greenhouse gas emissions. Arguably, more problematic from a climate change perspective are hydraulic fracturing operations themselves. These operations inevitably release gases from their wells, especially through venting and flaring. Thus, notable amounts of methane, as well as fine particulate matter and volatile organic compounds (VOCs), which can cause ground-level ozone, or smog, can be produced from hydraulic fracturing.

254. In this regard, gas leakage, especially methane leakage, may be particularly problematic. Although those concerned with climate change typically focus on emissions of carbon dioxide, "[p]ound for pound, the comparative impact of [methane] on climate change is over 20 times greater than CO_2 over a 100-year period."[304] The evidence on methane leakage from hydraulic fracturing operations remains somewhat mixed. Initially, the EPA concluded that the net environmental

299. 42 U.S.C. § 300h(d)(1) (2006).

300. 33 U.S.C. § 1362(6) (2006).

301. *See* U.S. Environmental Protection Agency, EPA Initiates Rulemaking to Set Discharge Standards for Wastewater from Shale Gas Extraction (Oct. 2011), *available at* http://water.epa.gov/scitech/wastetech/guide/upload/shalereporterfactsheet.pdf.

302. STRONGER is the State Review of Oil & Gas Environmental Regulations initiated by the U.S. EPA. *See* Regulatory Determination for Oil and Gas and Geothermal Exploration, Development and Production Wastes, 53 Fed. Reg. at 25,447 (decided July 6, 1988).

303. *See* U.S. Environmental Protection Agency, Study of the Potential Impacts of Hydraulic Fracturing on Drinking Water Resources: Progress Report (Dec. 2012), *available at* http://www2.epa.gov/sites/production/files/documents/hf-report20121214.pdf.

304. *Overview of Greenhouse Gases*, U.S. Environmental Protection Agency, http://epa.gov/climatechange/ghgemissions/gases/ch4.html (last visited Mar. 26, 2015); *see also Natural Gas—A Briefing Paper for Candidates*, Environmental Defense Fund, 5–6 (Aug. 10, 2012), *available at* http://blogs.edf.org/energyexchange/2012/08/10/natural-gas-a-briefing-paper-for-candidates/.

impact of hydraulic fracturing methane emissions is likely to be small,[305] but other studies have suggested that non-CO_2 emissions from these operations can make the lifecycle emissions of shale gas higher than those of coal. Still other studies conclude that shale gas does "not discernibly reduce the trajectory of greenhouse gas emissions" due to methane seepage during fracturing and drilling,[306] and suggest that previous studies had underestimated the amount of methane emissions.[307] In April 2012, EPA published finalized standards to reduce harmful air pollution from oil and gas production, which constitute the first federal air rules for natural gas wells that are hydraulically fractured.[308]

255. The Obama administration has issued a white paper warning of the dangers of methane leakage as part of its Climate Action Plan. The report notes that methane accounts for 9% of greenhouse gas emissions, and that, although they have decreased since 1990, methane emissions are likely to increase through 2030 if no additional action is taken.[309] In January 2015, the EPA outlined a series of steps that it plans to take to address methane and smog-forming VOC emissions from oil and gas production, including seeking to cut methane emissions 40% to 45% from 2012 levels by 2025.[310] The EPA is expected to issue proposed rules in this regard by 2016.

256. Another concern that has been raised in connection with shale gas development is community disruption, particularly in the eastern United States where oil and gas operations have not been common until recent years. Hydraulic fracturing operations can be extensive. Developing these sites requires bringing in trucks and

305. Carbon Pollution Emission Guidelines for Existing Stationary Sources: Electric Utility Generating Units; Proposed Rule, 79 Fed. Reg. 34830, 34862 (proposed June 18, 2014) (to be codified at 40 C.F.R. pt. 60).

306. Haewon McJeon et al., Research Letter, *Limited Impact on Decade-Scale Climate Change from Increased Use of Natural Gas*, 514 NATURE 482, 482 (Oct. 23, 2014); Robert W. Howarth, Renee Santoro & Anthony Ingraffea, Letter, *Methane and the Greenhouse-Gas Footprint of Natural Gas from Shale Formations*, 106 CLIMATIC CHANGE 679 (2011); *see also* Claire Moser, Nidhi Thakar & Matt Lee-Ashley, *Reducing Methane Pollution from Fossil-Fuel Production on America's Public Lands: A Needed Step to Combat Climate Change*, CENTER FOR AM. PROGRESS (Oct. 6, 2014), *available at* https://cdn.americanprogress.org/wp-content/uploads/2014/10/ReducingMethane.pdf.

307. A.R. Brandt et al., *Methane Leaks from North American Gas Systems*, 343 SCIENCE 733 (Feb. 14, 2014).

308. Press Release, U.S. Environmental Protection Agency, EPA Issues Updated, Achievable Air Pollution Standards For Oil and Natural Gas (Apr. 18, 2012), *available at* http://yosemite.epa.gov/opa/admpress.nsf/bd4379a92ceceeac8525735900400c27/c742df7944b37c50852579e400594f8f!OpenDocument (amended by Oil and Natural Gas Sector: Reconsideration of Additional Provision of New Source Performance Standards, 79 Fed. Reg. 79,018 (Final Rule) (Dec. 31, 2014) (to be codified at 40 C.F.R. pt. 60)).

309. WHITE HOUSE, CLIMATE ACTION PLAN: STRATEGY TO REDUCE METHANE EMISSIONS 1 (Mar. 2014), *available at* https://www.whitehouse.gov/sites/default/files/strategy_to_reduce_methane_emissions_2014-03-28_final.pdf.

310. Press Release, White House, FACT SHEET: Administration Takes Steps Forward on Climate Action Plan by Announcing Actions to Cut Methane Emissions, http://www.whitehouse.gov/the-press-office/2015/01/14/fact-sheet-administration-takes-steps-forward-climate-action-plan-anno-1 (Jan. 14, 2015).

other heavy equipment, and sometimes constructing new roads, drill pads, and gathering lines. These activities affect the immediate and surrounding areas. Their environmental impacts include air emissions, spill risks, land use changes, disruption to wildlife, noise, odors, and general changes in the life of these communities.

257. Concerns about such disruption, as well as other fears, have led citizens to seek bans to hydraulic fracturing in their communities. In New York, for instance, concerns about safe drinking water, seismic activity, respiratory health, and soil contamination led the state to ban hydraulic fracturing.[311] Unsurprisingly, industry has pushed back against such efforts, with some courts overturning local bans in favor of state-wide regulation favoring development.[312]

258. With the rise of hydraulic fracturing, the state of law addressing natural gas production in the United States is also rapidly evolving. There is no shortage of recommendations on how to regulate this technology, including growing responses at the state level and the various emerging federal efforts noted above. Scholars too have made extensive recommendations for improving regulation of hydraulic fracturing, including conducting environmental studies, developing spill prevention and response plans, adopting protections for surface water withdrawals and storage, and engaging in better siting of well pads.[313] Other scholars recommend more extensive use of liability mechanisms, such as bond requirements and insurance, to shift damages away from the public and onto producers themselves.[314] No matter what measures ultimately are adopted, the law governing natural gas extraction in the United States is certain to continue evolving in the years to come.

311. *See* Nicholas St. Fleur, *The Alarming Research Behind New York's Fracking Ban*, ATLANTIC (Dec. 14, 2014, 2:00 PM), http://www.theatlantic.com/national/archive/2014/12/the-alarming-research-behind-new-yorks-fracking-ban/383868/.

312. *See* Morrison v. Beck Energy Corp., No. 2013-0465, 2014 WL 687475 (Ohio Feb. 17, 2015).

313. Hannah J. Wiseman, *Risk and Response in Fracturing Policy*, 84 U. COLO. L. REV. 729, 799, 810 (2013).

314. David A. Dana & Hannah J. Wiseman, *A Market Approach to Regulating the Energy Revolution: Assurance Bonds, Insurance, and the Certain and Uncertain Risks of Hydraulic Fracturing*, 99 IOWA L. REV. 1523, 1523 (2014); SHEILA OLMSTEAD & NATHAN RICHARDSON, RESOURCES FOR THE FUTURE, MANAGING THE RISKS OF SHALE GAS DEVELOPMENT USING INNOVATIVE LEGAL AND REGULATORY APPROACHES 6, 7 (June 2014), *available at* http://www.rff.org/RFF/Documents/RFF-DP-14-15.pdf.

Chapter 2. Exploitation

§1. Natural Gas Industry and Markets

259. The natural gas industry in the United States is complex. Although traditional industry structure is comprised of three segments—production, transportation, and distribution—there are thousands of actors involved. For example, it is estimated that there are 6,300 natural gas producers in the United States, ranging from large integrated producers with worldwide operations (the "majors"), such as ExxonMobil, Royal Dutch Shell, or British Petroleum, to small operators with interests in as few as one well. Additionally, there are over 530 natural gas processing plants in the United States. Further, there are approximately 160 pipeline companies operating over 300,000 miles of pipelines in the United States. Of these, 180,000 miles are interstate pipelines. Over 123 companies store natural gas. Over 260 companies market gas and gas products. And natural gas is distributed to end users by over 1,200 natural gas local distribution companies (LDCs).[315]

260. The U.S. natural gas industry, then, is largely not vertically integrated. Production companies, however, are frequently sister companies to exploration companies. Nonetheless, sometimes, although not often, producers are affiliated with pipeline companies.

261. Gas producers explore, exploit, and process natural gas (and oil) that is discovered in the field for sale to gatherers or to pipelines. Gatherers build pipelines to collect natural gas from individual producers and sell it to other pipelines or to other specific customers. Natural gas pipelines, in turn, offer two services. They both sell and transport natural gas. If a pipeline sets one price for the sale and transportation of the gas, that is known as a bundled transaction. If, however, a pipeline sets one price for the gas and one price for the transportation service, this is known as an unbundled transaction. This dual function of pipelines—both offering transport services for natural gas and selling the gas itself—has presented regulatory problems that, as discussed below, have been some focus of energy law in the United States. The local distribution company, or LDC, is frequently a regulated utility selling gas to end-use customers. Currently in the United States, the distribution segment of the natural gas industry is undergoing restructuring.

262. The first U.S. natural gas well was developed in Fredonia, New York in 1821. The first gas utility in the United States opened its doors in Maryland in 1817 as the Gas Light Company of Baltimore. By 1859, there were nearly 300 manufactured gas plants in the country serving nearly 5 million customers for lighting. These small, privately owned, and largely unregulated manufactured local gas companies proliferated until the turn of the twentieth century, when local utility regulation began.

---

315. *Industry and Market Structure*, NaturalGas.org (Sept. 20, 2013), http://naturalgas.org/business/industry/.

263. Also at that time, two other developments occurred that profoundly affected the natural gas industry. First, natural gas pipeline technology improved and, therefore, gas was able to be transported long distances in interstate commerce. Second, electricity and coal began replacing gas in the residential, commercial, and industrial sectors.[316]

264. As pipeline transportation improved, the natural gas industry was also able to realize economies of scale, and with those economies of scale came industry consolidation, particularly in the pipeline segment. Additionally, during the 1920s and early 1930s, holding companies were organized to own and operate both electric and gas companies, including interstate pipelines. This industry structure gave rise to concerns over the exercise market power, particularly by the pipelines. In 1928, the Federal Trade Commission began investigating the extent of holding companies' market power. This investigation led to a 1935 report that recommended Congress regulate interstate natural gas transportation and sales.[317]

265. After World War II, the natural gas industry expanded considerably for several reasons: (1) abundant gas reserves were discovered, making gas available at favorable prices; (2) improvements in pipeline technology and storage capacity; (3) increased demand; and (4) increased investment in the interstate pipeline industry.[318] The largest segments of increased demand were from the residential sector, which could rely on natural gas as a clean-burning, efficient fuel for heating and cooking. The industrial sector also increased its demand because natural gas was competitively priced. Regardless of this expansion, the industry hit a substantial snag starting in the mid-1950s through the 1980s due to ineffective regulations that contributed to, if not created, a natural gas shortage that became most problematic in the 1970s. Ultimately, though, the natural gas markets were corrected through regulatory reform and restructuring, resulting in increased production and consumption and greater stability in natural gas markets.

266. In the late nineteenth century, and for most of the twentieth century, oil and natural gas were found and extracted together; such gas was called "associated gas." Because petroleum was believed to be the more valuable resource, natural gas was considered a nuisance byproduct and was often burned off, or "flared." Today, history repeats itself as natural gas is regularly flared in North Dakota as a nuisance byproduct of oil produced from the Bakken Shale formation. Nevertheless, today, most natural gas is non-associated gas: that is, gas not found with oil.

267. Two other types of natural gas should be noted. First, liquefied natural gas (LNG) is the process of discovering natural gas in one location and shipping it at low temperatures on specially designed tankers to another location. Once the LNG reaches its destination, it is reprocessed into its gaseous form. Second, synthetic

316. TOMAIN & CUDAHY, *supra* note 225, at 271.
317. Michael J. Thompson, *Natural Gas*, *in* 2 ENERGY LAW AND TRANSACTIONS, *supra* note 81, at § 50.02[1][c].
318. *Ibid* at § 50.02[2].

natural gas (SNG) is gas that is created by such processes as coal gasification or other chemical transformation. Today, SNG is too expensive to create viable gas markets.

268. Until 1986, domestic natural gas production and consumption were largely balanced. After that date, consumption began to outpace production, and imports of natural gas (93% from Canada and 7% overseas) began to increase. Then, beginning in 2001, production increased as domestic production provided over 93% of all natural gas consumed. Approximately 67% of total U.S. gas is produced in the states of Texas, Pennsylvania, Louisiana, Oklahoma, and Wyoming. In addition to onshore natural gas production, approximately 5% of U.S. natural gas is produced from formations in federal waters, mainly in the Gulf of Mexico.[319] Shale gas has been the main contributor to the increase in domestic natural gas production. Indeed, shale gas production itself is increasing dramatically, from 2.1 Tcf in 2008 to over 11.4 Tcf in 2013—out of a total of 24.3 Tcf produced that year.

269. Currently, then, the natural gas industry in the United States is vibrant and is helping push forward a transformation in the domestic energy landscape, particularly in how electricity is generated as shale gas replaces higher-priced coal. The strength of this market is largely due to the hydraulic fracturing and horizontal drilling employed to release shale gas. While the current natural gas market is healthy, that has not always been the case. Instead, the history and development of the natural gas industry reveals a series of successful and unsuccessful regulations that have shaped those markets and, consequently, the price of this resource.

270. This Chapter concentrates on the federal regulation of the production and transportation of natural gas. However, before turning to federal regulation, it is important to understand the state and local regulation that has played, and continues to play, a historically important role in the development of the industry.

§2. State Regulation

271. The first issue any regulatory scheme must address is who owns the natural gas. In the United States, ownership divides primarily on the type of land in question. Unlike many nations where the government owns title to mineral resources, ownership of minerals in the United States turns on ownership of the surface estate. For government-held lands, the government owns the mineral estate, including natural gas. The federal government owns one third of the land in the United States. Individual states in the U.S. also hold title to surface lands and the underlying minerals. While individual states exercise some control over offshore lands, usually up to 3 miles from the shore, the federal government controls offshore lands beyond that point. Much natural gas in the United States is also found on private lands,

319. *Natural Gas Explained*, *supra* note 284.

which means that private citizens own the natural gas underlying their land. Pursuant to the *ad coelum* doctrine, the surface owner of the land owns all of the various property estates up to the heavens and down to the center of the earth. Thus, on private lands in the United States, the landowner controls all mineral interests and may sell or lease them as he or she desires. On government owned lands, however, private developers lease mineral rights and exploit them in exchange for royalty payments. Consequently, the ability to explore and exploit energy resources such as natural gas in the United States depends first on acquiring an ownership interest in those resources.

272. Notably, most surface owners in the United States do not produce their own gas. Instead, they enter into contractual arrangements for either selling or leasing their mineral rights to natural gas producers. The bundle of laws that govern such leases is known as "oil and gas law" and is a subject unto itself.[320] Such laws affect the nature of ownership, the interpretation of what interests are being conveyed, the nature and duration of such grants, the value and interpretation of royalties, and the inclusion of implied covenants as well as the right to transfer property interests.[321]

273. Natural gas, like petroleum, is a migratory, or "fugacious," mineral. From the surface, it is impossible to locate the contours of the oil or gas reserves below the surface. Instead, the only way to determine whether oil or natural gas is located under a particular plot of land is to drill a well. Again, in the United States, the right to drill is owned by the public or private surface owner. This single feature of U.S. law is the origin point for domestic regulation of natural gas production.

274. Before the development of the modern natural gas industry, beginning in the late nineteenth century, issues of mineral ownership were addressed by state common law and then by state regulations. The most notorious of those common law rules is the "rule of capture." The United States inherited the rule of capture from England. As further detailed in Part III on *Petrol*, the capture of natural gas can be analogized to the capture of wild animals through the doctrine of *ferae naturae*, which simply means that whoever captures and has possession of wild animals owns them. Thus, under U.S. common law, the surface owner who drilled a well and recovered the natural gas from under her land (or, by changing the subsurface pressure of the reservoir, from under the land of an adjacent or nonadjacent landowner) became the owner of the natural gas.[322] As one commentator has summarized: "There is no liability for capturing oil and gas that drains from another's lands to a well on one's own land. The owner of mineral rights in a tract of land acquires title to the oil and gas produced from wells drilled on that land, though part of the oil and gas may have migrated from adjoining lands."[323] It soon became

320. *See, e.g.*, JOHN LOWE ET AL., OIL AND GAS LAW (2012); PATRICK H. MARTIN & BRUCE M. KRAMER, WILLIAMS & MEYERS, OIL AND GAS LAW (5th ed. 2013).
321. *See, e.g.*, OWEN L. ANDERSON & EUGENE O. KUNTZ, KUNTZ LAW OF OIL AND GAS (1998) (with annual supplements); RICHARD W. HEMINGWAY, THE LAW OF OIL AND GAS (1992).
322. *See, e.g.*, Barnard v. Monongahela Natural Gas Co., 65 A. 801 (Pa. 1907).
323. JOHN S. LOWE, OIL AND GAS LAW IN A NUTSHELL 12 (6th ed. 2014).

apparent, however, that the rule of capture doctrine did not fit well with the migratory nature of natural gas. Instead, the doctrine needed to be modified and limited in specific ways.

275. Strictly applied, the rule of capture leads to what economists call a "tragedy of the commons."[324] That is, by giving title to whoever first captures the gas, the rule of capture creates an incentive for each person to be the first mover, encouraging a rush to drill and overproduce the resource. Such a tragedy is not just a theoretical risk. It has borne out repeatedly in the U.S. oil and gas industry. The rush to drill occurred dramatically in the early years of the oil industry in East Texas and Oklahoma. In January 1901, for example, the Spindletop oil well hit a "gusher" that inaugurated the Texas oil boom. A normal well at that time might produce 50 barrels of oil per day. Spindletop, though, reached as much as 100,000 barrels per day. The move to find oil on surrounding plots was rapid. Land that had been as cheap as USD 10 per acre skyrocketed to USD 900,000 an acre. Boom times, however, did not last long. In 1902, Spindletop produced 17 million barrels of oil. However, as a result of the oil glut, prices dropped to an all-time low of USD 0.03 per barrel, whereas oil had been trading at about USD 2 per barrel.[325]

276. Two consequences follow from such a rush to produce. First, if too many mineral owners are drilling for natural gas, then well pressure is reduced and otherwise recoverable natural gas is left in the reservoir. Second, in the short term, natural gas will "flood the market," which can create dramatic price declines. Thus, another way of characterizing the tragedy of commons problem is to say that the rule of capture produced waste: "physical waste" when gas was left in the ground, and "economic waste" when unnecessary price drops were caused.

277. The first limitation, then, that courts began imposing on the rule of capture was to create liability for physical and economic waste. Courts initially began addressing liability for waste using another common law principle known as the doctrine of correlative rights. The case of *Elliff v. Texon Drilling Co.*[326] exemplifies the principle. Elliff was a property owner who owned land adjacent to property where Texon was drilling for oil and gas. Texon, however, was negligent in their operations, and one of its wells blew out, causing damage to oil and gas underneath Elliff's property (as well as Elliff's property itself). Elliff successfully argued that Texon was liable for damages for its lost oil and gas. Texon relied on the rule of capture as its defense. The Texas Supreme Court, however, rejected that argument, relying on the doctrine of correlative rights: Each property owner, the court said, has a right to drill to access underlying mineral resources, but each owner also bears an obligation to exercise that right without negligence. Thus, the doctrine of correlative rights cabined the rule of capture.

---

324. *See* Garrett Hardin, *The Tragedy of Commons*, 162 SCIENCE 1243 (1968).

325. *See* YERGIN, *supra* note 42, at 82–86 (1991); Mary G. Ramos, *Oil and Texas: A Cultural History*, TEXAS ALMANAC, http://www.texasalmanac.com/topics/business/oil-and-texas-cultural-history (last visited Mar. 27, 2015).

326. Elliff v. Texon Drilling Co. 210 S.W.2d 558 (Tex. 1948).

278. Courts also developed another limitation on the rule of capture by restricting the ownership of natural gas produced by someone else. In situations where one landowner has captured gas and injected it into a storage unit, a subsequent driller may not "capture" previously discovered and stored oil and gas.[327] This limitation has been applied in situations where one landowner has produced and stored natural gas, only to have it escape from its storage reservoir. "Generally, courts reason that oil or gas becomes personal property when produced, so that ownership is not lost by mere loss of possession."[328]

279. A third limitation that courts impose on the rule of capture arises in situations of enhanced recovery for oil and gas. Once a well is drilled and the initial pressure releases oil and gas to the surface, other oil and gas often remains in that reservoir. That oil and gas can be removed through enhanced recovery techniques. Such techniques involve injecting water or a mixture of water and carbon dioxide into the reservoir in order to increase the reservoir's pressure and thus release additional oil and gas. Some courts, employing theories of nuisance or trespass, have held that mineral owners who conduct secondary recovery operations are liable to adjoining land or mineral owners that are drained of oil and gas. Other courts, however, have held that the recovery may take place without liability if the adjoining owner has refused to participate in a "fair" proposal for the enhanced recovery.[329]

280. While common law developments limited the reach of the rule of capture, states soon learned that these rules were insufficient to avoid overproduction, declining prices, and waste. In response, beginning in the late nineteenth century, states adopted a variety of conservation laws that today form the basis of U.S. law governing oil and gas production and development. An early example of such a law prohibited well owners from allowing oil and gas to escape into the atmosphere for longer than a period of two days. This conservation measure was upheld by the U.S. Supreme Court as a means of protecting correlative rights.[330] The basic idea behind conservation laws is to avoid physical and economic waste through the reasonable development of oil and gas—that is, to seek full recovery from oil or gas reservoirs.[331] By 1970, virtually every producing state had such statutes on their books, albeit using a variety of techniques.[332]

---

327. Champlin Exploration, Inc. v. Western Bridge & Steel Co., Inc., 597 P.2d 1215 (Okla. 1979).
328. LOWE, *supra* note 320, at 15 (citing Pacific Gas & Elec. Co. v. Zuckerman, 234 Cal. Rptr. 630 (Cal. App. 1987)); *see also* Texas American Energy Corp. v. Citizens Fidelity Bank & Trust Co., 736 S.W.2d 25 (Ky. 1987).
329. LOWE, *supra* note 320, at 16.
330. Ohio Oil Co. v. Indiana, 177 U.S. 190 (1900).
331. *See, e.g.*, NORTHCUTT ELY, THE OIL AND GAS CONSERVATION STATUTES (ANNOTATED) (1933); NORTHCUTT ELY, OIL CONSERVATION THROUGH INTERSTATE AGREEMENT (1933); BRUCE M. KRAMER & PATRICK H. MARTIN, THE LAW OF POOLING AND UNITIZATION (3d ed. 2009).
332. Alan C. Wolf et al., *State Regulation of Oil and Gas Production*, *in* 1 ENERGY LAW AND TRANSACTIONS, *supra* note 81, at § 5.01.

281. As oil and gas extraction technology became more sophisticated, so too did state conservation laws. Thus, one basic modern approach to conservation is well-spacing rules that limit the number of wells that can be drilled in a particular area. By applying basic ideas about density, such spacing rules attempt to prevent over-production and maximize natural gas recovery. Often, these rules are set out in statutes, although some states allow a state agency to establish regulations. Directional drilling adds a complication to well-spacing requirements. Well-spacing requirements are easy to apply to typical wells that are drilled straight down. A directionally drilled well, however, has a surface location, a path that it takes, and a bottom-hole location. Thus, operators of such wells must obtain specific authorizations from surface owners in the well's path.[333]

282. Another regulation that attempts to preserve correlative rights, conserve natural gas, and prevent waste is a proration order. The proration order establishes a certain amount of gas that can be produced per day, with the intent of conserving gas by not exceeding market demand.[334] The amount of gas that can be taken from the particular pool or field is known as an "allowable." State commissions set a "maximum efficient rate" at which a well can produce without creating waste. Proration orders allocate allowables among producers in the field so that each producer will have an opportunity to receive "a just and equitable share" of that production.[335] Even though the natural gas industry is heavily federally regulated, state proration orders have been upheld as a constitutional way to prevent waste and protect correlative rights.[336]

283. A law that complements prorationing rules is called a ratable-take provision. Ratable-take provisions require purchasers of oil or gas from a common source to take gas at specific rates and amounts from producers, pursuant to a given allocation. Once a common pool of natural gas is identified, then, pursuant to a statute or regulation, a formula is set allocating specific amounts among purchasers who, in turn, are required to take that allowable amount from those designated producers without discrimination as to either price or amount. During the natural gas shortage, however, demand-prorationing worked differently. At that time, in the mid-1970s, purchasers set their demand and were required to allocate that demand among multiple producers. This version of ratable-take provisions was an attempt to "spread the pain" among producers, rather than have some producers go out of business. In other words, these provisions seek to limit waste not just by regulating the producers but also natural gas purchasers. They are based on the idea that "if a pipeline purchaser can take gas from only a few producers, those producers may effectively

---

333. *Ibid* at § 5.01[1].
334. *Ibid* at § 5.01[2].
335. *See, e.g.*, MISS. CODE ANN. § 53-3-9 (2013).
336. *See* Northwest Cent. Pipeline Corp. v. State Corp. Comm'n of Kan., 489 U.S. 493 (1989).

drain the field to the detriment of producers without a market outlet."[337] Ratable-take provisions, however, cannot interfere with federal regulation of gas being sold in interstate commerce.[338]

284. Conservation can also be accomplished through the imposition of gas/oil ratios. These ratios limit the amount of gas that can be produced relative to the production of oil. Such regulations are intended to prolong the life of a pool and increase the amount of oil that is ultimately recovered. This is because the gas in a field often creates the field's natural pressure; taking too much gas thus can limit the amount of oil that can be extracted from the reservoir. The ratio is crucial. If there is a high gas/oil ratio, then the amount of oil allowable will be reduced in proportion to natural gas. A lower gas/oil ratio, then, may more efficiently produce the oil in a given pool.[339]

285. While each of these tools is widely used among states, the most significant conservation statutes in the United States involve compulsory pooling and/or unitization. "Pooling and unitization modify the rights of parties and the nature of mineral interests and vary the method for allocating production, costs, and the rights thereto."[340] "Pooling" involves the consolidation of oil and gas rights in multiple tracts of land into a defined unit for the production of natural gas from a particular reservoir. "Unitization" is the consolidation of oil and gas interests into an area for the production of natural gas through the regulation of costs and operation. In other words, unitization affects not only rights but also methods of production. "Both pooling and unitization result in the consolidation of oil and gas rights into a single unit for cooperative development so that all parties within the unit will share in the production there."[341]

286. Because state law controls how pooling and unitization are conducted, these methods vary from state to state. Generally, however, the procedure for compulsory pooling or unitization involves the application by two or more tract owners in a common supply source. The regulatory authority then can issue an order for a compulsory pool or compulsory unitization, if it determines that such an order is necessary to prevent waste. The regulatory authority also considers how to allocate the just and equitable shares of oil and gas from the unit's pool or combination of pools.[342]

---

337. Wolf et al., *supra* note 332, at § 5.01[3].
338. It should be noted that state ratable take laws apply only to unregulated natural gas; they do not apply to natural gas sold in interstate commerce, which is under federal jurisdiction. *See, e.g.*, Northern Natural Gas Co. v. State Corp. Comm'n of Kan., 372 U.S. 84 (1963); Transcontinental Gas Pipe Line Corp. v. State Oil & Gas Bd. of Miss., 474 U.S. 409 (1986).
339. Wolf et al., *supra* note 332, at § 5.01[4].
340. *Ibid* at § 5.01[5].
341. *Ibid* at § 5.01[5] (citing Frederick Lawson, *Recent Developments in Pooling and Unitization*, 15th Oil & Gas Inst. 223–24 (1964)).
342. *Ibid* at § 5.02.

§3. Federal Regulation

287. This section presents an overview of federal regulation of the natural gas industry.[343] Understanding federal regulation of natural gas is important because, although it focuses on the transportation segment of the industry, for some time it also included regulation of wellhead sales, and thus directly implicated production and exploitation as well. Further, it is through its toehold in gas transport regulation that FERC has restructured the entire industry, thus opening up production and exploitation of natural gas in the United States.

288. As noted above, the United States began to consume gas first produced from coal, and then natural gas itself, for lighting purposes in the early and middle eighteenth century. At the time, there was no regulation of natural gas. Later in the century, gas companies began to divide service territories among themselves and agreed not to engage in price competition. Soon, municipalities recognized that the natural gas business had monopoly characteristics and began to regulate natural gas companies as public utilities. This meant that municipalities set gas rates.[344]

289. The natural gas industry, like the electricity industry, is a network industry. While there are multiple producers and millions of customers, producers and customers are connected through a pipeline, or "bottleneck." In network industries, the transportation segment of the industry is key. It effectively renders the entity that owns the bottleneck a natural monopoly. In natural gas, then, the development of the transportation segment of the industry gave cause for federal regulation, particularly when that transport became interstate in nature.[345]

290. Federal regulation of natural gas is based upon the Commerce Clause of the U.S. Constitution,[346] which gives the federal government authority over transactions in interstate commerce. Because the United States utilizes a federal structure, governmental authority is divided between federal and state governments. Early in the twentieth century, the Supreme Court ruled that states could not regulate the interstate sales of natural gas.[347] This ruling created a regulatory gap allowing

343. Federal law also governs much of the extraction of natural gas off the coasts of the United States. The law governing coastal extraction of oil and gas resources is detailed in the Section addressing Petrol.

344. Tomain & Cudahy, *supra* note 225, at 284.

345. *The History of Regulation*, NaturalGas.org (Sept. 20, 2013), http://naturalgas.org/regulation/history/. The U.S. Constitution gives the federal government authority to regulate interstate commerce. *See* U.S. Const. art. I, § 8, cl. 3; *see also, e.g.*, Gibbons v. Ogden, 22 U.S. 1 (1824); United States v. Lopez, 514 U.S. 549 (1995); United States v. Morrison, 529 U.S. 598 (2000). At the end of the eighteenth century, large oil and gas fields were discovered in the southwestern United States. At first, oil producers ignored the energy potential of natural gas and allowed it to escape into the atmosphere, or burned it off through flaring. State conservation regulations, as noted above, were eventually imposed to bring order to oil and gas markets, aiming to reduce waste and stabilize prices.

346. U.S. Const. art. I, § 8, cl. 3.

347. Missouri v. Kansas Natural Gas Co., 265 U.S. 298 (1924).

interstate natural gas companies to effectively set any price they wanted, free of state oversight. This was because, as a result of the ruling, states had the authority to regulate *intrastate* gas sales only—and could not set prices for *interstate* sales.

291. This ruling directly strengthened the monopoly position of the few major interstate pipeline companies that owned the transportation segment of the industry. Gas pipelines, as the only source of natural gas in many areas, had both monopoly and monopsony power. As purchasers of natural gas, they could exercise market power over producers. As sellers of natural gas, they could exercise market power over consumers. The absence of state regulation only exacerbated the problem. A "pipeline with the monopsony and monopoly power paid less than competitive wholesale prices for the natural gas it purchased, and it resold the gas for more than competitive retail prices."[348]

292. As a result of this regulatory gap and pipelines' exercise of market power, the Federal Trade Commission issued a report in 1935 finding that eleven holding companies controlled over 25% of interstate natural gas pipelines. These holding companies also controlled a significant portion of gas production, distribution, and electricity generation. The FTC's report first led to passage of the Public Utility Holding Company Act of 1935.[349] The purpose of PUHCA was to limit the ability of holding companies to exercise market power in natural gas and electricity industries. Because that law did not address the regulation of interstate natural gas sales, however, Congress passed the Natural Gas Act of 1938, which to this day forms the heart of federal regulation of the industry.[350]

293. The NGA gave the Federal Power Commission (now the Federal Energy Regulatory Commission) the authority to issue licenses for interstate pipelines, as well as to set rates for interstate natural gas sales and transportation services. Thus, the NGA focused attention on the monopolistic practices of interstate pipelines. However, the Act did not specifically authorize the exercise of ratemaking power over producers at the wellhead. Consequently, a pipeline could price discriminate in favor of itself if it was also a producer—or in favor of an affiliated producer. Concomitantly, a pipeline could price discriminate against a competitor.

294. Thus, from 1938 until 1947, the FPC exercised no jurisdiction over natural gas wellhead prices charged by producers. Instead, the prices that were charged by producers to pipelines were passed through to customers, without any regulatory protection. In 1947, however, the U.S. Supreme Court ruled that the FPC could exercise jurisdiction over the prices that producers charged to affiliated pipelines.[351]

348. TOMAIN & CUDAHY, *supra* note 225, at 286.

349. Public Utility Holding Company Act of 1935, Pub. L. No. 74-333, 49 Stat. 803 (codified at 15 U.S.C. §§ 79–79z-6) (repealed by the Energy Policy Act of 2005, Pub. L. 109-58, 119 Stat. 594 (codified in scattered sections of 42 U.S.C.).

350. Natural Gas Act of 1938, 15 U.S.C. §§ 717–717w.

351. Interstate Natural Gas Co. v. FPC, 331 U.S. 682 (1947).

Then, in 1954, in *Phillips Petroleum Co. v. Wisconsin*,[352] the Supreme Court ruled that the FPC could exercise jurisdiction over all producer prices.

295. The *Phillips Petroleum* decision was a watershed event for federal regulation of natural gas. The ruling's consequence was that the FPC began holding individual rate hearings for producer prices, which imposed a significant administrative burden on the agency. In 1960, President-elect John F. Kennedy asked James Landis, then dean of Harvard Law School, to undertake a study of regulatory agencies. The Landis report estimated that the FPC could not finish its 1960 caseload until the year 2043.[353] The FPC, in short, could not effectively exercise its jurisdiction over either pipelines or producers and, therefore, could not effectively set natural gas rates.

296. To address this caseload backlog, the FPC began to set rates for gas producing regions. First, the agency set area rates through adjudicatory hearings. These area rates were upheld as constitutional by the U.S. Supreme Court.[354] The adjudicatory hearings, however, were lengthy, and the ratemaking process did not adequately address all pending rate cases. The FPC then changed from ratemaking by adjudication to ratemaking by rulemaking. In other words, instead of evidentiary hearings, the agency proceeded by proposing a rule, receiving comments on the proposed rule, and finally issuing a ruling. Initially, this ratemaking by notice-and-comment rulemaking was done on a regional basis. Eventually, however, the FPC moved to setting national rates through rulemaking, a process that was also upheld as constitutional.[355]

297. Unfortunately, even with the more expeditious process of national ratemaking by rulemaking, FPC regulations caused severe disruption in national gas markets. Particularly problematic was the FPC's ratemaking methodology. Simply put, the FPC would set rates based on historic average costs. In the 1970s, however, due to fluctuations in world oil prices, world gas prices were higher than domestic prices. Consequently, there was a disconnect. Because rates were set on historic average (domestic) costs, producers had an incentive to sell their gas on the intrastate market where they could get the higher international price, rather than on the interstate market where they were limited to the lower domestic price.

298. Several consequences flowed from use of this historic-cost ratemaking methodology. First, there was a gap between domestic and world natural gas prices. As a result, domestic producers preferred to sell at the world price rather than the domestic price. However, if a producer had previously "dedicated" natural gas to interstate commerce, then those prices were set by the FPC and, as a result, were

352. Phillips Petroleum Co. v. Wisconsin, 347 U.S. 672 (1954).

353. James M. Landis, Report on the Regulation of Regulatory Agencies to the President-Elect 6 (1960).

354. *See* In re Permian Basin Area Rate Cases, 390 U.S. 747 (1968); Mobil Oil Corp. v. FPC, 417 U.S. 283 (1974).

355. *See* Shell Oil Co. v. FPC, 520 F.2d 1061 (5th Cir. 1975), *cert. denied*, 426 U.S. 941 (1976); American Pub. Gas Ass'n. v. FPC, 567 F.2d 1016 (D.C. Cir. 1977), *cert. denied*, 435 U.S. 907 (1978).

determined by the domestic, not world, market. At the same time, state regulators were setting natural gas prices at the world market price for gas sold in intrastate commerce, so intrastate production increased to take advantage of the higher prices. Interstate producers, then, had the incentive either to attempt to move their sales from the interstate to the intrastate market, or to keep gas off the interstate market until prices rose.[356]

299. Moreover, regulations limited producers' ability to easily move their gas from one market to another. Once a producer (or a merchant pipeline) dedicated its gas to interstate commerce, it could not remove the gas from that market without FPC permission. That permission had to come in the form of a certificate of abandonment issued by the agency.[357] Thus, both because of the incentive not to put gas into the interstate market and because of the regulatory limits on moving "dedicated" natural gas out of the interstate market, the nation experienced a natural gas shortage.

300. The FPC moved to address this shortage through a series of steps. First, the FPC decided to "pre-grant" abandonment certificates when it issued licenses to natural gas companies to engage in interstate commerce. The "pre-granted" abandonment certificates were included as conditions in companies' certificates of public convenience and necessity.[358] Second, the agency changed the test that companies had to meet to satisfy the abandonment requirements. Historically, when a jurisdictional company sought abandonment, it had to show its specific needs for such a request. Under the FPC's new test, however, an entity seeking abandonment could argue that the natural gas market as a whole would benefit from the pre-granting of abandonment certificates. Obviously, in the time of a natural gas shortage, this made it much easier for any given regulated entity to obtain an abandonment certificate.

301. In addition to liberalizing its abandonment policies, the FPC addressed the natural gas shortage by suggesting two curtailment policies. First, the agency proposed a pro-rata curtailment plan, which, as its name indicates, would reduce the share that each end user would receive in proportionate amounts. Second, the agency proposed, and adopted, an end-use curtailment plan, which delivered gas to priority customers such as schools, hospitals, and small residential users.[359] Because large industrial customers had fuel switching capability, their supplies were curtailed first.[360]

302. The natural gas shortage was severe and quickly became a matter of national concern. Congress responded by passing the Emergency Natural Gas Act

356. Thompson, *supra* note 317, at § 50.02[4][a].
357. 15 U.S.C § 717f (b); *see also* United Gas Pipe Line Co. v. McCombs, 442 U.S. 529 (1979); California v. Southland Royalty Co., 436 U.S. 519 (1978).
358. *See* FPC v. Moss, 424 U.S. 494 (1976).
359. Thompson, *supra* note 317, at § 50.02[4][b].
360. *See, e.g.*, City of Willcox v. FPC, 567 F.2d 394 (D.C. Cir. 1977).

of 1977,[361] under which President Carter declared a natural gas emergency and authorized the FPC to order the transportation of natural gas to the East Coast at prices similar to those set the world market. As a direct result of the shortage, the industry began to advocate for deregulation. Congress heard those calls and the next year adopted the Natural Gas Policy Act of 1978.[362]

303. President Carter touted the NGPA as a central element of his National Energy Act of 1978. The express intent of the Act was to eliminate the dual interstate and intrastate natural gas markets by aligning natural gas prices. As a first step, the Act set higher market prices for all "first sales" of natural gas in either the interstate or intrastate markets. Then, the jurisdiction of FERC—the FPC's successor—was removed from all natural gas, known as "old gas," that had not been dedicated to interstate commerce as of November 8, 1978. A series of other regulations addressing natural gas production also ensued. Federal price controls were imposed on the intrastate market. The Act imposed a formula for monthly price increases in the wellhead price of "new" natural gas, which was specifically defined in the statute. And, the ceiling price on the new gas was pegged to the price of refined oil, which was closer to the world price for natural gas.

304. Finally, the NGPA provided for the elimination of all price controls starting on January 1, 1985. During this period, the late 1970s and 1980s, federal efforts to eliminate the dual natural gas markets were successful—perhaps too much so, as the country then experienced a natural gas "bubble." With the bubble, producers encountered cash flow problems as relief was granted from long-term contracts. Consequently, drilling declined, and in order to stimulate drilling, Congress passed the Natural Gas Wellhead Decontrol Act to eliminate all wellhead price controls.[363] Wellhead price controls were phased out beginning July 27, 1989, and fully repealed as of January 1, 1993 in the Natural Gas Wellhead Decontrol Act of 1989.[364]

§4. Restructuring and "Deregulation"

305. In an effort to regulate the new market after the destabilization of the 1970s, FERC entered into a period of regulatory restructuring. The restructuring had two significant dimensions. First, as noted above, prices were decontrolled. Second, FERC applied open access policy to the transportation segment of the industry.

306. FERC's move to open access for natural gas transport was a long process. The market problems FERC faced were manifold. Gas producers and pipelines had entered into long-term contracts in an effort to: (1) stabilize supplies and reduce price volatility; (2) provide cash flow to producers; and (3) provide a reliable source of supply to pipelines. Long-term contracts were the norm in the industry because

361. Emergency Natural Gas Act of 1977, Pub. L. No. 95-2, 91 Stat. 4.
362. Natural Gas Policy Act of 1978, 15 U.S.C. §§ 3301–3432 (2012).
363. 15 U.S.C.§ 3331(f).
364. *Ibid.*

of the federal dedication and abandonment rules. Once gas was "dedicated" into interstate commerce, as noted above, it could not be removed without federal authorization. This not only encouraged long-term contracting but also meant that spot markets were unable to develop. Together, this meant that the natural gas market was quite constrained. The problem became worse when it became clear that the assumptions parties had made about price in their long-term contracts were inaccurate and, as prices fell, pipelines attempted to alleviate contractual obligations to purchase high-priced natural gas.

307. FERC responded to these problems through a series of rulemakings that sought to make cheaper natural gas more readily available. During the 1970s, a small spot market had begun to develop as curtailed purchasers were allowed to buy directly from producers. The NGPA, however, loosened those rules for "first sales" of gas, which expanded the spot markets. The development of spot markets, in turn, signaled the availability of cheap gas that needed to reach market. FERC thus sought to restructure the industry through a series of orders that aimed to make cheaper gas more available, culminating in a directive to open access to the industry's transportation segment.

308. One attempt to bring that cheaper gas to market occurred through FERC's Order No. 380, which sought to alleviate pipelines' obligations under "take-or-pay" contractual obligations. Take-or-pay contracts had emerged during the natural gas shortage because pipelines needed to have a reliable source of natural gas. In the 1970s and early 1980s, pipelines were willing to enter into contracts that required them either to "take" a specific volume of gas or to "pay" for it, regardless of whether they took the gas or not. When these contracts were negotiated, the take-or-pay provisions had the intent and effect of assuring pipelines' supply and of assuring producers' cash flow. This worked well when supplies were unreliable and prices volatile because long-term contracts can bring stability to a market. Once prices declined, however, problems quickly arose.[365]

309. This happened in the early 1980s when natural gas prices fell. The many interstate pipelines that had committed to take-or-pay contracts found themselves saddled with costly contractual obligations—by some estimates, between USD 5.6 billion and USD 9 billion by the end of 1986. Consequently, pipelines attempted to renegotiate contracts or to seek regulatory and judicial relief from their contractual obligations. In response, producers resorted to litigation, mostly successfully.[366]

310. In an attempt to provide price relief for pipeline customers, FERC issued Order No. 380, which eliminated the variable costs as part of a customers' minimum bill—in short, the take-or-pay contractual obligations that pipelines incurred by agreeing to those contracts. Order No. 380 thus relieved customers of the

365. Tomain & Cudahy, *supra* note 224, at 302–304.
366. Thompson, *supra* note 317, at § 50.02[5][a].

automatic obligation to pay for gas irrespective of whether they needed it, while leaving the pipelines' obligation to pay producers for that gas in place.[367]

311. While Order No. 380 provided some relief to customers, pipelines continued to feel constrained. In an effort to relieve some of the stress on pipelines, FERC initiated "special marketing programs" (SMP) from 1983 through 1985. The idea behind the programs was to direct natural gas to fuel-switchable industrial customers. As this gas was "released" to those customers, pipelines would reduce their take-or-pay obligations. Additionally, FERC used "blanket" certificates of abandonment, allowing customers to purchase gas directly from producers rather than from pipelines.[368] Under these rules, a producer and the pipeline effectively amended their contract to enable the producer to sell contract gas and credit the sale of the pipeline. The pipeline, in turn, would be relieved of the obligation to pay for the amount of gas sold by the producer, and the producer would receive money from another source. Thus, the SMP would relieve contractual obligations without injuring producers. FERC retained the right to review contracts and grant any necessary abandonments.

312. The SMP satisfied pipelines and producers. It did not, however, satisfy customers. Nor did it satisfy existing law. While pipelines were pleased by the reduction in their obligations, and producers were satisfied with receiving compensation, customers were still paying high-priced contract gas rather than cheaper gas that was available on the market. This led to legal challenges in court, where the SMP was invalidated and remanded to FERC.[369]

313. FERC continued to address contractual problems and to react to changing market conditions by loosening price, as well as entry and exit, controls to allow gas to flow more smoothly through the distribution system from wellhead to burner tip.[370] Partially in response to the failure of SMP under Order No. 380, FERC promulgated looser entry and exit requirements under Order No. 436.[371] The central

367. The order was upheld by the United States Court of Appeals. *See* Wisconsin Gas Co. v. FERC, 770 F.2d 1144 (D.C. Cir. 1985).

368. Thompson, *supra* note 317, at § 50.02[5][a].

369. Maryland People's Counsel v. FERC (MPC I), 761 F.2d 768 (D.C. Cir. 1985); Maryland People's Counsel v. FERC (MPC II), 761 F.2d 780 (D.C. Cir 1985); Maryland People's Counsel v. FERC (MPRC III), 768 F.2d 450 (D.C. Cir 1985). During remand, the SMP and blanket certificate programs expired.

370. One such effort was Order No. 451, which provided for incentive pricing. Under the order, producers would receive a certain price for "old gas" that was regulated under the Natural Gas Act, and they would receive the highest price for "old gas" under the NGPA. The rule was initially vacated by the U.S. Court of Appeals for the Fifth Circuit, but then was upheld by the United States Supreme Court. Mobil Oil Exploration & Producing Se. Inc. v. United Distrib. Cos., 498 U.S. 211 (1991). Further, FERC issued Order No. 500, attempting to respond to take-or-pay liability issues. Under that order, producers were required to credit pipelines' take-or-pay obligations against gas transported under open access rules. The regulations were also intended to provide for a sharing of

idea behind Order No. 436 was to separate the merchant and transportation roles of pipeline companies through a process called "unbundling." While fuel-switchable customers had access to various fuels, certain captive customers, as well as customers that found it difficult to switch fuels, had difficulty gaining access to the cheaper natural gas. Accordingly, under this order, pipelines would separate their merchant (buy-sell) and transportation (pipeline) functions and then price those functions accordingly. Thus, under Order No. 436, pipelines, on a voluntary basis, could offer transportation services to customers on a first-come, first-served basis. Interstate pipelines that chose to participate, however, were barred from discriminating against transportation requests in order to protect their own merchant services. The Order also required transportation rate minimums and maximums to be set, but within those limitations, pipelines were free to offer competitive rates to their customers. Order No. 436 worked a significant change in the industry. Although it was voluntary, all of the major pipeline systems participated.

314. Order No. 436 was comprised of several elements: (1) open-access and non-discriminatory transportation where, if demand exceeded capacity, the open-access pipeline would transport natural gas on a first-come, first-served basis; (2) rate regulation ensuring that prices were within a zone of reasonableness rather than based on historic average cost; (3) authorization for local distribution companies to switch their "contract demand" to transportation services, as well as the option to reduce contract demand itself, all in an effort to afford LDCs greater access to cheaper gas; and (4) expedited licensing for new pipelines and services that opted for open access status.

315. As with virtually every major federal regulatory rule, Order No. 436 was challenged in court. The order, however, largely survived this challenge, although the court remanded the rule to FERC for failure to adequately address take-or-pay burdens.[372] Specifically, in *Associated Gas Distributors v. FERC* (*AGD I*), the U.S. Court of Appeals for the D.C. Circuit upheld FERC's jurisdiction to issue open access provisions as long as the provisions were nondiscriminatory. The court also sustained FERC's ability to offer flexible rate treatment. This was critical. By allowing pipelines to set rates within a "zone of reasonableness," pipelines could offer discounts rather than being compelled to offer only cost-based rates. Additionally, the court upheld Order No. 436's allowing customers to unilaterally modify their

accrued take-or-pay liabilities and to avoid future take-or pay obligations. Order No. 500 was initially remanded, *see* Associated Gas Distribs. v. FERC (AGD II), 893 F.2d 349 (D.C. Cir. 1989); American Gas Ass's v. FERC, 912 F.2d 1496 (D.C. Cir 1990), but ultimately affirmed. *See* American Gas Ass'n, 912 F.2d at 1496.

371. Order No. 436, Regulation of Natural Gas Pipelines After Partial Wellhead Decontrol, 50 Fed. Reg. 42,408 (Oct. 9, 1985) (codified at scattered sections of 18 C.F.R.), FERC Stats. & Regs. ¶ 30,665, *modified,* Order Nos. 436-A, 50 Fed. Reg. 52,217 (Dec. 23, 1985) (codified at scattered sections of 18 C.F.R.), FERC Stats. & Regs ¶ 30,675 (1985); 436-B, FERC Stats. & Regs. ¶ 30,688; 436-C, 34 F.E.R.C. ¶ 61,404; 436-D, 34 F.E.R.C. ¶ 61,405 (1986); 436-E, 34 F.E.R.C. ¶ 61,403 (1986), *vacated and remanded,* Associated Gas Distribs. v. FERC, 824 F.2d 981 (D.C. Cir. 1987), *cert. denied,* 485 U.S. 1006 (1988).

372. Associated Gas Distribs., 824 F.2d at 981.

contracts with pipelines, enabling them to change a portion of their gas purchase obligation to lower-cost, contract demand gas.

316. FERC continued to address the dislocations of natural gas market in the wake of Order No. 436. Seven years later, in Order No. 636,[373] FERC ratcheted up the voluntary unbundling that Order No. 436 allowed and made it mandatory. Finding that its prior orders had not adequately moved the natural gas market toward competition, FERC sought to eliminate the natural monopoly aspects of the pipeline segment of the industry. In brief, Order No. 636, often referred to as the Final Restructuring Rule, stated that pipelines must separate their transportation and sales services so that pipeline customers had a choice in selecting sales, transportation, and/or storage services from any provider, in any quantity.

317. To effectuate these goals, Order No. 636 prohibited interstate pipelines from offering sales services on a bundled basis. Instead, pipelines were required to unbundle their transportation and sales services. Additionally, all customers were required to convert their previously bundled sales contracts into contracts for transportation and transportation services. Further, customers had to purchase natural gas independently—either from independent vendors or pipeline suppliers.[374]

318. Order No. 636 contained the following provisions: (1) mandatory unbundling of pipeline sales and transportation; (2) blanket certificates allowing pipelines to make unbundled sales at market-based rates; (3) mandatory open access transportation services that had to be provided "equal in quality" regardless of seller; (4) more expansive pre-granted abandonment authorizations, which applied to sales as well as transportation; and (5) policies allowing full recovery of transition costs incurred by pipelines. Further, Order No. 636 defined transportation to include storage, making storage also subject to the rule's open access requirements.[375] FERC recognized that industry restructuring would have a significant impact on pipelines and, therefore, authorized cost recovery for: costs of gas supply realignment; costs associated with the elimination of the purchase gas adjustment mechanism; costs associated with unneeded upstream capacity; and stranded facilities costs.[376]

319. Order 636, as ultimately implemented,[377] fine-tuned FERC's rate and pricing provisions and added scheduling and reporting requirements. Thus, although this rule dealt heavily with the transport segment of the industry, it had profound

373. Order No. 636, Pipeline Service Obligations and Revisions to Regulations Governing Self-Implementing Transportation; and Regulation of Natural Gas Pipelines After Partial Wellhead Decontrol, 57 Fed. Reg. 13,267 (Apr. 16, 1992) (codified at 18 C.F.R. pt. 284 (2012)), FERC Stats. & Regs. ¶ 30,939.

374. Jonathan D. Schneider, *Natural Gas Transportation*, *in* 4 ENERGY LAW AND TRANSACTIONS, *supra* note 81, at § 83.03[2][a].

375. TOMAIN & CUDAHY, *supra* note 35, at 304–11.

376. Schneider, *supra* note 94, at § 83.03[2][d].

377. Order No. 637-A, 65 Fed. Reg. 35,706 (June 5, 2000), FERC Stats. & Regs. ¶ 31,091 and ¶ 31,099 (2002); Order No. 637-B, 92 FERC Stats. & Regs. ¶ 61,202 (2000).

effects on the operation of the production side as well. Upheld by the courts,[378] Orders No. 436 and 636, together with the subsequent Order No. 712,[379] effectively restructured the natural gas industry, in part by moving pipelines toward common carrier status. That, in turn, increased competition while lessening regulatory obligations.

378. United Distrib. Cos. v. FERC, 88 F3d 1105 (D.C. Cir. 1996); Interstate Natural Gas Ass'n of America v. FERC, 285 F.3d 18 (D.C. Cir 2002).

379. 73 Fed. Reg. 37,058 (June 30, 2008) (codified at 18 C.F.R. pt. 284), FERC Stats. & Regs. ¶ 31,271 (2008). Order No. 712 continued deregulation of the natural gas industry by first removing the price ceilings for capacity sales of one year or less, thus allowing those prices to reach market clearing levels and encouraging distribution of gas to those who value it most highly. The rule also exempted other capacity releases to certain purchasers known as "asset management arrangements" and to state-approved retail access programs.

Chapter 3. Transport

320. In the United States, jurisdiction over natural gas transportation and rates is divided effectively between the federal government and the states. In brief, once gas is placed into interstate commerce, federal jurisdiction attaches. Thus, federal regulation controls transportation of natural gas at the large scale over bigger distances, while states maintain authority over smaller scale distribution of natural gas at the local level.

321. Federal regulation falls within the jurisdiction of FERC. FERC regulates the terms under, and rates at which, transportation service is provided. It also has eminent domain authority to facilitate the siting and construction of natural gas pipelines.

322. State regulation, particularly over rates, occurs through public utility commissions and predominantly affects the local distribution companies that deliver gas to end-users. LDCs receive a monopoly franchise in exchange for providing natural gas to all customers at just and reasonable rates. There are over 1,200 natural gas distribution companies in the United States, some of which operate intrastate and others that operate interstate. These companies operate more than 1.2 million miles of distribution lines.[380]

323. Because federal regulation controls natural gas transportation in the United States, this Chapter focuses on federal regulation. State regulation, which addresses natural gas distribution, is addressed in Chapter 4 *infra*.

§1. Natural Gas Pipeline Transportation

324. The Natural Gas Act of 1938 is the key statute governing transportation of natural gas in the United States.[381] As noted above, the NGA was passed in response to the monopoly characteristics exhibited by interstate pipelines. Initially, the NGA was administered by the Federal Power Commission, but today it is administered by FERC.

325. NGA § 1 establishes the scope of FERC's jurisdiction. That provision gives FERC authority to regulate "the transportation of natural gas in interstate commerce," as well as "the sale in interstate commerce of natural gas for resale for ultimate public consumption."[382] Additionally, Section 1 makes clear that federal jurisdiction does not extend to sales for local distribution or to facilities used for gathering natural gas. In other words, the NGA's objective is to regulate natural gas companies engaged interstate commerce as public utilities—interstate

380. Thompson, *supra* note 317, at § 50.03[4][b].
381. Natural Gas Act of 1938, 15 U.S.C. §§ 717–717w.
382. *Ibid* § 717.

pipelines—and not to regulate intrastate transportation or sales, nor LDCs or gathering facilities.[383] The restriction on regulating LDCs is known as the Hinshaw Amendment.[384]

326. The NGA imposes a number of regulatory requirements on jurisdictional natural gas companies. If the regulated company seeks to construct an interstate pipeline or other associated facilities, it must receive a certificate of public convenience and necessity from FERC after notice of an evidentiary hearing. The notice is issued to enable entities desiring to participate in a proceeding to file a motion to intervene. If "no issue of substance" is raised by the intervenor, then FERC, upon request of the applicant, can decide the matter without a hearing on the basis of the company's filings.[385] If an issue of substance is raised, however, then a full evidentiary, trial-type hearing, including cross-examination, takes place.[386]

327. To issue a certificate of public convenience and necessity, FERC must find that the applicant is willing and able to perform services in the public interest. Public interest considerations include environmental impacts, compliance with state licensing requirements, and economic benefits, as well as the interests of competing existing pipelines. More specifically, FERC certification procedures now mandate that the Commission examine four primary factors before issuing a certificate for new pipelines: (1) whether the applicant bears the full economic risk without imposing costs for new facilities on existing ratepayers; (2) effects on competing pipelines and their customers; (3) the effects of a project on landowners; and (4) the environmental effects of construction and operation of proposed facilities.[387] When the certificate is granted, an initial rate is established.

328. Just as the NGA requires a certificate for companies to enter the market, it also mandates that companies obtain a certificate in order to leave the market. As described in Chapter 3 *supra*, once natural gas has been dedicated to interstate commerce, it cannot be removed, or "abandoned," without federal acquiescence.[388] A natural gas company can request a certificate of abandonment when it desires to terminate service or desires to remove certain facilities from interstate commerce. FERC will grant abandonment when it determines that the available supply of natural gas for the relevant company has become depleted, that the continuation of service is unwarranted, or that abandonment is needed in the public interest.

329. FERC may impose reasonable conditions on the construction and operation of natural gas facilities. Under the NGA, FERC, after notice and hearing, may order a natural gas company to improve its transportation facilities or to establish physical

383. 15 U.S.C. § 717a–c.
384. James H. McGrew, Federal Energy Regulatory Commission 12 (2d ed. 2009).
385. 18 C.F.R. § 385.802.
386. Schneider, *supra* note 374, at § 83.0[1][b].
387. Certification of New Interstate Natural Gas Pipeline Facilities, 88 F.E.R.C. ¶ 61,227 (1999).
388. 15 U.S.C. § 717f(c).

connections to LDCs, as long as the order will not place an undue burden on the natural gas company or impair its ability to provide adequate service to its customers.[389]

330. In an effort to loosen regulation and open natural gas markets, FERC does allow certain construction activities to take place under a blanket certificate. Interstate pipelines that already have received a certificate of public convenience and necessity "may construct, acquire, operate, or replace certain eligible facilities, including mainline facilities, pursuant to blanket authority for the cost of the project if it does not exceed certain cost levels set out in the regulations."[390]

331. Fundamental concepts in the Natural Gas Act are premised on key pillars of the Federal Power Act. Perhaps most prominently, Sections 4 and 5 of the NGA parallel Sections 205 and 206 of the FPA. As described in more detail in Chapter 5 *infra*, these provisions dictate that jurisdictional sales of gas and electricity, respectively, must be made at "just and reasonable" rates—that is, at rates that ensure a reasonable rate of return to the gas company or utility but that do not impose too high a burden on customers. These provisions also mandate that such rates be nondiscriminatory, and that the rates charged be put on file with the Commission (originally the FPC, now FERC). If a party believes that a rate filed with FERC is not just, reasonable, or nondiscriminatory, it may intervene in the filing proceeding and challenge the rate. Likewise, if a party later believes that a rate on file with FERC is unlawful, it may bring an action under Section 5 of the NGA to challenge the rate.

332. The "just and reasonable" standard of Section 4 of the NGA (and of Section 205 of the FPA) is a flexible one. As the Supreme Court announced long ago in *FPC v. Hope Natural Gas Co.*, it is the result arrived at, not the methodology relied upon to get there, that determines whether a rate passes Section 4 muster.[391] Today, NGA Sections 4 and 5 apply perhaps most importantly to rates for transportation service on pipelines. Using the flexibility read into the just and reasonable standard by the Supreme Court, FERC today also allows parties to make jurisdictional natural gas sales using market-based rates, if the selling party can demonstrate that it lacks market power. The Energy Policy Act of 2005 substantially augmented FERC's ability to oversee gas sales at market-based rates. That law "granted FERC express authority to police the manipulation of wholesale natural gas markets, gave FERC discretionary authority to require dissemination of information that would improve the transparency of wholesale natural gas markets, and clarified FERC's exclusive authority to site LNG import and export projects."[392] The 2005 Act also gave FERC authority to use market-based rates for natural gas storage services, "and granted FERC authority to coordinate federal and state agency review of natural gas projects."[393]

---

389. 15 U.S.C. § 717f(a).
390. Schneider, *supra* note 374, at § 83.02[1][a].
391. 320 U.S. 591 (1944).
392. Joseph T. Kelliher & Maria Farinella, *The Changing Landscape of Federal Energy Law*, 61 ADMIN. L. REV. 611, 627 (2009).
393. *Ibid.*

333. The scope of FERC's transportation jurisdiction under the NGA has been broadly construed. According to the Supreme Court, gas "crossing the state line at any stage of its movement to the ultimate consumer" constitutes interstate commerce.[394] Once gas leaves interstate commerce, however, state jurisdiction attaches as long as state regulators do not interfere in ways that affect federal regulation. In other words, states may regulate LDCs and intrastate transportation, and FERC may not.[395]

334. The Natural Gas Policy Act of 1978 is the second major statute regulating the natural gas industry.[396] As described in greater depth in Chapter 3 *supra*, the NGPA was passed at a time of crisis in the domestic natural gas industry in an effort to eliminate the dual natural gas markets. One way that the NGPA facilitated that market liberalization is relevant to natural gas transportation. The Act authorized FERC to allow transactions between intrastate pipelines and interstate pipelines and distributors.

335. Specifically, Section 311 of the NGPA attempted to separate the transportation and merchant roles of interstate pipelines with the intent of lessening the monopoly power held by pipeline companies. This legislation gave FERC the authority to authorize: (1) any interstate pipeline to transport natural gas on behalf of any intrastate pipeline or local distribution company; (2) any intrastate pipeline to transport natural gas on behalf of any interstate pipeline or any LDC served by an interstate pipeline; and (3) any intrastate pipeline to sell gas to any interstate pipeline or any local distribution company served by an interstate pipeline.[397] The legislation also allowed these transactions to go forward without the scrutiny required under certification proceedings of the NGA. Additionally, intrastate pipelines engaged in such transactions are not subject to the Commission's general NGA jurisdiction.[398] However, under these provisions, intrastate pipeline rates must be reasonably comparable to the rates of interstate pipelines.[399]

336. In addition to regulation by FERC, natural gas pipelines are also subject to safety legislation through the Natural Gas Pipeline Safety Act, passed in 1968 and amended in 1976 and 1979.[400] This legislation requires the Secretary of Transportation to set minimum safety standards for gas transportation and for pipeline facilities. Additionally, the Secretary is required to set minimum safety standards for LNG facilities, including their location, design, construction, operation, and maintenance. Supplementary to this legislation, in 2006, Congress passed the Pipeline Inspection, Protection, Enforcement, and Safety Act,[401] which reauthorized Department of Transportation's regulatory authority over pipeline safety.

394. Maryland v. Louisiana, 451 U.S. 725, 755 (1981).
395. Schneider, *supra* note 374, at § 83.01[2][a]–[b], [3][a].
396. Natural Gas Policy Act of 1978, 15 U.S.C. §§ 3301–3432 (2012).
397. Schneider, *supra* note 374, at § 83.02[2].
398. 15 U.S.C. § 3371(a) (2006).
399. 15 U.S.C. § 3371(a)(2)(B).
400. 49 U.S.C. §§ 1671-1686.
401. Pub. L. No. 109-468, 120 Stat. 3486 (2006).

§2. Gas Pipeline Siting

337. The natural gas pipeline infrastructure in the United States is quite extensive. Currently, there are an estimated 2.6 million miles of interstate and intrastate pipelines in the United States, which together comprise the gathering, distribution, and transportation segments of the industry.

338. This system is subject to dual jurisdiction. Interstate transportation lines are regulated by the federal government, but intrastate transportation lines, plus gathering and distribution lines, are regulated by the states. The vast majority of this system, then, is subject to state regulation. There are about 2 million miles of distribution pipelines in the United States, which carry gas to end-users, including individual households, businesses, and manufacturing facilities.[402] There are also approximately 200,000 miles of gathering pipelines. There remains, then, over 300,000 miles of transmission pipelines, which transport gas longer distances in interstate commerce under FERC authority.[403]

339. Notwithstanding the expanse of this natural gas pipeline infrastructure, new discoveries and expanded production give rise to a need for additional pipelines. As noted earlier, U.S. oil and natural gas production has increased dramatically within the last decade. Hydraulic fracturing and directional drilling have enabled producers to reach more oil and gas from locations that were costly and inaccessible in the past. Therefore, to the extent that oil and natural gas are now being found in new locations, the transportation infrastructure is inadequate to properly service either resource. Quickly, this is becoming clear throughout the country. Oil, for example, increasingly is being transported by rail rather than pipeline, with the consequence of accidents and explosions that have caused significant environmental and human damage. New natural gas likewise has not always found ready transportation available to take it to market. Consequently, natural gas found in some fields, particularly in the Bakken play, is being flared rather than taken to market. Thus, some commentators have suggested that our transportation systems for both oil and natural gas are inadequate to handle increased development.[404]

340. Where a resource can be exploited economically, the natural response to discovery of new reserves is to build out additional infrastructure to move the commodity. In the case of oil and particularly natural gas, this typically has meant construction of pipelines. Despite the fact that oil and gas most often are naturally found together, their regulation in the United States has differed significantly. Those differences in regulation extend to pipeline regulation as well.

402. U.S. Gov't Accountability Office, GAO-13-221, Pipeline Permitting: Interstate and Intrastate Natural Gas Permitting Processes Include Multiple Steps, and Time Frames Vary 4 (Feb. 2013), *available at* www.gao.gov/assets/660/652225.pdf.

403. Paul W. Parfomak, Cong. Research Serv., R43138, Interstate Natural Gas Pipelines: Process and Timing of FERC Permit application Review (Jan. 16, 2015), *available at* https://fas.org/sgp/crs/misc/R43138.pdf.

404. Klass & Meinhardt, *supra* note 35, at 947.

341. A key reason for different regulatory regimes for oil and natural gas is that oil and natural gas have different physical characteristics. Because of its liquid form, oil can be transported by pipeline, truck, rail, or ship. Natural gas, however, unless liquefied, is most easily and efficiently transported via pipeline. Consequently, even if state oil pipeline regulation is restrictive, oil can be transported by other means, and thus there arguably is less need for federal regulation of oil pipelines than natural gas pipelines, particularly when it comes to siting and construction of new facilities. Thus, as described in greater detail in the part on Petrol, while FERC regulates interstate oil transportation rates, oil pipelines in the United States otherwise are largely regulated by the states, with some exceptions for federal oil pipeline safety control. Natural gas pipelines, however, are largely regulated by the federal government, and federal regulation of natural gas pipelines supersedes state regulation via the doctrine of preemption. In the context of U.S. law governing energy infrastructure generally, natural gas thus holds a rather unique position. The construction of no other energy infrastructure in the United States—oil pipelines, electricity transmission lines, or coal transportation—is controlled so heavily at the federal level.[405] In part because of its history, however, the siting of natural gas pipelines rests within federal control.

342. As described more fully above, federal regulation of the natural gas industry occurred precisely because of the monopolistic characteristics of, and the exercise of market power by, interstate pipelines. At the same time, natural gas companies also historically argued that state regulators, at the behest of the coal and railroad industries, were inhibiting competition by refusing to grant rights-of-way for pipelines. Thus, the Federal Trade Commission undertook a study of the natural gas industry, with a focus on transportation. The FTC report recommended that natural gas pipelines be considered "common carriers" subject to federal control of construction, operation, financing, and the transportation of natural gas. The Natural Gas Act of 1938 thus gave the FPC authority over sales for resale and transportation in interstate commerce—and, importantly, to regulate facilities used for such sales and transportation.

343. Despite the authority given to the FPC, natural gas pipelines, throughout the 1930s and 1940s, did not have eminent domain authority. State interests, then, could impose impediments to the siting and construction of natural gas pipelines. It is easy to imagine, for example, that a state would be reluctant to grant a right-of-way across its lands to a natural gas pipeline if that pipeline was not going to offer any of its gas for sale within that state. As a result of such "hold-out" problems, Congress passed legislation in 1947 granting pipelines eminent domain authority,

405. Rates for coal movements, which occur primarily by rail, are subject to federal regulation by the Surface Transportation Board, just as FERC regulates oil pipeline rates but not facility construction. Likewise, as detailed in Part I, *Electricity*, there have been some overtures to give the federal government additional control over the siting of interstate electricity transmission lines, but those efforts both have turned out inert in their application and were more limited than FERC's control over natural gas pipelines in their general thrust in any event. The Nuclear Regulatory Commission controls atomic materials from cradle to grave but does not regulate the siting or construction of the underlying infrastructure utilized to move nuclear fuel.

precisely to enable them to build out the nation's natural gas transportation infrastructure. This was seen as especially necessary at the federal level to allow pipelines to effectively carry out the certificates of public convenience and necessity that the FPC granted them and, thereby, to overcome state prohibitions and roadblocks.[406]

344. Interstate natural gas pipelines require FERC approval. The approval process for every new, extended, or modified pipeline requires a certificate of public convenience and necessity.[407] The certification process generally entails three phases. The first phase is the "pre-filing" phase. During this phase, pipelines, on a voluntary basis, can provide FERC with documentation for the purpose of facilitating and expediting the review process, through early coordination with FERC and other agencies and interested individuals and organizations.[408] Pursuant to the Energy Policy Act of 2005, FERC has been designated as the lead agency for coordinating all applicable federal authorizations, as well as for environmental impact statement (EIS) review under the National Environmental Policy Act. At this stage, other federal, state, and local agencies, as well as citizens and interest groups, can become involved in assessing pipeline applications. FERC has approval authority over a company's request for pre-filing.

345. During the pre-filing phase, the developer attempts to engage all stakeholders in the review process. Although developers can engage stakeholders, third parties do not become formal intervenors with legal rights at this stage of the certification process. Typically, the developer presents information regarding potential sites, pipeline route studies, and field surveys, and must begin the environmental review process. FERC consults with interested stakeholders, including government agencies, and holds public scoping meetings, in an effort to assess the scope of the project. FERC also conducts site visits to the proposed project. Most pipeline developers engage in the pre-filing process, but they need not.

346. The second stage consists of filing an application for the certificate of public convenience and necessity. Here, FERC must decide whether to issue the certificate "in the public interest." Considering the application, FERC assesses, among other things, the project's potential impact on competition, whether the pipeline will contribute to overbuilding, potential environmental consequences, and whether the pipeline is being subsidized by existing customers. The application must contain a description of the proposed pipeline, route maps, construction plans and schedules, as well as evidence of compliance with applicable statutory and regulatory requirements, such as needed permits. The applicant must also file environmental reports analyzing route alternatives, together with studies of potential environmental impacts as required by NEPA.

---

406. Natural Gas Amendments of 1947, Pub. L. No. 80-245, 61 Stat. 459 (codified as amended at 15 U.S.C. § 717f (2012)).
407. 15 U.S.C. § 717f(c)(1)(a) (2013).
408. 18 C.F.R. § 157.21 (2006).

347. The third phase is decisional. If FERC grants the certificate, the order will state terms and conditions of approval, including the authorized pipeline route, as well as any construction or environmental measures that must be taken to comply with the agency's order. The certification confers upon the developer eminent domain authority,[409] and the certification has the effect of preempting any state or local laws that duplicate or obstruct the federal certificate, such as siting or zoning requirements. Thus, a pipeline developer has the authority to acquire rights-of-way by law if they are not able to negotiate for them. In practice, eminent domain authority is used infrequently, as landowners are compensated through negotiations and ultimately protected by the Constitution.[410]

348. After the certificate is granted, parties to the proceeding may object to the order for any reason and may formally request a rehearing so that the FERC may reconsider its decision. Such requests must be made 30 days after the issuance of the final order, and that deadline is one that cannot be waived or extended. If there is a rehearing and the certificate is approved, the pipeline may proceed even if challenges have been filed in a federal court. The developer then provides FERC with outstanding information, and the agency can then issue a Notice to Proceed with Construction Activities and construction can begin. The developer then files weekly status reports with FERC documenting compliance until construction is completed.[411]

§3. Liquefied Natural Gas

349. LNG is simply ordinary natural gas that is been changed from its gaseous state to a liquefied state by reducing its temperature below -260° Fahrenheit. In addition to changing its physical state, the volume of LNG is 1/600 that of its gaseous state and, therefore, enables easier transportation, particularly by ship rather than pipeline. In its liquefied form, LNG can be more easily stored than in its gaseous state.

350. Although the liquefaction of natural gas was first patented in 1914 and facilities had been first constructed by 1944, it was not until the late 1960s that the United States began importing Algerian LNG. Due to domestic natural gas shortage, LNG seemed to be an attractive option. Unfortunately, by the end of the 1970s, LNG markets changed dramatically, precisely because LNG prices were tied to the price of oil, and oil prices were rising precipitously. Consequently, U.S. imports fell off significantly. In the 1990s, as gas prices receded, imports ticked up considerably.[412]

---

409. 15 U.S.C. § 717f(h).

410. The Constitution prohibits the taking of "private property . . . for public use, without just compensation." U.S. Const. amend. V.

411. Parfomak, *supra* note 403, at 6.

412. Brian O'Neill, *Liquefied Natural Gas (LNG), in* 4 Energy Law and Transactions, *supra* note 81, at § 56.01[1].

351. Since 2000, LNG imports have been volatile. In the early 2000s, there were multiple requests for approval of LNG terminals in the United States. While some of those terminals were built and LNG import capacity increased, by 2010 that capacity was unused. Excess import capacity was the direct result of domestic natural gas production, which proceeded apace beginning roughly in 2008, largely as a consequence of shale development. The boom in domestic natural gas production drove down prices considerably, which in turn significantly dampened demand for LNG imports. Thus, during the first decade of the twenty-first century, LNG imports varied dramatically, from a high 770.8 bcf in 2007 to a low of 157.8 bcf in 2012. FERC lists eleven existing LNG import terminals located along the Atlantic seaboard and Gulf Coast, as well as fourteen proposed export terminals.[413]

352. The NGA originally provided for FPC jurisdiction over natural gas imports and exports. Pursuant to the Act, "no person shall export any natural gas from the United States to a foreign country or import any natural gas from a foreign country without first having secured an order of the Commission ordering it do so."[414] To obtain such an order, the Act specified that the FPC needed to afford an opportunity for a hearing, after which it would make a finding whether exportation or importation would be consistent with the public interest. In 1977, however, the Department of Energy Organization Act transferred import and export authority to the Secretary of Energy.[415] The Energy Policy Act of 1992 then also prohibited FERC regulations from imposing unjust, unreasonable, unduly discriminatory, or preferential terms for natural gas imports and exports on the basis of national origin for any country with which the United States had a free trade agreement.[416] In the Energy Policy Act of 2005, Congress granted FERC the authority to site LNG terminals[417] and, in doing so, preempted state and local siting authority. It must be noted, however, that while preemption extends to siting, states may also exert rights under other applicable statutes such as the Coastal Zone Management Act (CZMA), the Clean Air Act, and the like.

353. Pursuant EPAct 2005, FERC has exclusive authority to approve or deny applications for the siting, construction, expansion, and operation of LNG terminals, with exceptions for other federal laws. FERC exercises its authority by holding hearings with notice to interested persons, including interested state agencies. FERC reviews plans for construction and operation of an LNG terminal and requires that the terminal also offer service to customers on an open access basis; that the project shall not result in subsidization of expansion capacity by existing consumers; that there will be no degradation of service to existing customers; and that there will not be any undue discrimination against existing customers regarding the terms and conditions of service from the facility.

413. *LNG*, FERC, http://www.ferc.gov/industries/gas/indus-act/lng.asp (last updated Mar. 6, 2015).
414. 15 U.S.C. § 717b(a).
415. 42 U.S.C. §§ 7151, 7172 (2012).
416. Energy Policy Act of 1992, Pub. L. No. 102-486, 106 Stat. 2776, tit. II. § 201(b) (codified at 15 U.S.C. § 717b (2012)).
417. 15 U.S.C. § 717b(e).

354. Volatile natural gas markets, combined with reduced domestic production in the 1990s and early 2000s, encouraged FERC to engage in relatively light-handed regulation of LNG terminals. In 2003, for example, FERC ruled that NGA § 7 applications were not required for new LNG terminals. The policy stated that new LNG terminals were not required to offer open access service under cost-based rates or to meet other tariff and rate schedules. Rather, rates, as well as terms and conditions for LNG terminal service, would be negotiated between the parties. FERC reasoned that such streamlining was in the public interest and better comported with market-based approaches to regulation.

355. LNG terminals are also subject to the requirements of the Deepwater Port Act, as amended in 2002.[418] This Act transferred regulatory authority over offshore LNG terminals from FERC to the Department of Transportation. The Act also facilitates the development of LNG by exempting natural gas deepwater ports from the requirement that they operate as common carriers. Thus, the licensing, citing, construction, and operation of deepwater ports are regulated exclusively under this Act. The Act also removes from FERC the jurisdiction to regulate pipelines connecting the LNG to the terminal.[419]

356. At the time of the passage of the Energy Policy Act of 2005, LNG was still seen as a desirable resource. The Act stated that FERC had "exclusive authority" to approve applications for the siting, construction, expansion, or operation of onshore LNG terminals.[420] As a way of streamlining the approval process, FERC was designated as the lead agency for coordinating all federal authorizations, including compliance with environmental legislation.

418. 33 U.S.C. §§ 1501-1524 (2012).
419. O'Neill, *supra* note 412, at § 56.01[1][2][c].
420. 15 U.S.C. § 717(b)(1).

Chapter 4. Distribution

357. Daily, close to 70 million customers in the United States rely on the natural gas distribution system to deliver gas to their homes and businesses. Natural gas users in the United States receive this fuel in two primary ways—directly from the mainline pipeline system (40% of consumption) and via local distribution companies (60% of consumption). Together, these customers consume approximately 20 Tcf of natural gas, constituting about 22% of the total energy consumed in the nation annually. Residential and business consumers receive almost all of their natural gas via the distribution system (99 and 98%, respectively), whereas electric generation receives virtually all of its natural gas directly from the mainline pipeline system (98%). Industrial consumers obtain natural gas both ways; they receive about 45% of their gas deliveries via the distribution system and 55% via mainline pipelines.[421] Distribution, then, plays a critical role in the delivery of natural gas in the United States, especially for residential and business consumers.

358. There are a variety of natural gas distributors. LDCs typically are either investor-owned or privately owned.[422] There are approximately 250 investor-owned LDCs serving roughly 61 million, mostly residential customers in the United States. The difference between investor-owned and privately owned LDCs is that the stock of an investor-owned LDC is publicly traded. Regardless of ownership, both forms of LDC are subject to state PUC regulation, including rate-setting or, where such programs exist, customer choice rules. Municipal governments can also serve as an LDC, and most have contracts with investor-owned or privately owned utilities for needed services. In addition to LDCs, intrastate and interstate pipeline companies can deliver natural gas to end-users directly; these end-users are mostly large-volume customers such as electricity producers.[423]

359. Local and state regulation of natural gas distribution sales to consumers is among the oldest forms of energy regulation in the United States. Initially, in the early nineteenth century, municipalities granted rights-of-way to gas companies for construction of pipelines and other facilities. Later in the century, to avoid duplication and waste, state regulation of gas pipelines took hold. In exchange for the right to construct pipelines and sell gas, gas companies agreed to a variety of conditions of service, including safety, security, taxes, and rate-setting. Today, for the most part, intrastate activities are regulated by state PUCs.[424]

421. U.S. Energy Info. Admin., *Distribution of Natural Gas: The Final Step in the Transmission Process* 1 (June 2008), *available at* http://www.eia.gov/pub/oil_gas/natural_gas/feature_articles/2008/ldc2008/ldc2008.pdf.

422. Some LDCs are municipally owned, or are run as cooperatives. *See Ibid* at 3.

423. *Ibid* at 3.

424. Thompson, *supra* note 317, at § 50.04[3][a].

360. LDCs, then, are regulated utilities within a specific geographic service area. The LDC links the transmission system to end-users through smaller-diameter distribution pipelines. These smaller pipelines operate at lower pressures than transmission pipelines, and the gas travels shorter distances to a greater number of customers. Because natural gas is essentially odorless, the LDC injects an odorant into the gas during the distribution process.

361. The transportation system delivers natural gas to the "city gate"—that is, the market center for pricing natural gas, particularly in large urban areas. The LDC takes ownership of natural gas at the city gate and then delivers it to individual customers. At this transfer point from transportation to distribution, the city gate station performs regulating and metering functions. The city gate station regulates the pressure of gas from high transmission pressures to intermediate and lower distribution pressures. Natural gas is also scrubbed and filtered to ensure low moisture and particulate content.[425]

362. In addition to serving as the conduit from transportation to distribution, LDCs also engage in a variety of other distribution and safety activities, including: (1) use of detection equipment along the distribution network to identify leaks and other safety problems; (2) public safety education programming; (3) customer service utilizing on-call technicians; (4) emergency preparedness for events such as natural disasters; and (5) "one call systems" that provide customers, contractors, and excavators a single phone number to call before commencing excavation or construction to make sure that pipelines and other buried facilities are not damaged during construction.[426]

363. The restructuring of the natural gas industry has affected LDCs as well as interstate pipelines. Under Order No. 636, pipeline companies had to provide open-access transmission and could no longer buy and sell natural gas. This regulatory change allowed large-volume users to buy natural gas on the competitive market rather than from a single-source pipeline. LDCs, consequently, lost some of their sales to large-volume customers.

364. In most instances, LDC authorization to serve is derived from a state-issued certificate of public convenience and necessity. If an LDC wants to initiate service, expand or modify its facilities and services, or offer new services, it submits an application to the state PUC justifying the need and evaluating the impact on existing services. If the PUC determines that the service is justified in the public interest, it can approve the certificate and impose the terms and conditions it deems

---

425. *See, e.g.*, Khondkar A. Saleque, *Gas Distribution Network Codes, Standards: Guidelines*, *available at* http://naruc.org/International/Documents/Session%203%20Gas%20Distribution%20Network.pdf (last visited Mar. 27, 2015) (PowerPoint for USAID and National Association of Regulatory Utility Commissioners).

426. *Natural Gas Distribution*, NATURALGAS.ORG (Sept. 20, 2013), http://naturalgas.org/naturalgas/distribution/.

necessary. The LDC then operates in a guaranteed service territory free from competition. This is particularly important because distribution costs typically constitute about half of the natural gas costs for households and small volume customers.

365. State PUC authority is generally broad. Although the scope of PUC authority varies somewhat from state to state, it is fairly typical for PUCs to have authority to (1) assess the valuation of a utility's plant and assets; (2) set rates by balancing investor and consumer interests; (3) determine rates of depreciation, as well as fix rates of returns on equity; (4) require uniform systems of accounts; (5) require the posting of rates and charges; (6) investigate and, if necessary, revise rates and services; and (7) review and approve securities issuance.[427] Additionally, following the federal government's move to open-access gas transportation, several states have instituted tariffs under which LDCs are obligated to transport customer-owned gas on a nondiscriminatory basis subject to reasonable operating conditions.

366. Also following federal initiatives, some PUCs now require unbundled utility service allowing purchases of natural gas or transportation services independently. Although historically LDCs only offered bundled services that combined the cost of transportation, distribution, and the gas itself into a single price to customers, in recent decades residential "consumer choice" programs have been offered in a move towards greater retail "unbundling" of natural gas sales. Many states offer programs from which customers may choose suppliers of natural gas while using the LDC for service and delivery. In order to further consumer choice, natural gas marketers have now entered the natural gas sector, leading to greater competition and innovation in the marketplace.

367. In exchange for a guaranteed service territory, the LDC must provide service to all customers within that area without discrimination. Regulators, in turn, set rates that afford the LDC a just and reasonable rate for its service while allowing it an opportunity to realize a reasonable profit on its investments. For the most part, LDC rates historically have been cost-based. That is, the idea was that a customer's rate was based upon the utility's costs. Because the LDC had an obligation to serve and that obligation included providing a system to meet the highest level of demand over the course of the year, it was to the customer's benefit that such plans were realized.

368. Cost-based ratemaking, however, has a downside. Because an LDC is rewarded for its capital investments, it has an incentive to build. Customers, however, cannot absorb fixed costs indefinitely. At some point, the cost of natural gas (including the cost of transportation) will either force them to reduce demand or find alternative energy sources. State PUCs, further, are hampered in one respect to rate-setting. Once FERC sets a rate for interstate sales, that rate is passed through to the LDC and then to the LDC's customers. In an era of deregulated first sales of

427. Thompson, *supra* note 317, at § 50.04[3][a].

natural gas, those deregulated rates are deemed to be "just and reasonable." Nevertheless, PUCs are not powerless. The PUC may not dictate a price that the distributor pays for certain transactions. The PUC may, however, instruct the LDC on the types of transactions it may enter. Consequently, many PUCs have imposed a requirement that the LDC engage in "least cost purchasing" policies. LDCs then can "schedule" the purchase of its requirements among available suppliers in an ascending order of prices.[428]

369. Ratemaking is discussed in more detail immediately below in Chapter 5. However, it is worth noting here that PUCs have begun to experiment with the idea of demand decoupling for natural gas, which is to separate the amount of gas sold from the number of customers served. The idea behind decoupling is to allow the LDC to reach its revenue requirement while either basing rates upon the number of customers served, or allowing rates to be adjusted between rate cases using actual revenues and revenue targets. From the LDC standpoint, demand decoupling can be beneficial because its revenues do not fluctuate with changes in the amount of sales. From a public benefit perspective, such programs have much to offer because they can promote efficiency. The idea is to remove the disincentive to the LDC to invest in efficiency programs: If an LDC has a revenue target, it can increase its profitability by increasing its efficiency.[429] Thus, just as regulatory approaches for natural gas distribution are moving toward competition in the form of customer choice programs, they also increasingly emphasize environmental goals through mechanisms such as demand decoupling.

---

428. *Ibid* at § 50.04[3][c].

429. *See, e.g.*, Regulatory Assistance Project, Revenue Regulation and Decoupling: A Guide to Theory and Application (June 2011), *available at* file:///Users/alexisjones/Downloads/RAP_RevenueRegulationandDecoupling_2011_04.pdf.

Chapter 5. Consumption

370. Natural gas in the United States, as noted above, is used across sectors. It is the fuel of choice for many purposes, but especially for home and business heating and, increasingly, for electricity production. As of 2014, the greatest use of natural gas in the United States was for electricity generation, at 8,120,925 MMcf.[430] Next was industrial consumption (7,658,676 MMcf), followed by residential use, primarily for heating (5,071,500 MMcf), and finally commercial use (3,458,754 MMcf).[431]

371. Natural gas consumption, of course, is heavily influenced by the fuel's price. Nonconventional techniques for extracting natural gas, including directional drilling and hydraulic fracturing, have driven natural gas prices down tremendously in recent years in the United States.[432] In 2008, for instance, wellhead prices of natural gas in the United States reached USD 7.97/Mcf. By 2012, that price had dropped to USD 2.66/Mcf.[433]

372. Regulation also clearly influences natural gas prices. Pricing of natural gas in the United States occurs in multiple layers and in multiple ways, and each of these is influenced by the government's choice to regulate—or not—that segment of the industry.

373. Government regulation of the natural gas industry has shifted and evolved over time in the United States, as further detailed in Chapter 3, but the most direct way is through rate regulation of natural gas sales. Initially, this occurred—and still occurs—both at the federal level (for transportation and for natural gas sales for resale in interstate commerce) and, importantly, at the state and local level (for retail sales).

374. Historically, rates were initially set by legislatures. When this was the case, courts would review ratemaking decisions and apply their own analysis. This meant that courts had to act as overseers of legislative and then administrative decisions—a time-consuming task for which they were not particularly well equipped. Eventually, then, courts decided to remove themselves from the ratemaking business;[434] today they largely defer to the ratemaking decisions of the regulators charged with setting utility prices.

375. Thus, natural gas prices today depend on the presence, or absence, of regulation in different sectors of the gas industry. As noted above, pricing at the gas wellhead has been deregulated, so that market functions unfettered by governmental

430. *Natural Gas Consumption*, U.S. Energy Info. Admin., http://www.eia.gov/dnav/ng/ng_cons_sum_dcu_nus_a.htm (last updated Feb. 27, 2015). An "Mcf" is one thousand cubic feet.

431. *Ibid.*

432. *U.S. Natural Gas Wellhead Price*, U.S. Energy Info. Admin., http://www.eia.gov/dnav/ng/hist/n9190us3a.htm (last updated Feb. 27, 2015).

433. *Ibid.*

434. *See* FPC v. Hope Natural Gas Co., 320 U.S. 591 (1944).

oversight. However, FERC still oversees transportation rates, as well as wholesale sales of natural gas. Finally, and perhaps most importantly, gas prices are controlled in the United States at the retail level by state regulators, usually state public service commissions or public utility commissions. In both contexts (wholesale and retail sales), however, the general move in the United States toward industry restructuring and competition-based regulation has made inroads into how gas prices are set today.

376. For decades, natural gas pricing was based on cost-of-service ratemaking. Cost-of-service ratemaking rewards prudently operated utilities. It effectively ensures that utilities' costs will be covered, and allows them to earn a reasonable return on their capital investment. In this way, utilities can attract investors while continuing to operate.

377. Cost-of-service ratemaking is based on the idea of the "regulatory compact." Under this compact, the utility is granted a guaranteed service territory and the right of eminent domain so that they can build infrastructure at fair market value. The guaranteed service territory means that the utility will have no competitors in a particular geographic region. In exchange for this protected market, a utility has the obligation to provide service to all customers, and that service will be based on reasonable rates set by the regulator.[435]

378. The need for the regulatory compact is based on two premises.[436] The first premise is that the activity being regulated is a public good (such as the supply of natural gas), and that the provision of that good in the public interest. The second premise is that market failures warrant regulation. In the case of natural gas pipelines, for instance, the market failure is the risk that the owner of a bottleneck facility will exhibit monopoly power, or that the characteristics of the industry are such that there is a natural monopoly—that is, that economies of scale could be realized from relying on a single provider, whereas any additional construction of pipelines would be duplicative and wasteful. Either condition serves as a justification for rate regulation.

379. Federal and state regulators long used cost-of-service ratemaking, which can be described by the following formula: $R = O + (V\text{-}d)r$. Each of these terms is significant. R constitutes the utility's revenue requirement, that is, the amount of revenue a utility is entitled to make by applying rates to all of its customers. O constitutes a utility's operating expenses, such as salaries, taxes, fuel costs, and the like. Operating expenses are recovered on a dollar-for-dollar basis, and the issue of whether any given item should be included as an operating expense is generally noncontroversial. Controversy, however, often arises regarding the rate base, which is represented as (V-d). V constitutes the value of the company generally, meaning

435. *See, e.g.*, Jersey Cent. Power & Light Co. v. FERC, 810 F2d 1168, 1189 (D.C. Cir. 1987).
436. Munn v. Illinois, 94 U.S. 113 (1876).

its capital investment minus d, its depreciation. Finally, r, the rate of return, is intended to be set at a risk level comparable to other similar industries.[437]

380. Once the cost-of-service is determined, regulators must segregate costs into functional groups: (1) transmission; (2) storage; (3) production and gathering; and (4) customer costs. These costs are then classified as either fixed or variable, and variable costs are further classified as a commodity charge or demand charge. In brief, the commodity charge constitutes the amount of natural gas that is consumed by an end-user. The demand charge constitutes an end-user's proportionate payment for the fixed costs of providing natural gas. These classifications and allocations provide a source of controversy. A utility would prefer to recover its fixed costs first, yet regulators have not always acceded to this desire. In the 1980s, for example, FERC adopted a "modified fixed-variable" (MFV) rate design, which limited the fixed costs that could be recovered in the commodity charge. Consequently, the commodity charge was lower, imposing some risk on the utility to meet its sales projections. The MFV method sought to make gas companies more competitive.[438] Later, as part of industry restructuring under Order No. 636, FERC concluded that the MFV formula inhibited competition and abandoned that method in favor of a straight fixed variable (SFV) methodology as the preferred rate design for pipelines' transportation services. Under SFV, the utility is able to recover all of its fixed costs in the demand component and all variable costs in the commodity component.

381. Traditionally, cost-of-service ratemaking served five functions: (1) capital attraction; (2) reasonably priced energy; (3) efficiency; (4) demand control; and (5) income transfer. Each of these functions is fairly straightforward. In order for the utility to stay in business and continue to meet demand, they must attract capital. Because natural gas is seen as a product in the public interest, it should be reasonably priced so that consumers can take advantage of it. Rates should be based upon prices available in competitive industries—in other words, in industries with similar risk profiles—so that efficiency is attained. And, finally, rates can be designed such that there will be income transfers from consumers to producers or among consumer classes.

382. Through rate design, regulators can encourage or discourage consumption. As energy markets were expanding and as infrastructure needed to be constructed, regulators often used declining block rates. Rates would be set based upon blocks of usage—for example, thousands of cubic feet of consumption. In the initial blocks, the price per Mcf was high, which allowed utilities to quickly recover their fixed costs. Then, as consumption increased, subsequent blocks would be gradually priced lower, thus encouraging consumption.

437. For a general discussion of ratemaking, *see* TOMAIN & CUDAHY, *supra* note 225, at 165–204; *see also* Thompson, *supra* note 317, at § 50.04[1][a].

438. Thompson, *supra* note 317, at § 50.04[1][a].

383. Natural gas rates, as noted, are set by state as well as federal regulators. The division between these two authorities follows jurisdictional lines. Federal regulators set wholesale rates for gas sold in interstate commerce, and state regulators, generally PUCs, set retail rates. Regardless of the dual authority, ratemaking follows similar principles as well as similar legal requirements. In brief, ratemaking authority aims to balance the interests of utility shareholders and the interests of consumers. Moreover, rates are intended to be set at a level that they would be set as if natural gas was sold in a competitive market.

384. The NGA establishes ratemaking authority for FERC, and many state statutes parallel the NGA in their general structure and thrust. Under the NGA, first, all rates and charges by a natural gas company in connection with either the transportation or sale of natural gas subject to Commission jurisdiction "shall be just and reasonable," and any rate that is not just and reasonable is unlawful. Further, jurisdictional natural gas companies are prohibited from granting any undue preference or advantage or from maintaining unreasonable differences in rates between localities or classes of service.[439]

385. While the initial rate is generally set at the time of the issuance of the certificate of public convenience or necessity, subsequent rates are set upon a filing made by a natural gas company and usually after a trial-type adjudicatory hearing. The burden is on the natural gas company to show that the proposed rates should be adopted in place of the existing rates, which are presumptively just and reasonable. The Commission can accept or reject the rate proposal. If it accepts the rate, it must articulate its reasons for doing so. The Commission may also allow those rates to go into effect subject to refund, if it wants to hold a hearing to determine whether the proposed rate is just and reasonable.

386. Once a rate is set, it becomes the lawful rate and cannot be changed without Commission authorization. This is known as the "filed rate doctrine." The purpose behind the doctrine is twofold: to allow customers to rely on the rates, and to prohibit the regulated company from charging rates other than those that are filed and publicized.

387. Upon complaint by a third party or on its own initiative, FERC is authorized to fix just and reasonable rates if existing rates are found to be unjust, unreasonable, unduly discriminatory, or preferential.[440] Under these provisions, the Commission bears the burden of proving that its proposed rate is just and reasonable.

388. FERC began modifying its rate designs in an effort to create rate structures that were more competitive. Cost-of-service ratemaking served the country well as long as energy industries were expanding. However, once the utility industry had

439. 15 U.S.C. § 717c.
440. 15 U.S.C. § 717d.

built out its infrastructure and reached certain technological plateaus, cost-of-service ratemaking was less effective because of its built-in bias to encourage utilities to make more capital investments (so that they can earn a return on an inflated rate base). Consequently, other ratemaking methods were encouraged and adopted.

389. PURPA perhaps most directly incentivized such changes. That statute encouraged state regulators to consider different rate formulas to promote conservation. The Act also required regulators to come up with new rate designs to better manage consumption.

390. One important form of this kind of innovative rate design is known as marginal cost pricing. In brief, marginal cost ratemaking better reflects the true cost of a utility's product. Under the cost-of-service method, utility rates were set based upon historical costs—typically represented by a historical "test year." Thus, costs that already had been incurred were the basis for setting rates. This created strange incentives for utilities because it sends the wrong signal: When a market is continuing to realize economies of scale, setting rates on the basis of average historic costs risks overcompensating utilities because their costs are declining. When market costs are rising, however, a utility suffers unnecessary losses if rates are set on a historic basis. The more sound economic approach is to set rates based on current costs—that is, on the marginal cost of producing the next unit.[441] Thus, under the marginal cost approach, a utility's costs are more accurately reflected and price signals to consumers are more accurate. There are several variations of marginal cost pricing, including time-of-day rates that reflect the actual cost at the time of use.

391. Other innovations in ratemaking include incentive rates that can be structured in various ways. If a particular utility, for example, is operating efficiently and saving costs, then they may be rewarded with a slightly higher rate of return. Similarly, to the extent that regulators want to encourage investments in technological innovations or capital improvements, rates can reflect such investments by including a portion or all of them in the rate base or, alternatively, by allowing a slightly higher rate of return.

392. Coincident with an effort to lessen the regulation of natural gas utilities, regulators also have allowed market-based rates. Under this approach, the regulator ensures that parties do not have power to manipulate market prices, and if they do not, they are allowed to freely bargain with others. The idea is that by policing market power rather than the price itself, the market will reach a more economically efficient price than regulators could themselves. FERC has not approved market-based transportation rates, but it has done so for gas storage.[442] FERC has also

441. In re Madison Gas & Elec. Co., 5 P.U.R. 4th 28 (Wisc. PSC 1974).

442. Order No. 678, Rate Regulation of Certain Natural Gas Storage Facilities, 71 Fed. Reg. 36,612 (codified at 18 C.F.R. pt. 284 (2006)), FERC Stats. & Regs. ¶ 31,220 (2006); Order No. 687-A, 117 F.E.R.C. ¶ 61,190 (2006).

allowed market-based rates for sales of gas where the selling party lacks market power.[443]

393. At the retail level, ratemaking for natural gas also changed toward the end of the twentieth century and at the beginning of the twenty-first century, when many states began experimenting with utility deregulation and restructuring. Although the bulk of states in the U.S. continue to provide natural gas at regulated cost-of-service rates, several states have adopted "customer choice" programs that allow consumers to choose the provider from whom they receive their supply. Such programs promote competition, and thus, change the way natural gas is priced. Only a few states, primarily in the Northeast, have adopted full natural gas unbundling. Other states, however, have adopted phased or pilot programs or have otherwise encouraged unbundling. According to the U.S. Energy Information Administration, as of 2009, states taking such approaches extended from coast to coast and border to border.[444]

443. Alternatives to Traditional Cost-of-Service Ratemaking for Natural Gas Pipelines and Regulation of Negotiated Transportation Services of Natural Gas Pipelines, 61 Fed. Reg. 4633, 4633 (Jan. 31, 1996); *see also* Elizabethtown Gas Co. v. FERC, 10 F.3d 866 (D.C. Cir. 1993).

444. U.S. Energy Info. Admin., *Natural Gas Residential Choice Programs*, (May, 17, 2010), http://www.eia.gov/oil_gas/natural_gas/restructure/restructure.html.

Part III. Regulatory Framework Concerning Petrol

394. Petrol, or oil as it is known in the United States, is a key input into the nation's energy system and economy. Oil, like natural gas and coal, is a fossil fuel. It is composed of ancient organic material, present as complex hydrocarbons, that was deposited millions of years ago in watery ecosystems. Crude oil is not easily usable for energy purposes in its natural form. It has to be processed, or "refined," before it yields the different oil-based commodities that are traded in global markets. Oil is measured in barrels, and one barrel equates to 42 gallons. A typical barrel of oil in the United States yields roughly 45 gallons of oil products: 19 gallons of gasoline, 11 gallons of diesel fuel, 4 gallons of jet fuel, 1 gallon of heavy fuel oil, and 7 gallons of other products such as asphalt, coke, waxes, and lubricating oils. Oil becomes a useful energy commodity, then, not by merely extracting it from the ground, but also by processing it, moving it, and delivering it to end users. This vast system includes different segments of the industry: (1) exploration and production; (2) pipelines, shipping, and storage; (3) refining; and (4) distribution and trading. Oil has a strong connection to the United States, in part because the earliest commercial production of oil began in Pennsylvania as an alternative to whale oil and town gas for lighting purposes,[445] and because the United States is the world's top oil consuming country.[446]

445. Daniel Yergin, The Prize: The Epic Quest for Oil, Money & Power 22–23 (3d ed. 2008).
446. U.S. Energy Info. Admin., *Monthly Energy Review* 60 (Apr. 2015).

Chapter 1. Production

395. Oil is a critically important energy source in the United States, especially for the transportation sector. Overall, the United States consumes more energy from oil—32% of total U.S. energy consumption—than from any other source.[447] Thus, how the United States produces oil also is incredibly important.

396. Oil resources in the United States can be classified along two different axes: onshore versus offshore, and conventional versus unconventional. The former distinction is obvious; onshore resources are located under land, and offshore resources are located under the oceans off the nation's coasts. The latter distinction bears more explanation. Conventional oil is found trapped in layers of sedimentary rock; it generally can flow naturally or be pumped to the surface.[448] Formations of conventional oil commonly are referred to as oil "fields" or "reservoirs." By contrast, unconventional oil includes two subcategories, namely, tar sands and shale oil. These resources are more tightly trapped in less porous rock formations and require relatively new technologies to be extracted.[449] Shale oil resources are sometimes called "plays."

397. While oil *resources* can be classified as either (1) onshore or offshore and either (2) conventional or unconventional, there are also several types of extraction *methods* used to produce oil in the United States. Three key recovery methods are used in conventional fields: primary production, secondary production, and tertiary production. Primary recovery is the production of oil that takes advantage of natural reservoir energy, with oil either flowing naturally to the surface or being pumped to the surface. In contrast, secondary recovery refers to the processes where energy is added to the reservoir to allow for the recovery of oil. Such energy is generally added by injecting water, air, or natural gas. And tertiary recovery, which is also known as "enhanced oil recovery" (EOR), involves trying to free the oil to flow more easily, such as through injecting heat; carbon dioxide; and detergents, microbials, or other chemicals into the well bores.[450]

398. In addition to these production techniques used for conventional oil resources, two new technologies offer recovery techniques that make the extraction of unconventional resources possible—hydraulic fracturing and deepwater drilling. Although hydraulic fracturing has been an available technology for some decades, its combination with horizontal drilling has enabled domestic production to increase

447. U.S. Energy Information Administration, *Oil: Crude and Petroleum Products Explained*, http://www.eia.gov/energyexplained/index.cfm?page=oil_use (last visited Dec. 9, 2014).

448. Davies et al., *supra* note 29, at 116.

449. Alberta Energy Regulator, *What Is Unconventional Oil and Gas?*, *available at* http://www.aer.ca/about-aer/spotlight-on/unconventional-regulatory-framework/what-is-unconventional-oil-and-gas; USGS, *USGS Releases First Continuous Oil and Gas Assessment for North Alaska Loop*, *available at* http://www.usgs.gov/blogs/features/usgs_top_story/usgs-releases-first-continuous-oil-and-gas-assessment-for-alaska-north-slope/.

450. John S. Lowe, et al., Cases and Materials on Oil and Gas Law 91 (4th ed. 2002); *see also* Anderson & Kuntz, *supra* note 321, at § 4.8.

rapidly in the United States. Hydraulic fracturing, together with horizontal drilling, enables producers to drill into shale resources, and then inject fluids and proppants, such as sand, to break and keep open the rock formation and release the mineral resource. Likewise, tar sands (also known as oil sands) require advanced production techniques. Tar sands are not pumped like oil, but rather, mined, either through strip mining or open pit techniques. The extracted resource is then processed to separate the valuable bitumen from the other components that make up the mined material (clay, water, and sand). The bitumen then has to be upgraded before it can be refined, and it is usually diluted with other light petroleum products before it can be transported by pipeline, because it is so viscous.

399. By way of example, because of hydraulic fracturing, North Dakota has shot up the charts in terms of importance for oil production in the United States. Less than ten years ago, that state was only ninth in oil production in the United States. Today, it is second. Moreover, extraction in the Bakken formation, which overlies the northwestern portion of North Dakota, is only likely to continue. "Estimates are that it will take 20 years to develop the 35,000 to 40,000 production wells needed to fully exploit the 'thermally mature' part of the Bakken shale, which is an area the size of West Virginia."[451]

400. In addition to hydraulic fracturing and horizontal drilling, deepwater drilling is another technology that is responsible for the increase in oil production in the United States in recent years. Significantly, deepwater production has changed the game for offshore oil extraction. In 1990, almost all offshore oil extraction in the Gulf of Mexico was shallow water production, or wells placed at 1,000 feet or less. By 2000, however, deepwater extraction, defined as wells between 1,000 and 5,000 feet, had vastly expanded, comprising roughly half the offshore oil production in the nation. Now, deepwater production in the Gulf greatly exceeds shallow water production, and ultra-deepwater production (greater than 5,000 feet) also has entered the mix.[452]

401. Oil is produced in many areas of the United States, but production is fairly heavily concentrated in several states. As of 2014, the top oil-producing states in the United States were, in order, Texas, North Dakota, California, Alaska, and Oklahoma. However, many U.S. offshore extraction sites are in the Gulf of Mexico, and if that region counted as a state, it would be second in the nation for oil production.[453] Overall, these five states comprise about two-thirds of the nation's energy production, with Texas alone accounting for 37% and North Dakota for 13%. The Gulf of Mexico region accounts for 16%.[454] The top 100 U.S. oil fields by reserves also are located largely in these areas. Moving west to east, these fields are found

451. *Ibid.*
452. U.S. Energy Info. Admin, *This Week in Petroleum* (May 26, 2010), http://www.eia.gov/oog/info/twip/twiparch/100526/twipprint.html.
453. U.S. Energy Info. Admin., *Petroleum Supply Monthly* tbl. 26 (Feb. 2015).
454. *Ibid.*

in Alaska, California, Utah, Wyoming, Colorado, New Mexico, Montana, North Dakota, Texas, Oklahoma, and the Gulf of Mexico.[455]

402. Oil production in the United States has varied over time. It peaked in 1970, at 3.5 million Mbbl (thousand barrels), then steeply declined until 1976, then slowly rebounded to just under 3.3 million Mbbl in 1985. From that time on, oil production in the United States generally declined until it bottomed out in 2008 at 1.83 million Mbbl. Since 2008, however, domestic oil production has been on a sharp uptick, reaching 3.16 million Mbbl in 2014.[456]

403. The renaissance of domestic oil production has important implications for the United States. In the 1970s, disruptions to the global crude supply were treated as crises domestically. Thus, the United States' ability to close the gap on net oil exports is seen as an issue of critical national importance at home. That is beginning to happen. In 2014 and 2015, the four-week average of U.S. net imports of crude oil and other petroleum products hit a lower mark than it had since 1991—at 4,686 Mbbl/day and 4,774 Mbbl/day, respectively—compared to the prior low of 5,210 Mbbl/day in 1991 and the high of 13,591 Mbbl/day in 2005.[457] In fact, the United States is now the third largest oil producer in the world, behind only Russia and Saudi Arabia, but closing the gap quickly. In 2010, Russia produced just under 9,700 Mbbl/day of crude oil, Saudi Arabia produced 8,900 Mbbl/day, and the United States produced just under 5,500 Mbbl/day. In 2014, however, Russia produced 10,107 Mbbl/day, Saudi Arabia 9,735 Mbbl/day, and the United States 8,653 Mbbl/day—more than either Africa, Europe, Central and South America, or Asia and Oceana produced as entire continents.[458]

---

455. U.S. Energy Info. Admin., *EIA Report Highlights Top 100 U.S. Oil and Natural Gas Fields* (Apr. 24, 2015), http://www.eia.gov/todayinenergy/detail.cfm?id=20952.
456. U.S. Energy Info. Admin., *Petroleum & Other Liquids: Crude Oil Production* (Apr. 29, 2015), http://www.eia.gov/dnav/pet/pet_crd_crpdn_adc_mbbl_a.htm.
457. U.S. Energy Info. Admin., *Petroleum & Other Liquids: 4-Week Avg U.S. Net Imports of Crude Oil and Petroleum Products* (Apr. 29, 2015), http://www.eia.gov/dnav/pet/hist/LeafHandler.ashx?n=pet&s=wttntus2&f=4.
458. U.S. Energy Info. Admin., *International Energy Statistics* (Feb. 2015), http://www.eia.gov/cfapps/ipdbproject/iedindex3.cfm?tid=5&pid=57&aid=1&cid=regions&syid=2010&eyid=2014&unit=TBPD.

Chapter 2. Exploitation

404. Oil in the United States is predominantly privately owned, in contrast to most countries in the world, where the state claims title to valuable mineral deposits. However, federal ownership of oil is also important in the United States, as production from federal lands has exceeded 20% of total U.S. production since the late 1990s (reaching a peak of 36% in 2010). Further, an estimated 43% of U.S. crude oil reserves are on federal land, which includes both onshore and offshore reserves.[459] State-owned lands also are an important source of oil, especially in the western United States.[460]

405. Oil is regulated at every stage in the United States, from its initial ownership and production through consumption. However, regulation is especially relevant during the exploitation phase in order to determine ownership rights, allow production, and provide for fair treatment of all involved parties. State common law and state statutes, along with federal environmental statutes, govern the production of onshore oil. Offshore oil production, by contrast, is regulated mainly at the federal level. This Chapter addresses oil development on private lands first, followed by a discussion of oil exploitation on federal lands.

§1. The Rule of Capture, Its Limits, and State Conservation Laws

406. Regulation of onshore oil reflects the underlying tension between property law's focus on ensuring private ownership rights and the public objective of protecting the resource from over-extraction and waste.[461] The regulation of oil production thus begins with basic property law. The fundamental starting place is to determine who owns the oil. In the United States, under the common law doctrine of *ad coelum*, possession of the surface generally further entitles landowners to own the heavens above as well as subsurface minerals to the core of the earth below.[462] Were this the only applicable property principle, the question of subsurface oil ownership would be simple enough; it would turn on who owned the surface above the oil itself.

459. Marc Humphries, *U.S. Crude Oil and Natural Gas Production in Federal and Non-Federal Areas* 1–2 (2014), *available at* http://energycommerce.house.gov/sites/republicans.energycommerce.house.gov/files/20140410CRS-US-crude-oil-natural-gas-production-federal-non-federal-areas.pdf; U.S. Energy Information Administration, *Analysis & Projections, Sales of Fossil Fuels Produced from Federal and Indian Lands, FY 2003 through FY 2013*, http://www.eia.gov/analysis/requests/federallands/ (last visited Dec. 9, 2014).

460. *See* Peter W. Culp et al, Trust Lands in the American West: A Legal Overview and Policy Assessment (2005), *available at* http://www.lincolninst.edu/subcenters/managing-state-trust-lands/publications/trustlands-report.pdf; Peter W. Culp et al, State Trust Lands in the West: Fiduciary Duty in a Changing Landscape 13 (2006), *available at* https://www.lincolninst.edu/pubs/dl/1151_State%20Trust%20Lands%20PF014.pdf.

461. *See* Davies et al., *supra* note 29, at 231.

462. *See* Del Monte Mining & Milling Co. v. Last Chance Mining & Milling Co., 171 U.S. 55 (1898).

407. However, the simple concept of *ad coelum* does not fully capture the way the common law has evolved to address modern resource extraction. Oil, in particular, presents two pertinent problems. It is both fugacious—it can move through the earth—and it is fungible—it is hard to determine from where any single barrel of oil originated.[463] Because oil can move and is not readily identified with the surface lands, landowners might be leery of drilling, for fear of being accused of draining their neighbor's oil and thus being subjected to liability.[464] In order to address this issue—and to encourage oil development—the courts departed from the *ad coelum* doctrine and developed the rule of capture.

408. Simply stated, the rule of capture is as follows: "[A] landowner in a common source of oil and gas supply is legally privileged to take oil and gas from his or her land, even though in so doing he or she may take some of the oil or gas from adjoining lands."[465] This is the baseline rule in the United States. As the Pennsylvania Supreme Court observed as early as 1889, "[T]he one who controls the gas—has it in his grasp, so to speak—is the one who has possession in the legal as well as the ordinary sense of the word."[466] When citing support for the rule of capture, some courts analogized oil to wild animals under the doctrine *ferae naturae*,[467] which affords ownership to the person who captures the animal.[468] Other courts have derived support for the rule of capture from a property owner's absolute right to do as she will with her property.[469]

409. If strictly applied, the rule of capture would allow—and encourage—landowners to drain oil from under adjoining landowners' property.[470] Even if a landowner did not intend to drain oil from an adjacent plot, they would still be incentivized to drill near the property line to prevent their neighbor from draining oil from under their own property.

410. This dynamic was exemplified by the race to drill in Spindletop, Texas in the early 1900s. In Spindletop, operators drilled as many wells as possible on each lease in order to extract oil as quickly as possible. Within a month of the first major oil strike, there were 214 wells crowded on one hill, with some companies drilling on sites just large enough for the drilling rig itself.[471] In fact, the drilling rigs were so close together that their legs intertwined.[472] The result was plain, and rapidly ensured. The Spindletop field went bust in only five years and yielded only 5% of the reserve—a stunning figure compared to reservoirs where production is properly

463. Lowe, *supra* note 323, at 11–12.
464. *Ibid*; *see* Kelly v. Ohio Oil Co., 49 N.E. 399 (Ohio 1897).
465. Summers § 3.7.
466. Westmoreland & Cambria Nat. Gas Co. v. De Witt, et al.,18 A. 724 (1889); *see also* Barnard v. Monongahela Natural Gas Co., 216 Pa. 362, 65 A. 801 (1907).
467. *See Westmoreland*, 18 A. at 724.
468. Pierson v. Post, 3 Cai. R. 175, 2 Am. Dec. 264 (N.Y. 1805).
469. *See Kelly*, 49 N.E. at 399.
470. Bruce M. Kramer & Owen L. Anderson, *The Rule of Capture—An Oil and Gas Perspective*, 35 Envtl. L. 899, 899 (2005).
471. Yergin, *supra* note 42, at 85–86.
472. Boyce House, Spindletop, 50 Southwest Hist. Q. 36, 41 (Jul. 1946).

controlled and managed, which typically yield 25% to 50% of the reservoir and up to 70% in ideal conditions.[473]

411. The low yield was a result of the rush to produce, which dissipated gas pressure in the reservoir. Moreover, this unbridled development led to boom-and-bust cycles both for the oil producers and for local communities. As production rapidly increased, the price of oil dropped to 3 cents a barrel—while water was selling for 5 cents a cup. Also, the nearby community of Beaumont grew from 10,000 to 50,000 residents in a matter of months and struggled with crime, highly inflated prices, and fraud.[474] To this day, the Spindletop experience stands as a reminder of the need for regulation of oil extraction beyond the rule of capture: Standing alone, the rule of capture creates the risk for economic and physical waste, both of which were exemplified in Spindletop. Market gluts caused by the rule of capture's natural incentive for over-extraction drive down market prices below efficient levels—economic waste. Likewise, oil left in the reservoir because of suboptimal extraction practices creates waste—physical waste.

412. Because of these risks, modern oil and gas law in the United States has limited the rule of capture in several ways. These limits include the restraints inherent in the common law rule itself, limits imposed by correlative rights, and the statutory limits imposed by state conservation statutes.

I. Inherent Limits on the Rule of Capture

413. First, the common law rule of capture itself has inherent limits. These limits are rooted in the policy behind the rule—"to encourage development of oil and gas resources for the benefit of society."[475] One such limitation is that the rule of capture does not apply to oil and gas that was captured and then injected back underground for storage.[476] Generally, courts have found that once the oil or gas is captured, it becomes personal property, and therefore ownership is not surrendered by loss of possession.[477] Thus, while oil or gas that is injected back into the well may be physically recovered, or captured, by another driller, they may not legally do so. This limitation is consistent with the policy of promoting resource development and protecting ownership rights.

414. Courts also have recognized a limitation regarding the use of secondary or tertiary recovery methods. Under this limitation, a landowner cannot use enhanced

473. John S. Lowe, et al., Oil and Gas Law 47 (4th ed., 2002); American Petroleum Institute, Introduction to Oil and Gas Production 6–7 (1983).
474. Yergin, *supra* note 42, at 85–86.
475. Lowe, *supra* note 323, at 14.
476. *See* Champlin Exploration, Inc. v. Western Bridge & Steel Co., Inc., 597 P.2d 1215 (Okla. 1979).
477. *See, e.g.*, Pacific Gas & Electric Co. v. Zuckerman, 189 Cal. App. 3d 1113 (Cal. App. 1987); Texas American Energy Corp. v. Citizens Fidelity Bank Trust Co., 736 S.W.2d 25 (Ky. 1987).

recovery methods to impair another landowner's recovery using primary production.[478] For instance, in *Tidewater Oil Co. v. Jackson*,[479] the Tenth Circuit held a producer who used a secondary recovery method liable for the loss of production on a neighboring lease. The defendant had injected "unreasonable and excessive" amounts of water within 100 feet of the plaintiff's wells, which were under primary production at the time. As a result, the plaintiff's wells were flooded, rendering them unproductive.[480] The court awarded the plaintiff damages based on the fact that the secondary recovery method employed by the defendant resulted in "the water flooding activities [which] were intentional and the consequences foreseeable," even if they were lawfully carried out to get the defendant's well to produce.[481]

415. Further, if an operator wants to conduct a secondary recovery operation that will affect the oil and gas under another's property, she must offer the affected landowner the opportunity to join in the secondary recovery operation and receive his fair share of the production.[482] However, if the surface owner could not have used primary production and chooses not to participate in the secondary recovery operation, the rule of capture could apply to protect the extracting party. As the Nebraska Supreme Court has observed, "[W]here the primary recoverable oil has been exhausted, all interested parties in the field must be offered an opportunity to join in any unitization project to recovery secondary oil on a fair and equitable basis, and if any interested party refuses to join he should not be permitted to capitalize on that refusal. To hold otherwise would discourage unitization and encourage rather than avoid waste."[483]

416. It must be noted, however, that state laws are not uniform regarding limitations on the rule of capture. In the recent case of *Coastal Oil & Gas Corp.*,[484] for example, the Texas Supreme Court held that subsurface hydraulic fracturing of a natural gas well that extended into another owner's property was not a trespass. Consequently, gas that was drained away through hydraulic fracturing from under another person's land belonged to the driller as a result of a traditional application of the rule of capture.

II. Correlative Rights

417. Second, in addition to its inherent limits, the rule of capture is also limited by correlative rights—the rights of all owners to their fair share of the resource. The correlative rights doctrine is illustrated by *Elliff v. Texon Drilling Co.*, where the Texas Supreme Court held that the rule of capture did not shield the defendant from liability for negligent drilling. In *Elliff*, the defendant, Texon, negligently drilled a

478. ANDERSON & KUNTZ, *supra* note 321, at § 4.8.
479. Tidewater Oil Co. v. Jackson, 320 F.2d 157 (10th Cir. 1963).
480. *Ibid.*
481. *Ibid* at 164.
482. ANDERSON & KUNTZ, *supra* note 321, at § 4.8; 34 Oil and Gas Reporter 245 (2013).
483. 34 Oil and Gas Reporter 245 (2013); ANDERSON & KUNTZ, *supra* note 321, at § 4.8.
484. Coastal Oil & Gas Corp. v. Garza Energy Trust, 268 S.W.3d 1 (Tex. 2008).

well that blew out and burned a substantial amount of oil from under the plaintiff's property.[485] The court ruled that while the rule of capture affords landowners broad rights to extract oil, such extraction must be exercised with due regard to the rights of others. "No owner should be permitted to carry on his operation in reckless or lawless irresponsibility," the court wrote, "but must submit to such limitations as are necessary to enable each to get his own."[486]

III. State Conservation Laws

418. Finally, the common law rule of capture has been significantly modified by state conservation legislation, which is enforced and administered by state oil and gas, resource, or conservation agencies.[487] These conservation laws were adopted because correlative rights and other inherit limitations on the rule of capture were not sufficient to address the dual waste problem that rule created. Accordingly, states began passing conservation statutes to protect their resources. Each state takes its own approach to drafting conservation legislation, so no two laws are identical. Overall, however, the effect of these conservation statutes, combined with the correlative rights doctrine, is to create a "fair share" doctrine, in which "each owner is entitled to a fair chance to capture oil and gas under his or her property."[488] One's fair share is normally based on the relationship between the surface acreage owned and the proportion of the reservoir under that acreage.

419. While the details and overall approaches that states have adopted to oil conservation vary, several general categories of regulation have emerged. These range from laws that dictate well construction specifications to more comprehensive approaches that seek to maximize long-term production and uphold environmental, health, and safety standards.[489] The key categories of typical state conservation laws include: (1) well-spacing rules; (2) production regulations; and (3) pooling and unitization requirements.

A. Well-Spacing Rules

420. Well-spacing laws are designed to prevent waste and to protect correlative rights by limiting the number of wells in a given area. Generally, well-spacing laws allow conservation agencies to set both the distance between wells and the distance

485. Elliff v. Texon Drilling Co., 210 S.W.2d 558 (Tex. 1948).

486. *Ibid.*

487. These agencies are generally delegated executive authority to enforce oil and gas statutes and typically have the authority to issue regulations implementing the statutes and to hold hearings and adjudicate particular disputes. *See, e.g.*, Railroad Comm'n v. Rowan and Nichols Oil Co., 310 U.S. 573, 580 (1940).

488. Lowe, *supra* note 323, at 23.

489. David B. Spence, *Federalism, Regulatory Lags, and the Political Economy of Energy Production*, 161 U. Pa. L. Rev. 431, 447-48 (2013).

between wells and property lines.[490] The spacing between wells is generally based on the area that can be efficiently drained by a single well in the reservoir at issue. By allowing conservation agencies to set the number and location of wells, these laws, in addition to seeking to prevent oil waste, aim to limit surface impacts by reducing the number of wells, and thus, their environmental impact.[491] For example, in Oklahoma, the legislature has explicitly granted the conservation agency broad authority to establish well spacing regulations "[t]o prevent or to assist in preventing the various types of waste of oil or gas prohibited by statute, or any wastes, or to protect or assist in protecting the correlative rights of interested parties . . . ".[492]

421. One key problem that well-spacing rules, and oil conservation statutes more generally, grapple with is how to limit waste. Each conservation statute defines "waste," but, because of the United States' federalist system of government, those definitions vary from state to state. One key place these laws diverge in their definitions is whether economic waste counts as waste. For instance, the Wyoming Supreme Court held that the definition of waste in the Wyoming oil conservation statute did not include economic waste, because the state legislature had rejected draft statutory language that included economic considerations in the definition of waste, in favor of alternate language.[493] Therefore, the conservation agency could not consider economic waste when establishing their well-spacing requirements.[494] In Oklahoma, the consideration of economic waste is balanced with other factors.[495] By contrast, Texas' conservation statute addresses the question much more specifically, defining waste to include "production of oil in excess of transportation or market facilities or reasonable market demand . . . ".[496]

422. Another problem addressed by well-spacing and oil conservation statutes is the scope of where drilling is allowed. Generally, state conservation agencies do not permit the drilling of wells beyond the boundary of a reservoir, on land without oil present. This rule prevents drainage of oil from productive lands to unproductive lands. Also, with the increased use of horizontal drilling and hydraulic fracturing, states have needed to rethink how the particular ways in which their statutes address well spacing.[497] Originally, well-spacing patterns were developed assuming that wells would be drilled vertically and thus drain a specific area. However, horizontal drilling, combined with hydraulic fracturing, has allowed operators to reach lands never contemplated in original well-spacing laws. States have addressed these

490. LOWE, ET AL., *supra* note 473, at 62.
491. *See Ibid* at 64.
492. OKLA. STAT. tit., 52 § 87.1(a).
493. Larson v. Oil & Gas Conservation Comm'n, 569 P.2d 87 (Wyo. 1977).
494. *Ibid.* However, the Wyoming legislature later amended their statute to allow the conservation agency to consider economic factors. LOWE, ET AL., *supra* note 473, at 75; WYO. STAT. ANN. § 30-5-104(d)(iv).
495. *See* OKLA. STAT. ANN. tit. 52, § 87.1.
496. TEX. NAT. RES. CODE ANN. § 85.046 (10).
497. *See* Browning Oil Co. v. Luecke, 38 S.W.3d 625, 634 (Tex. App. 2000).

technological developments in different ways. Many states apply their original well-spacing laws to horizontal drilling. Others, including North Dakota,[498] issue permits allowing horizontal wells to span several well-spacing units. And some, like Texas, have altered the way the unit lines are established in order to more closely approximate the area that is actually drained through horizontal drilling and hydraulic fracturing.[499]

423. Well-spacing laws also have to consider when they will not apply. Due to geological features or small tracts of land, conservation statutes generally allow exceptions to well-spacing regulations under certain circumstances. The basic assumption underlying well-spacing limits is that the oil reservoir is homogenous and that drainage of the field will be radial, that is, approximately circular.[500] When these assumptions do not hold, exceptions are usually made. For example, there may be a geological formation that does not allow oil to flow to the well site that would otherwise be permitted under the well-spacing regime. In such a case, the conservation agency may allow an exception to the well-spacing requirement to accommodate the landowner's need to recover his fair share of the resource. An exception may also be granted if the parcel is too small to allow for a well site that threatens the site's mineral interest because of potential drainage from surrounding parcels.[501] However, in this situation, the conservation agency may force the small tract owner to pool with an adjacent parcel, allocating the small tract owner a share of the production rather than allowing the small tract owner to drill himself.[502]

424. Exceptions also may be made to well-spacing rules to prevent economic waste. For instance, in *Exxon Corp. v. Railroad Commission*,[503] the Texas Supreme Court upheld an order by the Texas Railroad Commission granting the operator an exception to the statewide spacing rule based on economic factors. There, the operator had an existing well that was originally drilled to access a deep formation. The operator petitioned for an exception to the well-spacing rules to allow access to a shallower formation that would have been non-compliant with the well-spacing rules. The operator claimed, and the Commission found, that drilling a new well to access the shallower formation was economically infeasible.[504]

B. Production Limits: Allowables and Prorationing

425. A second common feature of conservation legislation, in addition to the well-spacing requirements, are laws regulating production. Production allowables, or prorationing rules, limit the production of oil on a daily, weekly, or monthly basis. They seek to achieve the most efficient rate of withdrawal from the

498. *See* N.D. ADMIN. CODE § 43-02-03-18.
499. *See* TEX. ADMIN. CODE tit. no. 16, § 3.86.
500. LOWE, *supra* note 323, at 28.
501. LOWE, ET AL., *supra note* 473, at 86.
502. *Ibid.* at 64.
503. 571 S.W.2d 497 (Tex. 1978).
504. *Ibid.*

reservoir—the rate that allows for the largest recovery. Thus, the goal of allowables, like conservation legislation in general, is to limit waste and to allow all owners to enjoy their fair share of the resource. Most commonly, allowables are based on surface acreage drained by a well but, depending on the statute, other factors may be considered, including the depth of the well and the open flow capacity of the well.[505] For instance, in Texas the allowable regulation is based on the date the field was discovered, the acreage assigned to the well, and the depth of the well.[506]

426. Allowables can also be aimed at preventing economic waste, such as large swings in the price of oil. This type of allowable is called market-demand prorationing and may limit production to "reasonable" market demand. Often, the reasonable market demand is determined by the state conservation agency based on evidence submitted at hearings. Such evidence may include estimates of demand from oil purchasers, along with historic records from the state agency, federal agencies, and industrial associations.[507] Such allowables can either limit production or seek to ensure it does not fall below an optimal level. As with many other aspects of oil conservation laws, each state has its own procedures for setting and enforcing the allowables, but it is common for states to do so through administrative orders specific to producers.

427. In this connection, the agencies address "overages"—pumping that exceeds an allowable—and "underages"—where a driller produces less than the set allowable, the violation of which can create administrative consequences and, in some cases, civil or criminal liability.[508] State conservation agencies often have discretion in how to treat overages and underages. For instance, they sometimes deal with overages and underages through the transfer of an underage from one well to another well in the same formation,[509] by requiring an operator to reduce production to make up the overage over a specific period of time, or by shutting in an overproduced well.[510]

428. In addition to allowables, another way that state conservation legislation addresses production limits is through gas-oil ratio requirements. Often, oil and gas are present together, and the gas can help provide the required reservoir pressure to allow for efficient oil production. Thus, rapid loss of gas—and, concomitantly, reservoir pressure—can result in otherwise recoverable oil remaining underground. To address this risk, conservation statutes often impose oil-gas ratio limits. These ratios are aimed at maintaining the pressure in the reservoir to allow for maximum oil recovery by forcing operators to stop production from a well when they exceed the

505. Lowe et al., *supra* note 473, at 90–91, 118–19.
506. Tex. Admin. Code tit. no. 16, § 3.45.
507. *See* Anderson & Kuntz, *supra* note 321, at § 76.1.
508. *Ibid.* For instance, a Michigan court held that violation of an allocation order subjects the violator to possible civil liability for conversion of oil from other producers in the field. *See* Wronksi v. Sun Oil *Co.*, 279 N.W.2d 564 (Mich. App. 1979).
509. Anderson & Kuntz, *supra* note 321, at § 76.1.
510. *Ibid*; *see, e.g.*, Murray v. R&M Well Servicing and Drilling Co., 297 S.W.2d 225 (Tex. Civ. App. 1956).

established ratio. Typically, the gas-oil ratio is set administratively by the state agency, under authority derived from its statutory mandate to prevent waste.[511] However, some states clearly delegate the authority to fix ratios. For instance, North Dakota has given their state agency the authority "to require the operation of wells with efficient gas-oil and water-oil ratios, and to fix these ratios."[512]

C. Unitization and Pooling

429. A third type of conservation legislation is unitization and pooling. Unitization refers to the joint operation of some or all of the wells over a specific reservoir. The goal of unitization, similar to other conservation legislation regulations, is to reduce waste by allowing maximum production from the formation over the long run.[513] Once a set of wells is unitized, it is operated at the unit level, and one operator—the "unit operator"—is tasked with implementing and managing operations.[514] Production from the unit is carefully coordinated and divided among the parties. The formula for dividing production can include such variables as "surface acreage of each tract, net acre feet of reservoir beneath each tract, the volume of oil originally in place beneath each tract, the amount of oil each tract produced, the projected primary recovery of each tract, and differences in porosity and permeability throughout the unit area."[515] Generally, unitization is voluntary, in that it requires approval of a certain percentage of those with an interest in the mineral reserve being unitized. However, once that percentage is met, those interest holders who did not approve the unitization can be forced into unitization.

430. Although the process of unitization varies from state to state, the procedure generally includes an application by an interested party, notice to all possibly affected mineral interest holders, followed by an administrative hearing to determine if unitization is advisable and, if so, to define the unit. For instance, Oklahoma's statute requires a petition, notice, and an administrative hearing where the Commission is required to make specific findings, including: (1) that unitization is "reasonably necessary in order to effectively carry on pressure maintenance or repressuring operations . . . or any other nonprimary production form of joint effort calculated to substantially increase the ultimate recovery of oil and gas from the common source of supply"; (2) that unitization is feasible and will prevent waste and is likely to increase production; (3) that the cost of unitization will not be greater than the economic benefit from the increased oil and gas recovery; and (4) that unitization is "for the common good and will result in general advantage of the owners of the oil and gas rights within the common source of supply . . . ".[516] Upon making these findings, the state agency issues an order creating the unit.

511. ANDERSON & KUNTZ, *supra* note 321, at § 76.2; *see also* Ex parte Wood, 93 P.2d 1058 (Cal. App. 1939); Denver Prod. & Ref. Co. v. State, 184 P.2d 961 (Okla. 1947).
512. N.D. CENT. CODE ANN. § 38-08-04.
513. *See* ANDERSON & KUNTZ, *supra* note 321, at § 77.3.
514. LOWE ET AL., *supra* note 473, at 106.
515. *Ibid.*
516. Unitized Management of Common Sources of Supply, OKLA. STAT. tit. 52, § 287.3.

§2. State Property Law

431. In addition to state conservation regulation, state property law also may limit how oil is extracted, even under the rule of capture. Theories of oil ownership vary from state to state. Generally, states view the ownership of oil as an interest in land. This interest then can be further divided into two categories: corporal (the right to physical possession) or incorporeal (the right to use the land).

432. States that adhere to a corporal theory of ownership rely on the idea of ownership-in-place, where the landowner owns all substances, including oil, that are found below his land. By contrast, states that adhere to the incorporeal theory rely on the idea of a right to capture rather than ownership per se.[517] Instead of owning the oil itself, the landowner merely holds the exclusive right to capture oil through operations on his land.[518] It is difficult to delineate states as falling into either category because the expression of the state's theory of ownership often varies from case to case, and sometimes even varies in the same opinion. However, twelve states have nominally adopted, or expressed an inclination to adopt, the ownership-in-place theory,[519] and nine states have nominally adopted, or expressed an inclination to adopt, the exclusive-right-to-take theory.[520] The distinction between ownership theories is largely academic, with courts commonly abandoning logical consistency for pragmatic solutions.[521]

433. A state's adoption of either ownership theory may impact the claims and remedies available to a party. For instance, incorporeal rights, under the exclusive-right-to-take theory, can be abandoned at common law, through an extended period of non-use and intent to abandon, while corporal rights, under the ownership-in-place theory, cannot. Thus, in California, which has adopted the exclusive-right-to-take theory, the state Supreme Court held that oil and gas rights were subject to abandonment, but it also has found the necessary intent to abandon lacking by a plaintiff's non-use for 47 years.[522] Such decisions illustrate that even where abandonment is available as a claim, it may be difficult to demonstrate. Also, at common law some remedies may only be available to those holding a possessory, corporal right. These include trespass, ejectment, and compulsory partition.[523]

434. Ownership of oil in the ground is also classified either as an interest in real property or personal property. The distinction turns on the duration of the interest and its possessory quality. If the ownership interest is presently possessory and can

517. Lowe, *supra* note 323, at 35–36.

518. Anderson & Kuntz, *supra* note 321, at § 2.4.

519. These states include Alabama, Arkansas, Colorado, Kansas, Michigan, Mississippi, Montana, Ohio, Pennsylvania, Tennessee, Texas, and West Virginia. *See Ibid.*

520. These states include California, Illinois, Indiana, Kentucky, Louisiana, New Mexico, New York, Oklahoma, and Wyoming. *Ibid.*

521. *See Ibid.*

522. Gerhard v. Stephens, 442 P.2d 692 (Cal. 1968).

523. Lowe, *supra* note 323, at 35–36.

last for a considerable time (life estate or fee estate), then it is classified as real property. Otherwise, it is viewed as personal property. It is also worth noting that when oil is severed from the land, it always becomes personal property. The classification of an oil interest as either an interest in real property or personal property may be important for a range of reasons, including intestate or testamentary rights and tax liabilities.[524]

435. Because U.S. law generally allows the division of property rights into different categories,[525] the scope of ownership interests for a given oil deposit also is relevant to its extraction. There are a variety of such ownership interests a property owner may hold. A landowner may hold a fee interest in a property with mineral deposits. Ownership in fee includes the ownership of all possible property rights for that parcel—the entire bundle of rights.

436. By contrast, ownership of the parcel may be severed from the mineral rights, separating the mineral interests from surface ownership. This separation of interests is quite common. Typically, ownership of the mineral interest includes the right to use the surface, in the form of an easement, in a manner that is reasonably necessary to extract the minerals. Activities by mineral owners that have been found "reasonably necessary" include the laying of pipes to gather produced minerals,[526] the building of roads for transportation of minerals and housing for employees,[527] the construction of salt-water disposal pits,[528] and the administration of seismographic tests.[529] The mineral interest also includes the rights to profits from the sale of minerals, as well as the obligation to pay the cost of mineral development, the right to lease or sell the mineral interest, and the right to benefits under an oil and gas lease. One can also hold a leasehold interest, commonly called a working interest, in minerals. This interest is very common and is one of the primary methods used for oil and gas extraction in the United States. Generally, any of these ownership interests can be transferred through conveyance, inheritance, judicial action, or adverse possession.[530]

§3. Oil and Gas Leases

437. Because mineral interests often are severed from ownership in the surface property, the oil and gas lease is at the center of much of oil and gas law and development in the United States. At its core, an oil and gas lease is a contract between the mineral owner and the exploration company. Consequently, it is governed mainly by contract law and property law, both of which are the domain of state

524. *Ibid.* at 38–40.
525. Property rights in the United States are said to be a "bundle of sticks," and there are many different sticks in the bundle.
526. Rostocil v. Phillips Petroleum Co., 502 P.2d 825 (Kan. 1972).
527. Livingston v. Indian Territory Illuminating Oil Co., 91 F.2d 833 (10th Cir. 1937).
528. Feland v. Placid Oil Co., 171 N.W.2d 829 (N.D. 1969).
529. Yates v. Gulf Oil Corp., 182 F.2d 286 (5th Cir. 1950).
530. Lowe, *supra* note 323, at 43–44, 67.

law.[531] Generally, the lessee—the oil company—aims to obtain the right to develop the leased property without a specific obligation to do so. In turn, it promises the lessor—the owner of the mineral estate, often the surface landowner—a percentage of revenues if oil is produced. In exchange, the lessor gives the lessee the right to maintain the lease as long as it is possible to economically produce oil.[532] In this exchange, the lessee and lessor have different interests. The lessee seeks to pay as little as possible for access to the mineral estate, and to develop the leasehold only if it becomes economic to do so. The lessor, on the other hand, wants to optimize any royalty payment she may receive. The lease, then, which is negotiated between these two parties, works to balance the lessee's interest with the lessor's.

438. An oil and gas lease is different than an ordinary real property lease. Because it includes the right to take oil and gas from the property, it generally does not run for a set term of years, but rather, terminates when oil and gas is no longer being produced in paying quantities. This is what is referred to as a determinable interest. An oil and gas lease must be in writing to avoid the statute of frauds; it must contain words of grant, a description of the property to be leased, and a designation of the parties. It also must be properly executed. As an interest in real property, oil and gas leases can be transferred and are assignable unless limitations in the lease so specify. However, most leases specifically allow for assignment. When a lease is assigned, the original lessee remains liable, as they are still in privity of contract with the lessor. However, the original lessee can require indemnification from the assignee. Single oil and gas leases are frequently transferred to allow for the consolidation of large blocks of leases, which are more efficient to explore and develop.[533]

439. Four key provisions are common to oil and gas leases in the United States: (1) the granting clause; (2) the habendum clause; (3) the delay-rental clause; and (4) the royalty clause. Although these provisions are common, there is significant variation in oil and gas leases, and there is no standard form across jurisdictions. Specific leases vary from region to region and transaction to transaction.

I. The Granting Clause

440. The granting clause defines the rights that the mineral owner grants to the lessee. The granting clause must address: "(1) What rights are given to use the land; (2) What substances are covered; and (3) What land and what interests are subject to the lease."[534] Implied within the grant to access mineral rights underlying a property is the right to reasonable use of the surface in order to locate, develop, and produce oil and gas.[535] In application, this implied right is treated as an implied

531. Lowe et al., *supra* note 473, at 125.
532. *Id.* at 183–85.
533. Lowe, *supra* note 323, at 34, 70–78, 377–79.
534. Lowe et al., *supra* note 473, at 128.
535. *See, e.g.*, Hunt Oil Co. v. Kerbaugh, 283 N.W.2d 131 (N.D. 1979).

easement and includes such activities as conducting seismographic tests, erecting oil storage tanks, and building roads.[536]

441. The implied easement to use the surface, however, is limited to use that is reasonably necessary to develop the mineral resource, and such efforts must be undertaken in a manner that gives due regard to the surface owner's right to use and enjoy the surface property. This work also must be in compliance with all applicable statutes, ordinances, and regulations.[537] Statutory and common law in any given jurisdiction define the scope of these implied rights. However, in an oil and gas lease, parties can—and often do—modify the implied right to use the surface land, delineating more specifically what rights for surface use the lease does or does not grant.[538]

II. The Habendum Clause

442. The habendum clause sets the term of the lease. Typically, it sets two terms: a primary term and a secondary term. During the primary term, the operator has the right to explore for, or drill, oil or gas but does not have any obligation to do so. Historically, the primary term was commonly ten years, and it still may be this long in unproven areas. However, it is more common now to have a one-to-five-year primary term.[539]

443. The secondary term kicks in with production. It extends the lease past the primary term once there is production; secondary terms typically run as long as the operator continues to produce oil or gas in paying quantities.[540] In effect, the secondary term creates a fee simple determinable, which will terminate when the operator fails to produce oil or gas in paying quantities. Common legal issues that arise surrounding the secondary term include: (1) how much "production" is necessary to extend the lease; (2) determining when production terminates; and (3) what constitutes "paying quantities" under the lease (that is, how to determine the lease's economic viability).[541]

444. To address these interpretation problems, courts have imposed a "reasonably prudent operator" standard on oil and gas leases. This standard looks to "whether or not under all the relevant circumstances a reasonably prudent operator would, for the purpose of making a profit and not merely for speculation, continue to operate a well."[542] This requirement to produce oil and gas in order to extend the

---

536. LOWE, *supra* note 323, at 196.
537. *See* Bruce M. Kramer, *The Legal Framework of Analyzing Multiple Surface Use Issues*, 44 ROCKY MTN. MIN. L. FND. J. 273 (2007); *see also, e.g.*, McFarland v. Taylor, 65 S.W.3d 468 (Ark. App. 2002).
538. LOWE, *supra* note 323, at 204.
539. LOWE ET AL., *supra* note 473, at 152.
540. *Ibid.*
541. LOWE, *supra* note 323, at 209–211.
542. *Ibid.* at 212 (citing Clifton v. Koontz, 325 S.W.2d 684, 691 (Tex.1959)).

lease term can put a substantial burden on the lessee. Even a lessee who is acting diligently and in good faith may find it difficult to meet these requirements.[543]

445. To protect themselves, lessees may insert a number of savings clauses that generally allow a lease to be extended through constructive production instead of actual production. One example of these clauses is the "operation clause." Operation clauses are designed to protect the lessee once drilling operations are in progress—that is, when the lessee has begun drilling during the primary term but has not yet produced oil or gas, which is what would activate the secondary term.[544] The operation clause, then, may be most determinative of a given lease's term.[545] An example of the definition of operations is "preparing the drillsite location or access road, drilling, testing, reworking, recompleting, deepening, sidetracking, plugging back, or repairing a well in search for, or in an endeavor to maintain, re-establish or enhance the production of oil or gas or both, whether or not in paying quantities"—that is, "preliminary actions usually associated with actual drilling on the premises in good faith, [which are] diligently pursue[d] to completion."[546] The inclusion of such clauses allows the lessee to complete the drilling operations that were commenced during the primary term.[547] It thus acts to bridge a possible gap in the habendum clause between the lease's primary and secondary terms.

III. The Delay-Rental Clause

446. Another common provision in an oil and gas lease is the delay rental clause. The delay rental clause avoids the implied covenant to drill through the payment of a delay rental, which is normally a nominal amount such as USD 1 per mineral acre.[548] Such clauses allow lessees to avoid the need to immediately produce oil from a given lease. The delay rental clause allows the lessee to simply pay the lessor instead of drilling during the primary term. That is, the delay rental clause is one way in which parties contract around the implied common law duty to drill.[549]

IV. The Royalty Clause

447. Finally, oil and gas leases include a royalty clause. This clause specifies the compensation that the lessor will receive in return for granting the right to extract oil and gas from their property. The lease normally states the royalty payment as "a

543. *Ibid.* at 203.
544. *Ibid.* at 250.
545. LOWE, *supra* note 323, at 249–250 2014.
546. *Ibid.* at 251; John S. Lowe, Pennsylvania Oil and Gas Lease, Paid Up Oil and Gas Lease, *in* OIL AND GAS LAW IN A NUTSHELL, *supra* note 323, at 477.
547. LOWE ET AL., *supra* note 473, at 220.
548. *Ibid.* at 153; LOWE, *supra* note 323, at 219; Warm Springs Dev. Co. v. McAulay, 576 P.2d 1120 (Nev. 1978).
549. LOWE ET AL., *supra* note 473, at 152–153.

percentage of production, or its value, or proceeds of its sale, free of the cost of production."[550] The standard royalty payment until the 1970s was 1/8 of production, value, or proceeds of sale. However, royalty payments of 1/6, 3/16, or 1/5 are more common today.[551] The royalty payment is sometimes based on a sliding scale of factors, such as the recovery of costs and minimum payment.

448. Failure to pay a royalty normally results in court action and damages to compensate the lessor. Courts, however, tend to be reluctant to terminate leases due to non-payment of royalties.[552] This is because the royalty payment is considered a covenant from the lessee to the lessor, and the standard remedy for a breach of an express covenant is damages. Also, courts generally reason that the lessor can be made whole through payment of the royalty plus any interest—a sufficient remedy under law. In *Cannon v. Cassidy*,[553] for instance, the Oklahoma Supreme Court refused to grant the lessor a cancellation of the lease for royalty non-payment over approximately eleven months. Because the lease did not specifically allow for cancellation, the court was reluctant to read that remedy into the contract. In so ruling, the court noted an overwhelming majority rule that a lessee's "failure to pay royalty as provided by the lease will not give lessors sufficient grounds to declare a forfeiture unless by the express terms of that lease they are given that right and power."[554]

V. Additional Key Lease Terms

449. Beyond these four core provisions—the granting, habendum, delay-rental, and royalty clauses—oil and gas leases often include a number of other terms that seek to achieve more specific objectives, or preserve flexibility within the context of a larger oil and gas development operation. Pooling and unitization provisions give the lessee the authority to include the lease in a pool, or unit. Without such a provision, the lessee would not have the right to willingly enter into a pooling or unitization agreement.[555] Further, pooling clauses may specify that production anywhere in the larger unit where the lease exists constitutes production for the lease itself. Such provisions limit royalty payments to the proportion of the lease to the total unit.[556]

450. A force-majeure clause allows the lessee to preserve the lease when circumstances beyond its control prevent it from operating or producing. The clause defines what constitutes a force-majeure, although construction of these clauses is often at issue in litigation.[557] Events covered by force-majeure clauses typically are

550. LOWE, *supra* note 323, at 301–302 (2014).
551. *Ibid.*
552. *Ibid.* at 326; *see also, e.g.*, Cannon v. Cassidy, 542 P.2d 514 (Okla. 1975).
553. 542 P.2d 514 (Okla. 1975).
554. *Ibid.* at 516.
555. LOWE, *supra* note 323, at 259–260.
556. *Ibid.* at 260; LOWE ET AL., *supra* note 473, at 222–223.
557. LOWE ET AL., *supra* note 473, at 228.

unforeseen, extraordinary, and "beyond the control of the asserting party, such as a flood, war, riot, or act of God, that delays performance under the contract."[558]

451. Shut-in royalty clauses allow the lessor to maintain a lease if there is a well that has been shut-in for economic reasons. Without this clause, an operator may lose a lease when it drills a well that is capable of production but that it chooses not to use to produce oil immediately because of market conditions. The shut-in royalty clause thus allows the lessee to pay the lessor a royalty payment in order to keep the lease. This clause often is drafted to include only gas, because oil is easier to store during poor market conditions. However, out of caution, leases sometimes include shut-in clauses for both oil and gas.[559]

452. Cessation of production clauses deal with the determinable nature of the secondary term of the lease: If production ceases, the lease automatically ends. This can be a harsh result for the lessee. The common law allowed for a temporary cessation of operations without forfeiture of the lease, but would terminate the lease if there was a permanent cessation.[560] Courts thus faced the question whether the cessation was "temporary" or "permanent," and would consider a variety of factors to make this determination, including duration, cause, and the lessee's efforts to restore production.[561] Today, instead of leaving this issue to the courts, many oil and gas leases set a specific period of cessation that terminates the lease—generally between 60 and 180 days.[562]

VI. Implied Covenants in Oil and Gas Leases

453. In addition to oil and gas leases' express language, courts have developed a set of implied covenants to protect the lessor's interests. These covenants derive from the unequal sophistication and bargaining power of the lessee and lessor, the fact that the lessee typically drafts the lease, and the tension between lessee's inherent interest in controlling the pace of development and the lessor's interest in relatively rapid production to yield faster royalty payments. There is a debate whether the implied covenants are implied in fact—representing the unexpressed intention of the parties—or implied in law—not contemplated by the parties but inserted by the courts to correct the imbalance in bargaining power. The most important implication for the distinction is that, if the covenants are implied in fact, they can be written out of the contract completely. In practice, most states allow for alteration

558. Jessica S. Hoppe & William S. Wright, *Force Majeure Clauses in Leases*, PROBATE & PROPERTY, at 8, 9 (2007).
559. LOWE, *supra* note 323, at 281, 285.
560. *Ibid.* at 288–89.
561. *Ibid*; *see also* Saulsberry v. Siegel, 252 S.W.2d 834 (Ark. 1952).
562. LOWE, *supra* note 323, at 289. In addition to the savings clauses outlined above, the lessee may also include a series of administrative clauses to protect its interest. These may include warranty clauses, subrogation clauses, equipment removal provisions, and notice of assignment clauses.

of the implied covenants through specific contract language. States vary, however, in how specific the contract language must be to override an implied covenant.[563]

454. Underlying all of the implied covenants is the reasonably prudent operator standard. The presumption under the law is that the lessee will conduct itself as a reasonably prudent operator would under the same or similar circumstances. This standard generally has been interpreted to impose a duty on the lessee to act in good faith, to act competently, and to act with due concern for the lessor's interests.[564]

455. Courts have developed several other covenants to help ensure that lessors' interests are adequately protected. These include: (1) the covenant to test; (2) the covenant to reasonably develop; (3) the covenant to further explore; (4) the covenant to protect against drainage; (5) the covenant to market; and (6) the covenant of diligent and prudent operations. Each of these covenants historically has been critical to oil exploitation in the United States. However, it is important to note that as lessors become more sophisticated and negotiate express covenants to protect their interests, these implied covenants have become less important. The remedy for a breach of an implied covenant is generally money damages. The majority rule is that the court may only cancel a lease if damages completely fail to compensate the plaintiff.[565]

456. The covenant to test requires the lessee to drill a test well on the leased property within a reasonable amount of time. However, today, parties typically write around this covenant using the delay rental clause discussed above.[566]

457. The covenant to reasonably develop requires that an operator that has found oil continue to develop the lease in a reasonable manner.[567] For instance, the lessee must commence preliminary operations within an agreed upon period of time and undertake that work in "good faith" with due diligence.[568]

458. The covenant to further explore dictates that once a lessee has secured production on a lease and extended its term, it must continue to explore for additional mineral resources if reasonably prudent to do so. Violation of this covenant can be more difficult for the lessor to prove, because it requires the lessor to demonstrate that additional exploration can reasonably be expected to be successful.[569]

---

563. *Ibid.* at 330–34; *see also* Heritage Resources Inc. v. NationsBank, 939 S.W.2d 118 (Tex. 1996) (finding that rather broad language modified an implied covenant); Rogers v. Westerman Farm Co., 29 P.3d 887 (Colo. 2001) (finding that the same language was not enough to limit the implied covenant).
564. Lowe, *supra* note 323, at 334–35.
565. *Ibid.* at 346–47, 375.
566. *Ibid.* at 339.
567. *Ibid.* at 339–40.
568. John S. Lowe, et al., Oil and Gas Law 253–54 (6th ed. 2014).
569. *Ibid.* at 348, 351.

459. The covenant to protect against drainage requires the lessee to act if the mineral resources are threatened by drainage from drilling on an adjacent property. This covenant may require the lessee to drill even when it would otherwise be in the lessee's discretion—for example, if the lessee is currently in the primary term or holds the lease through delay rentals. To some extent, the need for this covenant has been diminished over time, as modern drilling and spacing rules reduce the threat of drainage and thus the need to drill to protect against it.[570]

460. The covenant to market requires the lessee to use due diligence to market oil and gas at a reasonable price and in a reasonable time. Most problems with the covenant to market involve gas, because it is harder to store and transport. Further, most oil leases stipulate in-kind payment. Under in-kind payment, the landowner takes possession of the oil before it is sold to a third party. Therefore, the lessor has increased flexibility and can manage the marketing of the oil as it sees fit.[571]

461. The final covenant, the covenant of diligent and proper operations, overlaps with many of the covenants outlined above. Four claims that generally arise under this covenant include that the lessee: (1) damaged the property through negligence or incompetence; (2) prematurely abandoned a well capable of economic production; (3) failed to use advance production techniques where appropriate; or (4) failed to seek favorable regulatory action.[572]

§4. Oil Leasing on Federal Lands

462. In the United States, the law for exploitation of oil on federal lands is also important, as oil and gas located on federal lands constitute a vital source of energy in the country. Oil reserves on federal lands in Alaska and the Outer Continental Shelf (OCS) already account for significant domestic oil production, and are thought to contain even more resources.[573] Generally, the law for oil and gas on public lands is patterned after the law that applies on private lands.[574] However, there are some important differences as to federal land both for onshore and offshore oil exploitation. Also, federal lands in Alaska are a special case and are managed under the Alaska Claim Settlement Act of 1971 and the Alaska National Interest Lands Conservation Act of 1980.[575]

463. Onshore oil extraction on federal lands is managed through the Department of the Interior's Bureau of Land Management. While private law governing oil and gas extraction serves as the model for federal law, the government does not

570. *Ibid.* at 354.
571. *Ibid.* at 364–65.
572. *Ibid.* at 374–75.
573. Lowe et al., *supra* note 473, at 837.
574. *Ibid.* at 838.
575. *Ibid.* at 843–844; *see also* 43 U.S.C. §§ 1601–1629h; 16 U.S.C. §§ 1301–1303, 3111–3126, 3141–3151, 3161–3173, 3181–3183, 3191–3215, 3231–3233.

have as much freedom to negotiate the terms of their oil and gas leases as do private parties. Rather, many of the terms are set by law, and those laws specifically address what, in the private context, is handled by implied covenants.[576] Further, an oil and gas lease with the government does not convey a "dominant estate" and all of the associated rights and easements; rather, the government retains greater rights in its surface lands than do many private landowners who lease their mineral rights for oil exploitation.[577]

464. Historically, oil and gas were classified as "locatable" minerals under the Oil Placer Act of 1897 and, therefore, were subject to claim under the General Mining Law of 1872.[578] Under that statute, one could lay a claim to locatable minerals on public land and develop those mineral deposits without payment to the government.[579] Discovery and development were the key steps to ownership. The U.S. Court of Appeals for the Ninth Circuit described the process:

> Under the general mining law enacted in 1872, individuals were encouraged to prospect, explore and develop the mineral resources of the public domain through an assurance of ultimate private ownership of the minerals and the lands so developed. The system envisaged . . . was that the prospector could go out into the public domain, search for minerals and upon discovery establish a claim to the lands upon which the discovery was made. This required location of the claim, which involved staking the corners of the claim, posting a notice of location thereon and complying with the state laws concerning the filing or recording of the claim . . . .
> [T]he claimant thus had the present and exclusive possession for the purpose of mining, but the federal government retained fee title and could protect the land and the surface resources from trespass, waste or from uses other than those associated with mining. The claimant could apply for a patent to the land under 30 U.S.C. § 29, and, upon meeting the statutory requirements, would be granted a patent which usually conveyed the full fee title to the land.[580]

465. The goal of this process, of course, was to promote development of public lands.[581] In a burgeoning West, where government resources were scarce and the public lands needing management and supervision were vast, the government aimed to promote private development, entrepreneurial spirit, and general expansion westward—a legal embodiment, one could say, of the nation's view of its "Manifest Destiny."

576. *Ibid.*
577. *Ibid.* (citing 2 G. COGGINS, PUBLIC NATURAL RESOURCES LAW § 23.01[1] (1992)).
578. *Ibid.* at 839.
579. *Ibid.*
580. United States v. Curtis-Nevada Mines, Inc., 611 F.2d 1277 (9th Cir. 1980).
581. The Statutes at Large of the United States of America, vol. 29. Washington: Government Printing Office, 1897, "An Act to authorize the entry and patenting of lands containing petroleum and other mineral oils under the placer-mining laws of the United States" (Feb. 11, 1897).

466. Soon, however, this all changed. In the early 1900s, oil became strategically important to the United States Navy, a fact that became particularly acute with the onset of World War I. Thus, in 1909, on the advice of the Director of the Geological Survey, President Taft issued a temporary order withdrawing over 3 million acres of public lands in California and Wyoming from private oil development under the 1872 law.[582] As the Director had noted, the government was effectively giving away oil reserves that it would then need to turn around and repurchase from private developers.[583] Congress ratified the President's actions with the Picket Act of 1910,[584] and the United States Supreme Court also upheld the President's executive withdrawal.[585]

467. Subsequently, in 1920, Congress passed the Mineral Leasing Act (MLA), which reclassified oil from a "locatable" mineral to a "leasable" mineral. As a leasable mineral, federal oil reserves no longer afforded the right to public land development through location, but instead, had to be leased, with royalty payments made to the government.[586]

468. Under the MLA, then, a six-point system of governance was established for mineral leasing on federal lands. First, unlike under the 1872 law, private citizens no longer had a right to mine on federal land, but rather, had to obtain government approval to do so. Second, the United States would recoup an economic return for the severance of its resources. Third, the implementing agencies had discretion to determine which leases would proceed. Fourth, leases would be for fixed terms. Fifth, the government could cancel leases if a lessee did not act with due diligence to develop its lease. And, finally, the government would include within the lease terms conditions to protect the surface land.[587]

469. The MLA process remained in place for many years. Under this law, a bifurcated method for oil and gas leasing on federal lands developed. Leases under this law could be issued on either a competitive or a noncompetitive basis. Competitive leases were used for "known geologic structures" (KGSs)—areas where it was known that mineral resources were available.[588] Noncompetitive leases, on the other hand, were used for non-KGS lands. Thus, two different systems were used to lease federal lands for oil and gas exploration. And, for leases that fell within the category of noncompetitive leases, two subsystems were employed: (1) so-called "over the counter" leases, where the first person to file an application for a tract of land would receive it; and (2) a "simultaneous" system, often used for expiring leases, where numerous parties would submit applications, which were all deemed

582. United States v. Midwest Oil Co., 236 U.S. 459, 467 (1915).
583. *Id* at 466–467.
584. Ch. 421, 36 Stat. 847 (1910).
585. *Midwest Oil Co.,* 236 U.S. at 459.
586. LOWE ET AL., *supra* note 473, at 839.
587. *See* CHARLES F. WILKINSON, CROSSING THE NEXT MERIDIAN: LAND, WATER, AND THE FUTURE OF THE AMERICAN WEST 53–54 (1992).
588. 43 C.F.R. § 3100.0–5 (1987).

to have been submitted "simultaneously," and then a computer-based "lottery" system would determine who would receive the lease.[589]

470. Several problems emerged with these systems. First, non-KGS lands often proved to have extremely valuable resources on them. In one case, for instance, the Interior Department leased 33,000 acres under the noncompetitive system for USD 1 per acre, only to turn around the following year and lease another 24,000 acres in the same area using competitive bidding for USD 1,705 per acre.[590]

471. Second, the Interior Department construed the definition of KGS quite narrowly, "so that many lands that were believed to be valuable for oil and gas were found not to be in a KGS and thus outside the competitive leasing process."[591] Eventually, this led Secretary of the Interior Cecil Andrus to halt noncompetitive oil and gas leasing in 1980, but his successor, James Watt, immediately lifted the moratorium in 1981. The result was predictable. Lessees quickly resold leases purchased cheaply from the government for windfall profits, such as the fourteen noncompetitive leases issued in Wyoming, which the lessees immediately turned around and sold for millions of dollars.[592]

472. Third, parties began to game the noncompetitive leasing system. Under applicable law, a party could only make one application for a noncompetitive lease. Thus, to increase the chances that a lease would be issued to a given party, filing services soon proliferated. "For a fee, these services would choose the tracts that were supposedly the most valuable and would then file applications on those tracts on behalf of each of its customers . . . . At least some of these filing services were run by unscrupulous operators who made false promises to customers while filing on virtually worthless tracts of land."[593] The confluence of these problems, plus environmental and other concerns, soon led Congress to amend the MLA for oil and gas leasing on federal lands, just as it did for coal.[594]

473. One early amendment was the Mineral Leasing Act for Acquired Lands of 1947.[595] This law gave other land management agencies (in addition to the BLM) some say in the leasing process for lands under their jurisdiction, including the ability to impose conditions on such leases.[596] The Act applies only to lands that the United States acquired after 1947, and thus does not apply to large swaths of federal lands in the western United States. It mainly affects eastern U.S. Forest Service

589. 43 C.F.R. §§ 3111.1, 3112 (1987).
590. James Rasband et al., Natural Resources Law and Policy 1155 (2d ed. 2009).
591. *Ibid.*
592. *Ibid.*
593. *Ibid.*
594. *See* the Part on Coal.
595. Mineral Leasing Act for Acquired Lands of 1947, 61 Stat. 913 (codified as amended at 30 U.S.C. §§ 351-360).
596. Rasband et al., *supra* note 590, at 841. "No mineral deposit covered by this section shall be leased except with the consent of the head of the executive department, independent establishment, or instrumentality having jurisdiction over the lands containing such deposit . . . ." 30 U.S.C. § 352.

lands, and requires the BLM to obtain Forest Service consent before leasing such lands for mineral development.[597] As most oil reserves on federal land are located in the western United States, this law does not affect a majority of federal oil leases.[598]

474. Even more critical was the Federal Onshore Oil and Gas Leasing Reform Act of 1987 (FOOGLRA).[599] This law, while keeping the main structure of the MLA in place, introduced a number of innovations and reforms to oil and gas leasing, thus establishing the modern leasing system for these resources on federal lands.[600]

475. First, while FOOGLRA maintained the competitive/noncompetitive system of bidding, the Act also substantially modified it. Now, all leases for oil on public lands must be first offered under a competitive leasing system; only if the government does not receive a minimum bid under the competitive bidding system can the parcels be released to noncompetitive leasing.[601]

476. Second, the law eliminated the KGS system of lease classification. Thus, competitive leases "are conducted by oral bidding,"[602] and the winning bids are given to the "highest bid from a responsible bidder which is equal to or greater than the national minimum acceptable bid . . . ."[603] The Secretary of the Interior determines the minimum bids by regulation. By contrast, noncompetitive leases "are issued to the first qualified applicant upon payment of a nonrefundable USD 75 fee."[604]

477. Third, lessees pay both minimum rental payments plus royalties to the federal government. FOOGLRA set the minimum rental payment at USD 1.50 per acre for a lease's first five years and at USD 2 per acre for the lease's duration. Minimum oil royalty payments under the law are 12.5%.[605] Currently, the Department of the Interior applies a royalty fee of 12.5% on both competitive and noncompetitive leases.[606] However, FOOGLRA left in place existing lease terms. Competitive leases thus have a five-year primary term but continue in effect "so long [thereafter]

---

597. *See* Kerr-McCree Corp. v. Hodel, DCDC 1987.
598. *The 10 Most Oil Rich States*, USA TODAY, *available at* http://www.usatoday.com/story/money/business/2013/08/03/the-most-oil-rich-states/2613497/.
599. Federal Onshore Oil and Gas Leasing Reform Act of 1987, Pub. L. 100-203, 101 Stat. 1330-256 (codified at 30 U.S.C. §§ 195, 226-3).
600. RASBAND ET AL., *supra* note 590, at 840.
601. 30 U.S.C. § 226(b)(1)(A).
602. RASBAND ET AL., *supra* note 590, at 1156.
603. 30 U.S.C. § 226(b)(1)(A).
604. RASBAND ET AL., *supra* note 590, at 1156.
605. 30 U.S.C. § 226(b)(1)(A).
606. U.S. Dep't of the Interior: Bureau of Land Mgmt., *Qs and As About Oil and Gas Leasing*, http://www.blm.gov/wo/st/en/prog/energy/oil_and_gas/questions_and_answers.html. However, there are some exceptions, such as reduced royalties on certain oil leases with declining production and using a sliding scale on older leases, and reinstated leases. *Ibid.*

as oil and gas is produced in paying quantities."[607] Unit and pooling plans are encouraged to minimize waste and maximize yield.

478. Fourth, FOOGLRA specifies that certain lands are unavailable for oil and gas production.[608] This includes national monuments, national parks, wildlife refuges, federal lands within wilderness study areas, and lands within Indian reservations.[609]

479. Thus, today, oil and gas exploitation proceeds on federal lands under a general three-phase process. Lands are first made available by the government for leasing, and competitive bidding commences. Next, the winning bidders begin the exploration process. Then, if exploration reveals promising tracts, development occurs. Before actual drilling can commence, however, either for exploration or for development, the lessee must obtain a drilling permit from the relevant BLM field office. FOOGLRA further mandates that such permits may not be approved until it is ensured that proper surface management, including reclamation, will occur. This may include the posting of a bond or other surety to protect the area.[610]

480. Finally, it must be noted that in leasing oil and gas on public lands, the federal government must follow applicable environmental regulations. These include the National Environmental Policy Act, the Endangered Species Act, and the National Historic Preservation Act (NHPA). We address these laws in more detail in Part V, *Interaction between Energy Law and Environmental Law*. These laws are important to note because they both affect the leasing process, often making it more time-intensive as NEPA imposes important procedural requirements, and may limit where leasing may be performed. For instance, the presence of a listed species under the ESA may be a significant roadblock to oil and gas development.

§5. Offshore Oil and Gas Exploitation

481. Offshore oil exploitation in the United States is managed under a distinct legal regime that is dominated by federal rather than state law. This regime has several parallels to the leasing system used for onshore federal oil and gas leasing, but also some significant differences.

607. 30 U.S.C. § 226(e).
608. 30 U.S.C. § 226-3.
609. *Ibid.* The Indian Mineral Leasing Act of 1938 authorized the Secretary of the Interior to approve the leasing of allotted and unalloted lands, and in 1982 the Indian Mineral Development Act authorized Indian tribes to enter into agreements for oil and gas development in addition to leases under the IMLA. This law allows an individual Indian who owns a restricted or beneficial interest in mineral resources to include those resources in a tribal mineral agreement, if the parties agree and the Secretary of the Interior determines that such participation is in the individual's best interest. Intermountain Oil and Gas BMP Project, Univ. of Colo. Law Sch., *Indian Law*, http://www.oilandgasbmps.org/laws/tribal/.
610. 30 U.S.C. § 226(g).

482. An initial key question is the scope of federal jurisdiction over offshore oil and gas leasing. The United States has claimed a 200-mile offshore exclusive economic zone surrounding the country. In the middle of the twentieth century, the U.S. Supreme Court determined that the federal government had control over the entire outer continental shelf, including shallow coastal waters.[611] However, in the Submerged Lands Act of 1953, Congress modified this determination and initially spelled out federal jurisdiction over offshore lands with oil deposits. Under the Act, states were given jurisdiction over submerged lands within three nautical miles of their shorelines—with the exception of Texas and the west coast of Florida, which received jurisdiction up to nine nautical miles off the coastline.[612] Beyond these narrow coastal zones, the federal government retained jurisdiction over mineral development.

483. Also in 1953, Congress adopted the Outer Continental Shelf Lands Act (OCSLA),[613] which authorized federal leasing of oil and gas deposits in the outer continental shelf. Subsequently, in 1978, Congress amended OSCLA,[614] and that law, combined with the Federal Oil and Gas Royalty Management Act of 1982,[615] now combine to form the basic structure for federal offshore oil and gas leasing in the United States. The Federal Oil and Gas Royalty Management Act was passed to "ensure that offshore federal lands have proper accounting and enforcement mechanisms."[616] Likewise, the 1978 amendments to OSCLA sought to ensure that "the sale of a lease [is] a distinct stage of the OCS administrative process, carefully separated from the issuance of a federal license or permit to explore, develop, or produce gas or oil on the OCS."[617]

484. Thus, under OSCLA as amended in 1978, offshore oil and gas exploitation in federal waters proceeds under four separate stages. The first stage is leasing program preparation. OCSLA requires the Department of the Interior to prepare a five-year plan for proposed outer continental shelf leases. In doing so, the agency is required to seek comments from interested federal agencies and states, and to respond in writing to the comments received from states. Interior then submits the plan to the President and Congress, along with comments from the states.[618]

485. The second stage is the sale of OCS leases. In this stage, Interior solicits bids for, and then issues, OCS leases.[619] Before the agency can conduct the sale,

611. United States v. California, 332 U.S. 19 (1947); United States v. Louisiana, 339 U.S. 699 (1950); United States v. Texas, 339 U.S. 707 (1950).
612. Davies et al, *supra* note 29, at 839.
613. Outer Continental Shelf Lands Act, 67 Stat. 462 (codified as amended at 43 U.S.C. 1331 et al.).
614. Outer Continental Shelf Lands Act Amendments of 1978, Pub. L. 95-372, 92 Stat. 629 (codified as amended at 43 U.S.C. § 1801 et al.).
615. Federal Oil and Gas Royalty Management Act of 1982, Pub. L. 97-451, 96 Stat. 2447 (codified as amended at § 30 U.S.C. 1701 et al.).
616. Davies et al, *supra* note 29, at 839.
617. Sec'y of the Interior v. California, 464 U.S. 312 (1984).
618. 43 U.S.C. § 1344.
619. 43 U.S.C. § 1337(a).

however, it must go through several procedural steps. These include ensuring compliance with the National Environmental Policy Act's environmental review process and the Endangered Species Act; requesting comments from the governors of affected states as to the size, timing, and location of the sale; and addressing any governors' comments.[620] "Interior is required to accept these [governors'] recommendations if it determines they strike a reasonable balance between the national interest and the well-being of the citizens of the affected state."[621] Once these procedural steps are complete, Interior may proceed with the lease sale, typically conducted via sealed bids. Those leases, however, only give the lease winners the right to conduct preliminary activities, not to actually extract oil.

486. The third phase is exploration. Under this stage, lessees submit exploration plans to the agency, which they cannot implement until they have agency approval.[622] These plans must include a certification of compliance with the Coastal Zone Management Act,[623] which provides for the management of the United States' coastal resources (including the Great Lakes). The agency thus reviews the plan to ensure that it will not cause significant harm to the marine, coastal, or human environment.[624] OCSLA also "expressly provides for federal disapproval of a plan that is not consistent with an applicable state management plan unless the Secretary of Commerce finds that the plan is consistent with CZMA goals or in the interest of national security."[625] If Interior disapproves an exploration plan because it will cause environmental harm, it may also cancel the lease, though compensation may be due to the lessee.[626]

487. The final stage under OCSLA is actual development and production of the mineral resource.[627] Again, however, the lessee must submit another plan to Interior before it can proceed. And, again, the agency must forward this plan to affected states for comment and review, and "the governor's recommendations must be accepted . . . if they strike a reasonable balance between local and national interests."[628] The agency must communicate its reasons for accepting or rejecting such states' comments. The development plan also must be consistent with the state's approved coastal management plan under CZMA. If it is not, the state can veto the plan, and that veto "can be overridden by the Secretary of Commerce."[629] As with the exploration plan, Interior can also disapprove a development plan if it will

620. 43 U.S.C. § 1345.
621. *Sec'y of Interior*, 464 U.S. at 338.
622. 43 U.S.C. § 1340.
623. Coastal Zone Management Act of 1972, Pub. L. 89-454, 86 Stat. 1280 (codified as amended at 16 U.S.C. § 1451 et al.).
624. 43 U.S.C. §§ 1334 & 1340.
625. *Sec'y of Interior*, 464 U.S. at 339; *see* 43 U.S.C. § 1343(c)(2).
626. 43 U.S.C. § 1340(c)(1).
627. 43 U.S.C. § 1351.
628. *Sec'y of Interior*, 464 U.S. at 340; *see* 43 U.S.C. § 1351.
629. *Sec'y of Interior*, 464 U.S. at 340; *see* 43 U.S.C. § 1351(d).

significantly harm the environment, and disapproval for that reason can lead to cancellation of the lease.[630]

488. A key issue in the implementation of OSCLA is to determine what lands are available for oil and gas development. Early on in the program, Congress withdrew millions of acres from possible development—what became known as congressional OCS moratoria. First, in 1982, Congress responded to Interior Secretary James Watts' plan to lease almost a billion acres from 1982-87 by prohibiting offshore leases off parts of California. "By 1989, succeeding appropriation bills had extended the moratoria to more than 181 million acres off the coasts of California, the North Atlantic, and the eastern Gulf of Mexico (near Florida). This 'off limits' acreage equaled more than twice the acreage that had been leased in the entire history of the OCS program."[631] Following these congressional moratoria, President George H.W. Bush issued his own moratorium on leasing in California, Florida, Oregon, Washington, and the Georges Bank area in New England. Then, in 1998, President Clinton extended this moratorium through 2012. When President George W. Bush took office, he left in place the OCS moratoria on the United States' East and West Coasts.[632]

489. In 2008, the congressional moratoria expired, and President George W. Bush "lifted the Executive Order against certain OCS leasing."[633] Currently, no nationwide moratoria are in place, but President Obama has withdrawn all marine sanctuary areas from oil and gas leasing, plus Bristol Bay, off Alaska, through June 30, 2017. Also, part of the central and eastern portions of the Gulf of Mexico are under restriction until 2022 pursuant to the Gulf of Mexico Energy Security Act of 2006.[634]

490. For years, the Minerals Management Agency, within the Department of the Interior, was charged with implementing the OCSLA. However, following the *Deepwater Horizon* disaster in 2010, the Obama administration renamed this agency the Bureau of Ocean Energy Management, Regulation and Enforcement. The administration then replaced BOEMRE with three separate agencies: the Office of Natural Resources Revenue (ONRR); the Bureau of Ocean Energy Management (BOEM); and the Bureau of Safety and Environmental Enforcement (BSEE). ONRR falls under the jurisdiction of Interior's Office of Policy, Management and Budget. BOEM has responsibility for implementing OCSLA leasing and plan administration, resource evaluation, economic assessment, and renewable energy

630 43 U.S.C. § 1351(h)(2)(C).
631 Fred Bosselman et al., Energy, Economics and the Environment: Cases and Materials 308 (3d ed. 2010).
632. *Ibid.* at 309.
633. *Ibid.* at 310.
634. Gulf of Mexico Energy Security Act of 2006, Pub. L. 109-432, 120 Stat. 3000 (codified at 43 U.S.C. § 1331).

development. BSEE is charged with implementation of safety and environmental regulations, including oil spill response and offshore regulatory programs.[635]

491. As we move through the second decade of this century, United States offshore oil production continues to shift, just as it has considerably over the last two decades. The Oil Pollution Act of 1990, adopted in response to the *Exxon Valdez* disaster off the coast of Alaska, continues to play an important role. That law imposes requirements on shipping vessels for oil, mandates remediation plans, and creates a liability and compensation scheme. Increasingly, however, some commentators have asked whether the OPA goes far enough,[636] particularly as production from deepwater drilling—drilling at depths greater than 500 feet—continues to increase.[637] As companies push the limits of deepwater drilling, safety concerns have also grown. The 2010 *Deepwater Horizon* disaster put a particularly bright spotlight on these issues.[638] That explosion, which killed eleven crewmembers and eventually gushed an estimated 4.9 million barrels of oil into the Gulf of Mexico, still casts a shadow over offshore oil development in the United States.[639] Nonetheless, deepwater drilling remains a powerful force in U.S. oil exploitation. How it may further shift oil exploitation in the United States remains to be seen.

---

635. Bureau of Ocean Energy Mgmt., *The Reorganization of the Former MMS*, http://www.boem.gov/About-BOEM/Reorganization/Reorganization.aspx (last visited May 1, 2015).
636. Robin Kundis Craig, *Legal Remedies for Deep Marine Oil Spills and Long-Term Ecological Resilience: A Match Made in Hell*, 2011 B.Y.U. L. REV. 1863, 1863 (2011); Leila Monroe, *Restructure and Reform: Post-Bp Deepwater Horizon Proposals to Improve Oversight of Offshore Oil and Gas Activities*, 5 GOLDEN GATE U. ENVTL. L.J. 61 (2011); David Hults, *Environmental Regulation at the Frontier: Government Oversight of Offshore Oil Drilling North of Alaska*, 44 ENVTL. L. 761 (2014).
637. *Ibid.* at 826.
638. For a full discussion, *see* Hari M. Osofsky, *Multidimensional Governance and the Bp Deepwater Horizon Oil Spill*, 63 FLA. L. REV. 1077 (2011).
639. *Ibid.* at 827; United States Coast Guard, UNITED STATES COAST GUARD, ON SCENE COORDINATOR REPORT *DEEPWATER HORIZON* OIL SPILL (2011), *available at* http://www.uscg.mil/foia/docs/dwh/fosc_dwh_report.pdf.

Chapter 3. Transportation

492. Oil moves from the extraction point to a refinery through some combination of pipeline, tanker, and rail. There are an estimated 2.5 million miles of pipelines that transport oil throughout the United States.[640] Tankers transported 46 trillion barrels of oil around the world in the first decade of the twenty-first century.[641] However, rail is becoming a more important transportation method for oil as pipeline capacity is maxed out. In the United States, this is particularly true for unconventional extraction in areas not served by sufficient pipelines, namely, shale and tar sands formations. Transporting oil in this way involves various levels of risk to human safety and the environment.

493. Pipelines are often the most economical way to transport oil, and thus, the most prevalent transport method. The first oil pipeline was constructed in Pennsylvania in 1865 and undercut the cost of transporting oil by horse. Currently, the U.S. pipeline system transports more than 10 billion barrels of oil each year.[642] "In 2009 the United States transported roughly 70% of all petroleum products, or approximately 11.3 billion barrels, using interstate pipelines."[643]

494. To understand the regulation of pipelines, it is useful to recognize several underlying realities: Pipelines are often a bottleneck between oil producers and consumers. The pipeline system is controlled by a limited number of companies. And, it is expensive to site and build new pipelines.[644]

495. Because oil pipelines are bottleneck facilities, there is a risk that companies will charge excessive fees for transportation, thus driving up prices for consumers. To combat this risk, federal regulation of oil pipelines aims at ensuring fair rates. Oil pipeline rates are regulated by FERC under the Interstate Commerce Act of 1887[645] and the Energy Policy Act of 2005.[646] FERC's mandate under the ICA is to ensure "just and reasonable rates" for pipeline transport of oil, akin to what it does for natural gas pipelines and electricity transmission lines.

496. Federal jurisdiction of oil pipelines is not, however, exclusive. State and local governments, through land use planning regulations, regulate the siting of oil pipelines. Eminent domain power for oil pipelines varies by state, with some jurisdictions allowing companies to use eminent domain for all pipelines, others not

640. Davies et al., *supra* note 29, at 609.
641. *Ibid.*
642. *Ibid.* at 655.
643. *Ibid.* at 652.
644. *Ibid.* at 611, 652–55.
645. Interstate Commerce Act of 1887, 24 Stat. 379 (codified as amended at 49 U.S.C. § 501 et al.). The Hepburn Act of 1906, 34 Stat. 584 (codified at 49 U.S.C. § 10102 et al.), classified interstate pipelines as common carriers, making them subject to the ICA.
646. Energy Policy Act of 2005, Pub. L. 109-58, 119 Stat. 594 (codified as amended at 42 U.S.C. § 15801 et al.).

allowing use of eminent domain at all, and still others allowing the use of eminent domain in limited circumstances—such as when the pipeline is a public utility.[647] Interestingly, the siting of interstate oil pipelines does not proceed under federal law, unlike natural gas pipelines, which obtain eminent domain authority from FERC under the Natural Gas Act of 1938.[648] The resulting patchwork of state regulations makes siting an interstate or transboundary oil pipeline difficult. A recent example of the complexity of the siting process is the Keystone XL pipeline, which requires approval by both Canada and the United States, as well as the states through which it runs.

497. Pipelines have proven a relatively safe method for oil transport compared to tanker or rail. The Pipeline Hazardous Materials Safety Administration (PHMSA), which is part of the Department of Transportation, regulates pipeline safety in the United States. PHMSA operates under an array of statutes, including its enabling acts—the Natural Gas Pipeline Safety Act of 1968 and the Hazardous Liquid Pipeline Act of 1979.[649] Under these statutes, the PHMSA regulates "key aspects of interstate pipeline safety: design, construction, operation and maintenance, and spill response planning."[650] PHMSA regulation of pipeline safety are technical, comprehensive, and complex. Post-September 11, the Department of Homeland Security regulates pipeline security.[651]

498. Historically, oil was transported mainly by pipeline, with rail transport playing only a minor role. Pipelines long have been the preferred method of transport in the United States because they are effective, can be constructed relatively easily, and are cheaper than the railroads. In recent years, however, as domestic oil production has increased, there has not been enough pipeline capacity, particularly in emerging areas of production. Accordingly, producers are increasingly turning to the railroads to transport their oil. Today, 10% of the oil produced in the United States moves by rail, compared with only 1% in 2009,[652] and that figure varies by region, with some areas much higher. In the Bakken Shale, for instance, around two-thirds of oil is transported by rail because there are not adequate pipeline connections.[653] Increased transportation by railroad is troublesome from a safety and environmental standpoint. Derailments occur and can have devastating consequences. For example, in 2013, a freight train carrying oil from the Bakken

---

647. Alexandra B. Klass & Elizabeth J. Wilson, *Interstate Transmission Challenges for Renewable Energy: A Federalism Mismatch*, 44 ENVTL. L. REP. NEWS & ANALYSIS 10705 (2014); *see also* Christopher J. Barr, *Growing Pains: FERC's Responses to Challenges to the Development of Oil Pipeline Infrastructure*, 28 ENERGY L.J. 43, 49–51 (2007).
648. Natural Gas Act of 1938, 52 Stat. 821 (codified as amended at 15 U.S.C. § 717–717z).
649. 49 U.S.C. §§ 60101–60140; Hazardous Liquid Pipeline Act of 1979, Pub. L. No. 96-129, 93 Stat. 1003.
650. DAVIES ET AL, *supra* note 29, at 683; *see also* PAUL W. PARFOMAK, CONGRESSIONAL RESEARCH SERVICE, KEEPING AMERICA'S PIPELINES SAFE AND SECURE: KEY ISSUES FOR CONGRESS 5 (2013).
651. DAVIES ET AL., *supra* note 29, at 683–84.
652. *Ibid*; *see also* Jad Mouawad, *Bakken Crude, Rolling Through* Albany, N.Y. TIMES, Feb. 27, 2014.
653. DAVIES ET AL., *supra* note 29, at 683–84.

formation derailed in the town of Lac-Megantic in Quebec, Canada. The resulting spill and explosion killed 42 people and destroyed more than 30 buildings.[654]

499. The Federal Railroad Administration (FRA), which is also part of the Department of Transportation, regulates railroad safety. Federal laws largely preempt state law in this area, although FRA inspectors work with state safety inspectors to ensure compliance.[655] The FRA regulates several aspects of railroad safety, including track safety, crossings, equipment, operating practices, and movement of hazardous materials. Under the Rail Safety Improvement Act of 2008,[656] the FRA promulgated regulations "requiring railroads to achieve specific track failure rates on their rails."[657] While the FRA regulates rail safety, including coordinating with states to conduct inspections, the National Transportation Safety Board investigates railroad incidents.

500. Some oil in the United States also moves by tanker—roughly 8.5%.[658] While necessary to move oil internationally and from remote locations, oil spills from tankers pose a serious environmental and public health threat, as demonstrated by the *Exxon Valdez* spill, which released 11 million gallons of oil into the Prince William Sound in Alaska.[659] As noted, in response to the *Exxon Valdez* disaster, Congress passed the Oil Pollution Act of 1990.[660] The OPA established a regulatory framework for oil spill prevention, funding and coordination of cleanup efforts, and liability.

501. Specifically, the OPA introduced five key changes that now dictate environmental consequences of oil spills in the United States. First, the OPA requires the President to create a National Contingency Plan for worst-case spill scenarios. Second, the law requires tankers, offshore facilities, and some onshore facilities to prepare oil spill response plans. Without such a plan, they cannot transport or handle oil. Third, the OPA mandated that all oil tankers, with certain exceptions, utilize double hull designs in order to help prevent the risk of oil spills. Fourth, the OPA instituted a strict liability regime for spilling oil into waters of the United States, and broadened the scope of damages possible for such spills. Potential damages include: natural resource damages, economic loss damages, lost revenues from property destruction, and costs of public services for spill response. However, the

654. Mark Memmott, *42 Bodies Recovered After Quebec Disaster; 5 People Missing*, NPR, Jul. 22, 2013, *available at* http://www.npr.org/blogs/thetwo-way/2013/07/22/204479603/42-bodies-recovered-after-quebec-disaster-5-people-missing.
655. Davies et al., *supra* note 29, at 684.
656. Rail Safety Improvement Act of 2008, Pub. L. 110-432, 122 Stat. 4848 (codified at 49 U.S.C. § 20156 et al.).
657. Davies et al., *supra* note 29, at 685.
658. U.S. Energy Info. Admin., *Refinery Receipts of Crude Oil by Method of Transportation*, *available at* http://www.eia.gov/dnav/pet/pet_pnp_caprec_dcu_nus_a.htm.
659. Davies et al., *supra* note 29, at 688.
660. Oil Pollution Act of 1990, Pub. L. 101-380, 104 Stat. 484 (codified as amended at 33 U.S.C. § 2701 et. seq.).

law also established liability caps, typically based on the tonnage of oil being transported. Finally, the OPA created an oil spill liability trust fund, which was funded by a five-cent-per-barrel and, subsequently, eight-cents-per-barrel tax.[661]

661. JONATHAN L. RAMSEUR, CONGRESSIONAL RESEARCH SERVICE, OIL SPILLS IN U.S. COASTAL WATERS: BACKGROUND, GOVERNANCE, AND ISSUES FOR CONGRESS (2010).

Chapter 4. Distribution

502. The vast majority of petroleum use in the United States is for transportation in the form of gasoline. Therefore, the main distribution network for petroleum products in the United States is retail fueling stations. To arrive at these fueling stations, crude oil first goes to a refinery, where it is processed into gasoline and other petroleum products. It is then shipped by pipeline to storage terminals close to distribution centers. Finally, it is trucked to fueling stations, where it is stored in underground storage tanks until it is purchased by consumers.[662] Currently, there are approximately 153,000 fueling stations in the United States.[663]

503. Increasingly, independent business owners operate these fueling stations, as large oil companies are exiting the retail business. Particularly since 2007, large oil companies have been selling off their directly-owned-and-operated fueling stations, in order to focus on oil production and refining. As of June 2013, less than 0.4% of the fueling stations were owned by one of the five major oil companies. Small business owners operate most fueling stations and, of those, most are convenience stores selling fuel as well as other products, such as snack food, beverages, magazines, and other items. Thus, today, about 80% of the fuel purchased in the United States is acquired from convenience store fueling stations. Further, these convenience store fueling operations are mainly single-store operations.[664] There has also been growth in gasoline sold by big-box grocery stores, such as Costco and Sam's Club.[665]

504. Fueling stations are mainly regulated at the state and local level. These laws, moreover, are not expressly aimed at petroleum regulation. For example, at the local level, fueling stations are often regulated through generally applicable land use planning and zoning laws. These regulations vary by municipality but may be aimed at issues such as neighborhood compatibility, aesthetics, and traffic concerns.[666] Also, the business operations inherent in fueling stations may be regulated at the state level. For instance, if a gas station is set up in a franchisee/franchisor relationship, state laws designed to protect the franchisee from the superior bargaining power of the franchisor oil company often apply.[667] Further, independently owned fueling stations will often carry the brand of a major oil company. These relationships may be governed by a branding contract. This contract may address such issues as the volume requirements for both the gasoline that the oil company will provide and the amount the station owner will be required to sell; the specific

662. U.S. Energy Info. Admin., *Gasoline Explained: Where Our Gasoline Comes From*, *available at* http://www.eia.gov/energyexplained/index.cfm?page=gasoline_where (last visited Feb. 4, 2015).

663. NAT'L ASS'N OF CONVENIENCE STORES, 2014 RETAIL FUELS REPORT (2014), *available at* http://www.nacsonline.com/YourBusiness/FuelsReports/GasPrices_2014/Documents/2014NACSFuelsReport_full.pdf.

664. Fifty-eight % of these convenience stores are single-store operations.

665. *Ibid.* at 29. In March 2013, they sold 12.6% of total fuels purchased in the United States.

666. See, e.g., Riverside California Municipal Code 19.410, *available at* http://www.riversideca.gov/municode/pdf/19/article-7/19-410.pdf.

667. See Virginia Petroleum Products Franchise Act, *available at* https://leg1.state.va.us/cgi-bin/legp504.exe?000+cod+TOC59010000002000020000000.

branding requirements concerning signage and logos; and the wholesale price requirement.[668] In the United States, contract law is generally the domain of the states, and it can vary from jurisdiction to jurisdiction.

505. The major federal regulation in this area is the Petroleum Marketing Practices Act of 1978 (PMPA).[669] Similar to the state legislation discussed above, the PMPA is aimed at protecting petroleum franchisees from a franchisor's unfair termination of the franchisee/franchisor relationship.[670] The Act was passed in response to abusive practices by oil company franchisors that built repressive terms into their franchise agreements. Often, the franchisor would demand high leases for their land, set unrealistic minimum gallonage requirements, require dealers to purchase items such as tires or batteries from the franchisor at inflated prices, and retain the power to terminate the agreement immediately if the dealer failed to meet the agreement's requirements.[671]

506. Specifically, the PMPA limits the terms of gasoline distribution franchise relationships. It preempts state laws on the issue, and dictates that franchisors cannot terminate a franchise except in accordance with the Act. In attempting to protect franchisees, Title I enumerates limited reasons why a franchisor can cancel an agreement, including the franchisee's failure to carry out the agreement's terms in good faith, or if the trademark for the gasoline is no longer available. Likewise, the law limits the reasons why a franchisor can opt not to renew an agreement, including all those for which it can terminate the agreement plus some others, such as the franchisee receiving numerous customer complaints. The law creates a cause of action to enforce its terms in federal court. Title II of the Act mandates disclosure of fuel octane ratings.

507. In addition to the PMPA, federal environmental regulations are relevant to petroleum distribution in the United States, particularly to the extent they address the storage of gasoline in underground tanks and the remediation of sites contaminated with gasoline spills.[672] Leaks from underground storage tanks, including those containing gasoline, pose a threat to drinking water supplies and are costly to remediate.[673] Many old tanks were particularly problematic because they were made of bare steel, and thus, would corrode underground over time.

---

668. NATIONAL ASSOCIATION OF CONVENIENCE STORES, 2014 RETAIL FUELS REPORT 32 (2014), *available at* http://www.nacsonline.com/YourBusiness/FuelsReports/GasPrices_2014/Documents/2014 NACSFuelsReport_full.pdf; *see also* Chevron Branded Jobber Petroleum Products Agreement, Mar. 15, 2005, *available at* http://www.sec.gov/Archives/edgar/data/1361709/0001193125061093 68/dex1015.htm.
669. Petroleum Marketing Practices Act of 1978, Pub. L. 95-297, 92 Stat. 322 (codified as amended at 15 U.S.C. §§ 2801–2807, 2821–2824, 2841.
670. 15 U.S.C. §§ 2801–2807.
671. Comment, *Retail Gasoline Franchise Terminations and Nonrenewals under Title I of the Petroleum Marketing Practices Act*,1980 DUKE L.J. 522, 524–25.
672. U.S. Envt'l Protection Agency, *Underground Storage Tanks: Regulations Pertaining to Underground Storage Tanks*, *available at* http://www.epa.gov/oust/fedlaws/cfr.htm.
673. GROUND WATER PROTECTION COUNCIL, GROUND WATER REPORT TO THE NATION: A CALL TO ACTION 7-1,12, *available at* http://www.gwpc.org/sites/default/files/GroundWaterReport-2007-.pdf.

508. Congress has addressed the leaking underground storage tank (LUST) problem through several actions. In 1984, it added Subtitle I to the Solid Waste Disposal Act, which required EPA to create a regulatory program to deal with LUSTs. Congress also banned installation of unprotected steel tanks and piping, effective 1985. Subsequently, in 1986, Congress amended Subtitle I to create the Leaking Underground Storage Tank Trust Fund, which helps pay for cleanups of leaking tanks by responsible parties as well as at sites that require an emergency action or where the responsible party is not known or is unwilling to respond to the leak. EPAct 2005 then expanded the use of LUST Trust Fund and provided further specifications for inspections, operator training, and cleanup of leaks with oxygenated fuel additives. The American Recovery and Reinvestment Act of 2009 then assigned USD 200 million to the LUST Trust Fund, the vast majority of which EPA allocated to states to conduct cleanups.

509. To implement these measures, the EPA created the Office of Underground Storage Tanks. The EPA has issued both technical and financial responsibility regulations with detailed requirements for the design, construction, and operation of underground storage tanks.[674] EPA works with states and territories to ensure proper cleanup and inspection of LUSTs. EPA lists roughly 571,000 active underground storage tanks at approximately 205,000 sites that are regulated by the LUST program. Since the LUST program began, over 1.8 million tanks have been properly closed, and of the roughly half million releases reported since, 85.8% of them have been cleaned up. Between October 2013 and September 2014, 92,883 on-site inspections were conducted under the LUST program. States, territories, and third-party inspectors conducted all but 359 of these; the remainder were conducted by EPA or by credentialed tribal inspectors in Indian Country.[675]

674. 40 C.F.R. § 280; based on statutory authority from 42 U.S.C.A. § 6991.
675. U.S. ENVT'L PROTECTION AGENCY, SEMIANNUAL REPORT OF UST PERFORMANCE MEASURES: END OF FISCAL YEAR 2014, *available at* http://www.epa.gov/oust/cat/ca-14-34.pdf.

Chapter 5. Consumption

510. As noted, oil is used in the United States for two key purposes: most dominantly, for transportation, and secondly, for industry. Only a very small percentage of oil consumption is for electricity generation. Approximately 71% of oil goes to transportation in the United States, with 23% going to industrial use, 5% to residential use, and only 1% to electricity generation.[676] These trends are not expected to change. Rather, petroleum demand in the transportation sector is projected to increase by 80 to 130% by 2050.[677] Road transport, or vehicle use, makes up a majority of the transportation sector's consumption of oil.[678] In the United States, approximately 6 million new cars are sold each year, and the total fleet is composed of 131 million cars and 100 million light trucks. These vehicles have an average fuel economy of 23 miles per gallon.[679]

511. In the United States, regulation of oil consumption historically has focused not on directly restricting consumption (*i.e.*, restricting vehicle use),[680] but rather, on addressing the two major issues created by consumption: (1) the impact of emissions on public health and (2) fuel economy. Increasingly, this regulation also targets petroleum combustion's contribution to climate change. As vehicles burn oil (in the form of refined gasoline), numerous air emissions result, including carbon dioxide (a greenhouse gas); carbon monoxide, particulate matter, and volatile organic compounds (all of which can contribute to smog); nitrogen oxides; and sulfur dioxide.[681] Each of these pollutants has important environmental consequences. Ozone and particulate matter both impact human health. Particulate matter has been linked to "increased hospitalizations due to lung and respiratory symptoms, compromised cardiac function, student absences from school, and increased mortality."[682] A significant amount of the global emissions of carbon dioxide comes from the transportation sector—an estimated 23%.[683] These emissions are projected to increase significantly under current government policies. Thus, a system of regulation in the United States addresses these problems to different degrees.

512. The Clean Air Act regulates both tailpipe emissions of vehicles and the content of fuels.[684] Specifically, Section 202(a) of the Clean Air Act requires the EPA to regulate tailpipe emissions by setting regulatory emissions standards.[685]

---

676. U.S. Energy Info. Admin., *Annual Energy Review* (Sep. 27, 2012), *available at* http://www.eia.gov/totalenergy/data/annual/pecss_diagram.cfm.
677. Davies et al., *supra* note 29, at 510–11.
678. World Energy Council, Global Transport Scenarios 2050 8 (2011).
679. Davies et al., *supra* note 29, at 546–47.
680. The exception is the 1970s when, in response to the global oil crises, the government implemented a complex system of oil price controls that have since been abandoned. Also adopted at this time were gasoline rationing, daylight savings time, and a national speed limit.
681. World Energy Council, *supra* note 678, at 512-13.
682. *Ibid.* at 513.
683. *Ibid.*
684. 42 U.S.C. § 7521.
685. *Ibid.*

Section 209 creates a national standard by preempting states from setting individual emissions standards, although there is an exception for California, which was setting emission standards before the Clean Air Act was passed, as well as any other states in non-attainment of national ambient air quality standards (NAAQS) that choose to adopt California's standards.

513. The 1990 amendments to the Clean Air Act also required refiners to produce gasoline including 2% oxygen by weight, or what the statute referred to as reformulated gasoline (RFG).[686] The idea was simple; gasoline with higher levels of oxygen burns cleaner, thus reducing emissions, particularly of carbon monoxide, nitrogen oxides (NO_x), and volatile organic compounds (VOCs). RFG thus was required to be used in large metropolitan areas that were in non-compliance with certain NAAQS. In addition, some areas, including Boston and Washington, D.C., voluntarily opted into the program. Implementation was more complicated. EPA's regulations effectively allowed two different oxygenates to be used to satisfy the requirement—MTBE or ethanol—but they quickly faced battles in court.[687] Nevertheless, by 2000, "more than 30% of the gasoline sold in the U.S. was reformulated, and about 87% of that gasoline contained MTBE."[688]

514. MTBE, however, quickly was criticized because it can be tasted at extremely low levels in water, and because it can spread quickly and thus exacerbates water contamination by gasoline. Given the large number of leaking underground storage tanks in the United States, a number of jurisdictions began banning MTBE, and MTBE manufacturers failed in their lawsuits seeking to overturn those bans.[689] Then, in the Energy Policy Act of 2005, Congress eliminated the oxygenate requirement from the Clean Air Act, effectively removing this incentive for refiners to use MTBE. In that Act, Congress also imposed a renewable fuel standard (RFS), effectively promoting ethanol as an oxygenate in U.S. gasoline supplies.[690]

515. The Clean Air Act's RFG program and tailpipe emissions limits have had some success. They generally improved air pollution levels, with ozone precursors nationally decreasing by about 18% from 1970 to 1998 and one-hour ozone levels declining by 20% and eight-hour levels decreasing by 12% during this same period. After 1990, however, these emissions either leveled off or increased. In fact, from 1990 to 1999, one-hour and eight-hour ozone levels increased in the Mid-Atlantic, Midwest, and North Central regions of the country. This trend can be attributed largely to vehicle use. The contribution of vehicles to overall NO_x emissions was down 2% points from 1970 to 1998, and the contribution to overall VOC emissions

---

686. 42 U.S.C. § 7545(k)(2)(B).
687. *See, e.g.*, Am. Petroleum Inst. v. EPA, 52 F.3d 1113 (D.C. Cir. 1995).
688. Thomas O. McGarity, *MTBE: A Precautionary Tale*, 28 Harv. Envtl. L. Rev. 281, 285 (2004).
689. Oxygenated Fuels Ass'n v. Davis, 331 F.3d 665 (9th Cir. 2003); Oxygenated Fuels Ass'n v. Pataki, 158 F. Supp. 2d 248, 254 (N.D.N.Y.2001); *cf.* In re MTBE Prod. Liab. Litig., 175 F. Supp. 2d 593, 612 (S.D.N.Y.2001). *But cf.* Davis v. EPA, 348 F.3d 772 (9th Cir. 2003).
690. For a critical view of the RFS, *see* Arnold W. Reitze, *Biofuels—Snake Oil for the Twenty-First Century*, 87 Or. L. Rev. 1183 (2008).

was down 12%. That was the success of the Clean Air Act. But consumer behavior erased these gains. "[I]increased population levels, increased activity levels, and population dispersal all have contributed to a more than doubling since 1970 of the number of vehicles on the road and the vehicle miles traveled (VMT). Thus, although per-vehicle tailpipe emissions have decreased by as much as 90% since 1970, VMT is doubling roughly every twenty years."[691]

516. In addition to direct regulation of tailpipe emissions and the RFG and RFS programs, the government has also forced the redesign of automobiles to increase fuel efficiency. The National Highway Traffic Safety Administration, within the Department of Transportation, "regulates fleet-wide automobile efficiency standards known as Corporate Average Fuel Economy (CAFE) standards."[692] Congress established CAFE standards through the 1975 Energy Policy and Conservation Act in response to the oil crisis of 1973 and 1974.[693] CAFE standards are different for passenger vehicles and light-duty trucks, which include many SUVs, vans, and pickup trucks. For passenger vehicles, "CAFE standards began at 18.0 miles per gallon (mpg) for [model year (MY)] 1978 and increased to 27.5 mpg in MY 1985. NHTSA reduced the fuel economy requirements 26.0 mpg in MYs 1986 through 1988, because a drop in gasoline prices had changed consumer demand. . . ."[694] For model year 1989, "CAFE standards increased to 26.5 mpg, and increased again to 27.5 mpg for passenger cars from MY 1990 through MY 2010."[695]

517. CAFE standards for light-duty trucks are less than for other vehicles. For these vehicles, the standard was between 20.0 to 20.6 mpg from MY 1986 through MY 1995,[696] and it was then limited by Congress at 20.7 for MY 1996 through 2004. "After the legislative freeze ended, NHTSA set more stringent light-duty truck CAFE standards . . . [at] 21.0 mpg in MY 2005, 21.6 mpg in MY 2006, 22.2 mpg in MY 2007, 22.5 mpg in MY 2008, 23.1 in MY 2009, and 23.5 in MY 2010."[697] As a result, cars and light trucks in the United States achieved an "actual combined fuel economy of about 25 mpg from 1980 to 2004."[698]

518. In 2006, the Department of Transportation NHTSA adopted new, more aggressive CAFE standards for SUVs, vans, minivans, and pickup trucks, seeking to achieve an approximate 24 mpg for these fleets. That rule was found unlawful by

691. Michael P. Vandenbergh, *From Smokestack to SUV: The Individual as Regulated Entity in the New Era of Environmental Law*, 57 VAND. L. REV. 515, 628 (2004).
692. DAVIES ET AL. *supra* note 29, at 548.
693. Energy Policy and Conservation Act of 1975, Pub. L. 94-163, 89 Stat. 871 (codified as amended at 42 U.S.C. § 6201 et al.).
694. Arnold W. Reitze, Jr., *Controlling Greenhouse Gases From Highway Vehicles*, 31 UTAH ENVTL. L. REV. 309, 314 (2011); *see* Public Citizen v. Nat'l Highway Traffic & Safety Admin., 848 F.2d 256 (D.C. Cir. 1988).
695. Reitze, *supra* note 694, at 314.
696. Ctr. for Auto Safety v. Nat'l Highway Traffic Safety Admin., 793 F.2d 1322 (D.C. Cir. 1986).
697. Reitze, *supra* note 694, at 314–15.
698. *Ibid.* at 315.

the U.S. Court of Appeals for the Ninth Circuit the next year,[699] but another decision by the U.S. Supreme Court quickly reset the stage for fuel economy standards in the United States. In *Massachusetts v. EPA*,[700] the High court ruled that carbon dioxide is a pollutant under the Clean Air Act, and thus the EPA was required to regulate the emission of this pollutant from motor vehicles, if the agency found that greenhouse gas emissions endanger the environment.[701]

519. As a result, in 2007, the CAFE program received a reboot. In that year, in addition to the Supreme Court's decision, Congress enacted the Energy Independence and Security Act of 2007, which required the NHTSA to increase fuel efficiency for passenger cars and light trucks to 35 mpg by 2020.[702] The legislation also required the NHTSA to study the feasibility of increased fuel efficiency standards for heavy-duty vehicles and issue those standards if they were found feasible.[703]

520. In 2010, the Obama administration took executive action through the EPA and the NHTSA. These agencies promulgated rules establishing the first greenhouse gas emission standards in the United States for new passenger cars, light-duty trucks, and medium-passenger vehicles. The regulations established CAFE standards for model years 2012 through 2016.[704] Specifically, the standards limited emissions to a combined average emission level of 250 grams of CO_2 per mile in model year 2016—which is equal to a fuel efficiency of 35.5 mpg if the standard is met through fuel efficiency alone. The aim of these regulations is to speed up the congressionally mandated requirement of 35 mpg by four years. The administration also issued fuel efficiency standards for medium and light-duty engines and vehicle model years 2014 through 2018.[705]

521. Violation of CAFE standards carries significant risks. There is a civil penalty of USD 5 for every 0.1 mpg a fleet exceeds the standard, multiplied by each manufactured vehicle in the fleet. However, big manufacturers can usually pass the standard because they have so many vehicles across which they can spread their efficiencies. Thus, total CAFE fines in 2008 was just under USD 13 million, nearly all of which was paid by small foreign importers and manufacturers.

522. In addition to CAFE penalties, U.S. law also imposes a gas-guzzler tax for vehicles with low fuel economy ratings.[706] These begin at USD 1,000 for vehicles

699. Ctr. for Biological Diversity v. Nat'l Highway Traffic Safety Admin., 508 F.3d 508, 513 (9th Cir. 2007).
700. Massachusetts v. EPA, 549 U.S. 497 (2008).
701. 549 U.S. 497 (2007).
702. Energy Independence and Security Act of 2007, Pub. L. 110-140, 121 Stat. 1492 (codified as amended at 42 U.S.C. § 17001 et al.).
703. *Ibid.*
704. 40 C.F.R. Parts 85, 86, and 600; 49 C.F.R. Parts 531, 533, 536.
705. 40 C.F.R. Parts 85, 86, 600, 1033, 1036, 1037, 1039, 1065, 1066, and 1068 (2015); 49 C.F.R. Parts 523, 534, and 535 (2015).
706. Energy Tax Act of 1978, Pub. L. No. 95-618, 92 Stat. 3180; Pub. L. No. 101-508, 104 Stat. 1388 (1990) (codified at 26 U.S.C. § 4064).

with fuel economies between 21 and 22.5 mpg, and reach as high as USD 7,700 mpg for vehicles with fuel economies less than 12.5 mpg. This tax, however, does not apply to trucks. "Nevertheless, tax receipts from the gas-guzzler tax totaled over USD 172[] million in 2008."[707]

707. Reitze, *supra* note 694, at 316.

Part IV. Regulatory Framework Concerning Coal

Chapter 1. Production

523. The United States holds the largest estimated recoverable reserves of coal in the world.[708] For that reason, the United States is often referred to as "the Saudi Arabia of coal."[709] Globally, it is the second largest producer of coal, and until recently, it was the largest producer in the world, until China took that spot in 2011.[710] It is estimated that the United States has enough recoverable reserves of coal to last more than 200 years, based on current production levels.[711] Coal is produced in 25 states in the United States, spread across three major coal-producing regions: (1) Appalachia; (2) the Intermountain West (heavily in the Wyoming Powder River Basin); and (3) a region that stretches up from Texas northward into the Midwest.

524. Nevertheless, while coal reserves are widespread in the United States, a few states dominate production. In 2013, approximately 70% of domestic production was in just five states: Wyoming, West Virginia, Kentucky, Pennsylvania, and Illinois. Of these, Wyoming currently is the largest producer of coal in the United States. In 2013, Wyoming alone produced 39% of coal extracted in the United States, or 388 million short tons. The next closest state was West Virginia, with 11% of U.S. coal production, or 113 million short tons.[712]

525. In 2013, U.S. coalmines produced just under 1 billion short tons of coal, their lowest output level since 1993.[713] More than 90% of this coal was used by U.S.

708. U.S. Energy Info. Admin., *What Is the Role of Coal in the United States?,* (June 2, 2014), *available at* http://www.eia.gov/energy_in_brief/article/role_coal_us.cfm [hereinafter EIA, *Coal*]; Steven Chu, *Carbon Capture and Sequestration*, SCIENCE 1599 (Sep. 2009).
709. *The Republican Debate*, N.Y. TIMES (Jan. 24, 2008), http://www.nytimes.com/2008/01/24/us/politics/24text-debate.html?pagewanted=print&_r=0; Chris Mayer, *The Saudi Arabia of Coal*, BUSINESS INSIDER (May 25, 2011), http://www.businessinsider.com/the-saudi-arabia-of-coal-2011-5.
710. *See* EIA, *Coal*, *supra* note 708.
711. *Ibid; see also* Sharon Burke and Christine Parthemore, *A Strategy for American Power: Energy, Climate and National Security*, SOLARIUM STRATEGY SERIES 5(Jun. 2008), *available at* http://citeseerx.ist.psu.edu/viewdoc/download?doi=10.1.1.357.9005&rep=rep1&type=pdf#page=7.
712. EIA, *Coal*, *supra* note 708.
713. U.S. ENERGY INFO. ADMIN., MONTHLY ENERGY REVIEW 92 (Apr. 2015).

power plants to generate electricity. While coal has been the largest source of electricity generation in the United States for over 60 years, only 39% of the electricity produced in the United States in 2013 was generated from coal. The remaining 61% was produced by natural gas (27%), nuclear power (19%), petroleum (1%), and renewable resources such as hydro, wind, and solar power (13%).[714]

526. Coal is not uniform. Coal varies by its origin, which means that coal from different locations has different characteristics, including its sulfur, mercury, and heat content. Four general categories, or ranks, of coal are recognized. From highest to lowest heat content (generally), these ranks are: anthracite, bituminous, sub-bituminous, and lignite.[715] In the United States, bituminous coal mines dwarf those for other types of coal; bituminous coal mines comprise roughly 90% of the coal mines nationally. Likewise, bituminous and sub-bituminous coal account for roughly 97% of all coal production in the United States.[716]

527. In the United States, bituminous coal is sometimes referred to as "eastern coal," because it is produced in the mountains and foothills of Appalachia and in the Midwest.[717] Although bituminous coal typically has a higher heat content than sub-bituminous coal, it also contains a comparatively higher sulfur content, which has negative environmental consequences when it is combusted. The higher sulfur content produces sulfur oxide (SO_x) gases at combustion, a precursor to acid rain.[718] For this reason, the 1990 Amendments to the Clean Air Act[719] introduced new measures to address air pollution from coal plants using higher sulfur content, perhaps most notably, including an emissions trading program.[720] In part because of the additional cost imposed by the 1990 Amendments on facilities that use higher sulfur coal, many plants have since switched to lower sulfur western coal, which reduces their costs of complying with air quality limits.[721] Lower sulfur sub-bituminous, or western, coal is primarily found in the United States' West,

---

714. EIA, *Coal*, *supra* note 708.
715. *Ibid*; *see also* Mark T. Brown, Gaetano Protano & Sergio Ulgati, *Assessing Geobiosphere Work of Generating Global Reserves of Coal, Crude, Oil, and Natural Gas*, ECOLOGICAL MODELING, (Feb. 2011).
716. EIA, *Coal*, *supra* note 708.
717. JAMES G. SPEIGHT, THE CHEMISTRY AND TECHNOLOGY OF COAL 179 (3d ed. 2013); *see also How Coal Works,* UNION OF CONCERNED SCIENTISTS, http://www.ucsusa.org/clean_energy/coalvswind/brief_coal.html#.VOJ6wClNkRY (last visited Feb. 16, 2015).
718. U.S. Envt'l Protection Agency, *Causes of Acid Rain*, http://www.epa.gov/region1/eco/acidrain/causes.html (last visited Feb. 16, 2015).
719. Air Quality Acts of 1967, Pub. L. No. 90-148, 81 Stat. 485 (codified as amended at 42 U.S.C. §§ 7401–7626 (1990).
720. 42 U.S.C. §§ 7651–7651o.
721. David McDermott, *Coal Mining in the U.S. West: Price and Employment Trends*, BUREAU OF LABOR STATISTICS, *available at* http://www.bls.gov/mlr/1997/08/art2full.pdf (Aug. 1997); *see also* B.D. Hong & E.R. Slatick, *Carbon Dioxide Emissions Factors for Coal,* U.S. ENERGY INFORMATION ADMINISTRATION, *available at* http://www.eia.gov/coal/production/quarterly/co2_article/co2.html (last accessed Feb. 17 2015).

especially in the Powder River Basin of Wyoming and Montana.[722] Although sub-bituminous coal typically has a lower sulfur content than bituminous coal, it also tends to have a lower heat value, which means that more sub-bituminous coal must be mined and shipped to produce the same energy as a single ton of bituminous coal.[723]

528. Roughly half of the coal extracted in the United States is mined underground (bituminous coal), and most of that production is in the East. Coal also is extracted in the United States from surface mines, which unearth large coal seams (sub-bituminous coal) not as deep down in the earth's crust. Most of these mines are in the West. These mines use quite different extraction techniques, which themselves vary for both underground and surface mines.

529. There are three basic techniques that underground coal mining operations use. First, "conventional" subsurface mining techniques typically use "room and pillar" mining. In this method, coal is removed from the subsurface, creating giant underground "rooms," which are then supported by "pillars" of remaining rock. Second, some mines use a "continuous" extraction technique. In this method, a large, earth-moving machine with spinning teeth rips coal from the underground seam, and then continuously transports the extracted coal to the surface. Finally, longwall mining is a newer technique. In this method, giant machines continuously sheer the face of the coal seam, often in strips as long as 1000 feet, and then convey the coal to the surface. For this method, the roof is temporarily supported in the working area, and then the mine is allowed to collapse on itself behind the working zone.[724]

530. Surface mining in the United States also utilizes three key techniques for coal extraction. These are: *first*, contour mining, which occurs along sloping terrain; *second*, area mining, in which miners extract a large "box" of coal and overburden from a delineated area, and then replace the overburden into the affected area once the coal has been removed in order to remediate the area; and, *third*, mountaintop mining, where, as the name implies, a large amount of overburden is removed from mountaintops to allow for the mining of coal seams from the top-down.[725] Of these, mountaintop mining has received the most attention in the United States of late, largely because of its aesthetic and environmental effects but also because of environmental justice concerns since the mining often takes places in economically

722. U.S. Dep't of Energy, *Fossil Fuel Energy Guide: Coal*, http://energy.gov/sites/prod/files/2013/04/f0/HS_Coal_Studyguide_draft1.pdf (last visited Feb. 17, 2015); U.S. Geological Survey, *USGS Estimates 162 Billion Short Tons of Recoverable Coal in the Powder River Basin*, *available at* http://www.usgs.gov/newsroom/article.asp?ID=3518&from=rss#.VOOf6SlNkRY (Feb. 26, 2013); U.S. Energy Info. Admin., *Subbituminous and Bituminous Coal Dominate U.S. Coal Production*, *available at* http://www.eia.gov/todayinenergy/detail.cfm?id=2670 (Aug. 16, 2011).

723. Otto C. Kopp, *Subbituminous Coal*, ENCYCLOPEDIA BRITANNICA, http://www.britannica.com/EBchecked/topic/570576/subbituminous-coal (last visited Feb. 17, 2015).

724. DAVIES ET AL., *supra* note 29, at 108–09.

725. *Ibid.* at 109–10.

poor, rural areas.[726] Mountaintop mining is most common in Appalachia; such mines often can take off 200-600 feet of a mountain or ridge.[727]

531. Power plants throughout the United States rely on coal to produce electricity, making coal one of the nation's most important energy resources, even as the proportion of electricity produced from coal-fired generation recently has dipped below 40% for the first time in nearly four decades.[728] In many states, coal-fired generation dominates electricity production, with some states obtaining upwards of 80% of their electricity from this resource.[729] Further, because the United States is so rich in domestic coal reserves and is currently a net exporter of coal,[730] it is often suggested that coal is critical to national energy security.[731] Indeed, in the wake of the terrorist attacks of September 11, the National Research Council observed that one benefit of coal is that it is much less vulnerable to terrorist attacks than other energy resources, such as oil, natural gas, nuclear, and hydropower, whose distribution chains and production facilities might be more vulnerable targets for terrorist attacks than massive coal mines dispersed throughout the country.[732]

---

726. Claudia Copeland, Congressional Research Service, Mountaintop Mining: Background on Current Controversies (2015); M.A. Palmer et al., *Mountaintop Mining Consequences*, Science, (Jan. 8, 2010), *available at* http://www.filonverde.org/images/Mountaintop_Mining_Consequences_Science1[1].pdf.
727. Davies et al., *supra* note 29, at 109–10.
728. U.S. Energy Info. Admin., *Coal's Share of Total U.S. Electricity Generation Falls Below 40% in November and December* (Mar. 9, 2012), *available at* http://www.eia.gov/todayinenergy/detail.cfm?id=5331; U.S. Energy Info. Admin., *What Is U.S. Electricity Generation by Energy Source?*, http://www.eia.gov/tools/faqs/faq.cfm?id=427&t=3 (last updated Mar. 31, 2015).
729. U.S. Energy Info. Admin., *Electric Power Monthly*, (Feb. 2014), *available at* http://www.eia.gov/electricity/monthly/current_year/february2014.pdf.
730. U.S. Energy Info. Admin., *Annual Coal Report*, (Jan. 2015), *available at* http://www.eia.gov/coal/annual/pdf/acr.pdf.
731. Tom Donilon, National Security Advisor to the President, *Remarks at the Launch of Columbia University's Center on Global Energy Policy* (Apr. 24, 2013).
732. Nat'l Acad. of Sciences, Nat'l Research Council, *Making the Nation Safer: The Role of Science and Technology in Countering Terrorism* (2002).

Chapter 2. Exploitation

532. The legal issues that arise from the exploitation of coal in the United States can be divided primarily into three categories. The first two categories, in turn, depend on the type of land from which the coal is extracted. Mining on private lands raises questions of property and contract law, traditionally the province of state common law and, today, some statutes. By contrast, coal mining on public lands tends to be governed by statutory law, most notably federal leasing and mineral extraction statutes, but also state law where state public lands are at issue. Second, Congress has adopted national environmental statutes that govern coal extraction regardless of source, whether on private or public lands. And, third, Congress has passed legislation for the protection of mineworkers' health and safety.[733] We focus here on the legal issues related to mining on private and federal lands, and on general environmental regulation of coal mining, particularly mountaintop mining.[734]

§1. Coal Mining on Private Land

533. Two major issues dominate the legal questions raised by mining coal on private land. They are (1) what duties, if any, are owed to the surface owners when coal is extracted from the subsurface of private lands, and (2) who owns valuable byproducts of coal production, specifically, methane gas, when the parties fail to account for such questions in their contractual arrangements to allow the coal extraction to take place.

I. Surface Owners' Rights

534. Unlike oil and gas, coal rights are easy to apportion between the surface owner and the entity engaging in coal extraction, because coal is not a transitory resource. Thus, the common pool problems that arise in oil and gas exploitation, including the need for the law to incentivize extraction (*e.g.*, via the rule of capture) and to limit waste (*e.g.*, through well-spacing, conservation, and forced pooling and

733. *See* the Federal Coal Mine Safety Act, 30 U.S.C. §§ 451–483; Federal Coal Mine Health and Safety Act of 1969, (codified in various provisions of U.S.C. titles 15 and 30); significant amendments to the 1969 Act in the Federal Mine Safety and Health Amendments Act of 1977, 30 U.S.C. §§ 801 et seq.; Black Lung Benefits Act of 1972, Pub. L. No. 91-173, 83 Stat. 792 (codified as amended in scattered sections of 30 U.S.C.; and, the Coal Industry Retiree Health Benefits Act of 1992, 26 U.S.C. §§ 9701–9722. However, some of these laws have been repealed or superseded.

734. *See* Federal Mine Safety and Health Act of 1977, Pub L. 91-173, 83 Stat. 742 (codified at 20 U.S.C. 801-964). For more on the regulation of mine safety regulation, *see generally* Michael T. Heenan & C. Gregory Ruffennach, *National Institute of Occupational Safety and Health: Limits of Authority in Rulemaking Under the Federal Mine Safety and Health Act of 1977*, 7 J. Min. L. & Pol'y 171 (1992); R. Henry Moore, *The Doctrine of Judicial Deference and the Independence of the Federal Mine Safety and Health Review Commission*, 107 W. Va. L. Rev. 187 (2004); Wesley Kobylak, *What Are Protected Activities of Miners Under § 105(c)(1) of the Federal Mine Safety and Health Act of 1977 (30 U.S.C.A. § 815(c)(1))*, 67 A.L.R. Fed. 554 (originally published in 1984).

unitization requirements) do not exist for coal mining. Until it is mined, coal remains in one physical location on the property, and the owner of the surface estate may easily and clearly deed ownership of the underlying mineral rights to a different party. The primary legal tool used to govern coal extraction is thus primarily private law: contracts in the form of leases entered into by the property owners (lessor) and the entity that wishes to extract the coal (lessee). Despite this, leases governing coal extraction are not without legal dispute.

535. Arguably, the key issue that arises in the United States in the extraction of coal on private lands is what duties, if any, the extracting entity owes to the land's surface owner. Often, the parties themselves decide what constitutes reasonable use. Coal rights are generally leased rather than purchased from the surface owner, so the lessor (surface owner) and the lessee (the mining company, which acquires ownership of the coal by extracting it) may define the rights conveyed in the lease in any manner they choose through arms-length negotiation. Modern coal deeds and leases thus typically lay out the obligations of each party in regard to how the coal will be extracted, and what impact that extraction can have on the surface land. These include: (1) the method of mining (underground, or surface); (2) the extent to which the lessee has the right to use the land's surface, including for mine openings, transport, and access; (3) the extent to which the mine operator is liable for subsidence of the land; and (4) whether and under what conditions the lessee has an obligation to explore, develop, and extract the coal within a set period of time.[735] These issues, as with other kinds of mineral leases, raise numerous financial questions for both parties in how they structure their agreement.

536. Notwithstanding the ability of parties to mutually agree upon the surface impacts that coal mining can have, legal issues inevitably arise. Specifically, three key issues tend to pervade the relationship between surface owner (lessor) and mine operator (lessee): the mine's impact on the surface property, its impact on adjoining properties, and the scope of the grant to mine coal beneath the surface.

A. *Impact on Surface Land*

537. The question of the impact on the surface results from the interplay of competing property rules. On one hand, the general rule is that "the severed mineral estate is deemed dominant over the surface estate."[736] It is thus a basic principle of mineral law that the holder of the mineral estate may reasonably use the surface to

735. *Model Form of Mining Lease, Form—K,* Model Mining Development Agreement Project, International Bar Association: Mining Law Committee, *available at* http://www.mmdaproject.org/presentations/Form%20K%20-%20MODEL%20FORM%20OF%20MINING%20LEASE.pdf (last visited Feb. 22, 2015).

736. Marla E. Mansfield, *Coal*, *in* HICKEY ET AL., ENERGY LAW AND POLICY FOR THE 21ST CENTURY, *supra* note 109, at 9-1, 9-10; *see, e.g.*, Squires v. Lafferty, 121 S.E. 90, 91 (W. Va. 1924) (finding that an owner of mineral rights has "the right to use the 'surface' of the land in such manner and with such means as would be fairly necessary for the enjoyment of the mineral estate").

extract the minerals.[737] As the West Virginia Supreme Court has observed, "[A]ny use of the surface by virtue of rights granted by a mining deed must be exercised reasonably so as not to unduly burden the surface owner's use. . . . In order for such a claim to be successful, it must be demonstrated not only that the right is reasonably necessary for the extraction of the mineral, but also that the right can be exercised without any substantial burden to the surface owner."[738] This should make sense. The rule allows for mining to proceed, but it also means that the mining operation may be liable to the surface owner for negligently damaging the surface property or for unreasonableness in its use of the underlying mineral estate.[739] On the other hand, another general rule is that the surface owner retains the right to use and enjoy the surface property in every way for which she has not granted away rights. This also should make sense. As a basic principle of contract law, the mineral estate only goes so far as the parties agreed.

538. The difficulty comes, then, when the mining operation damages the surface estate.[740] This is what might be framed as coal mining's "split estate problem"—that is, the issues that arise from splitting the surface estate from the mineral estate, and thus, creating conflicting interests for different parties holding property interests in the same piece of land.[741]

539. A core version of these issues comes from subsidence, or the sinking or downward movement of the earth that can occur when supporting minerals are extracted from beneath the surface.[742] This is what is called the right to "subjacent" support—the right to support the surface land along a horizontal plane. In the United States, the "right to subjacent support is an absolute right: the mineral estate owner would be liable for subsidence even if its mining operations were not negligent."[743] Thus, surface owners that incur subsidence damage may sue the leasing coal company either for an injunction or for damages.[744] As with all other property rights, however, the surface owner may waive its right to seek redress for subsidence damage by contracting it away in the lease.[745]

737. LOWE, *supra* note 323, at 39.

738. Buffalo Min. Co. v. Martin, 165 W. Va. 10, 18 (1980).

739. In fact, to protect surface owners, some states require mines to post bonds or other security before commencing operations. *See, e.g.*, KAN. STAT. ANN. § 49-615; IDAHO CODE ANN. § 47-1512.

740. *See generally* Wendy B. Davis, *Out of the Black Hole: Reclaiming the Crown of King Coal*, 51 AM. U. L. REV. 905 (2002).

741. For more on these issues, *see, e.g.*, Timothy W. Gresham & Monroe Jamison, *Do Waivers of Support and Damage Authorize Full Extraction Mining?*, 92 W. VA. L. REV. 911 (1990).

742. *See, e.g.*, Keystone Bituminous Coal Co. v. DeBenedictis, 480 U.S. 270 (1987).

743. Mansfield, *supra* note 736, at 9-11.

744. *See, e.g.*, Stonegap Colliery v. Hamilton, 89 S.E. 305 (Va. 1916); Breeding v. Koch Carbon, Inc., 726 F. Supp. 645 (W.D. Va. 1989); Large v. Clinchfield Coal Co., 387 S.E.2d 783 (Va. 1990); McGurn v. Reichel, 268 S.W. 399 (Mo. App. 1925).

745. *See, e.g.*, Ball v. Island Creek Coal Co., 722 F. Supp. 1370 (W.D. Va. 1989); Vandyke v. Island Creek Coal Co., No. 91-3022, 1991 U.S. App. LEXIS 27680, at *1 (4th Cir. Nov. 22, 1991).

B. Impact on Neighboring Lands

540. A second issue that arises from subsurface mining's effect on property rights is its impact on neighboring lands.[746] This is known as the right to "lateral" support—property rights along a vertical plane. Generally in the United States, "[r]emoval of minerals cannot make adjoining property subside because of sideward and downward movement."[747] Typically, however, this rule is limited to protecting the adjoining land in its natural condition; if the adjoining land has been modified and that is the cause of the damage, a cause of action for harm to lateral support cannot prevail.[748] Likewise, damages for this cause of action generally are limited to actual harm to the property, although damages may be recovered irrespective of whether there is intervening land between the mined property and the damaged property.[749]

C. Scope of the Lease's Grant

541. The third issue is the extent to which a mining lease might imply the right for the operator to do damage to the surface land. Parties to a mineral lease can expressly waive claims for damages to the surface land, or explicitly give the lessee the right to create subsidence.[750] Absent express waiver, however, the question becomes whether there is an implied waiver in the lease allowing surface damage to occur. Generally, courts have been hesitant to imply such a waiver from the right to mine alone, but rather, have found that the language creating it "must be clear, unequivocal, and free from ambiguity."[751] However, several courts have held that a

---

746. *See, e.g.*, Livingston v. Moingona Coal Co., 49 Iowa 369 (1878).

747. Mansfield, *supra* note 736, at 9-11; *cf., e.g.*, Myer v. Hobbs, 57 Ala. 175 (1876); Green v. Berge, 38 P. 539 (Cal. 1894); Abrey v. City of Detroit, 86 N.W. 785 (Mich. 1901); Stearns v. City of Richmond, 14 S.E. 847, 848 (Va. 1892); XI Props., Inc. v. RaceTrac Petroleum, Inc., 151 S.W.3d 443 (Tenn. 2004).

748. *See, e.g.*, Western Coal & Mining Co. v. Randolph, 89 S.W.2d 741 (Ark. 1936); Valley Smokeless Coal Co. v. Manufacturers' Water Co., 144 A. 740 (Pa. 1928). Most courts have not extended this natural condition limitation to the right of subjacent support. *See* 5 POWELL ON REAL PROPERTY 312 (1968).

749. *See, e.g.*, Ohio Collieries Co. v. Cocke, 140 N.E. 356 (Ohio 1923); Home Brewing Co. v. Thomas Colliery Co. 117 A. 542 (Pa. 1922).

750. Mason v. Peabody Coal Co., 51 N.E.2d 285 (Ill. App. 1943); Paull v. Island Coal Co., 88 N.E. 959 (Ind. App. 1909); Madden v. Lehigh Valley Coal Co., 61 A. 559 (Pa. 1905); Chartnetski v. Miner's Mills Coal Mining Co., 113 A. 683 (Pa. 1921); *see also* Comment, *The Common Law Rights to Subjacent Support and Surface Preservation*, 38 MO. L. REV. 234, 237–42 (1973).

751. Gresham & Jamison, *supra* note 741, at 920. The exception is West Virginia. In *Griffin* v. *Fairmont Coal Co.*, 53 S.E. 24 (W. Va. 1905), that state's appellate court ruled that a mineral lease allowing the lessee to extract all coal impliedly waived the right to subjacent support — although that ruling was later limited by other decisions. *See* Hall v. Harvey Coal & Coke Co., 108 S.E. 491 (W. Va. 1921); Winnings v. Wilpen Coal Co., 59 S.E.29 655 (W. Va. 1950); Stamp v. Windsor Power House Coal Co., 177 S.E. 2d 146 (W. Va. 1970); Simmers v. Star Coal & Coke Co., 167 S.E. 737 (W. Va. 1933).

mineral lease's waiver of liability for damages resulting from the mining activity implies a waiver for damages from a loss of subjacent support.[752]

542. Perhaps the most controversial area where this implied waiver question arises, however, is with modern mining technologies. The question generally is whether older mineral deeds, sometimes called broad form deeds, which conveyed rights widely to mine, still have import when the type of mining technology used changes. Thus, a number of courts have addressed whether these deeds extend to the introduction of longwall mining. Generally, these courts have found the deeds to allow such mining, even when parties have argued that the mining technology could not have been contemplated in the contract because it did exist at the time of the deed.[753] By contrast, courts have varied on whether broad form deeds also allow for surface mining, namely, strip mining. The Kentucky courts, for instance, have gone back and forth on the issue, first finding that broad form deeds allowed strip mining,[754] then later allowing recovery for damages,[755] and finally reversing course and disallowing strip mining under such deeds.[756] Other courts, however, have rejected this approach, finding that mineral deeds allowing subsurface mining cannot, without more, give operators the right to strip mine.[757] To grapple with this problem, some states—including Colorado, Kentucky, Montana, Tennessee, and Wyoming—also have passed statutes to foreclose strip mining based on common mineral rights agreements entered into before that practice was used.[758]

II. Coalbed Methane

543. Irrespective of how mineral deeds are construed as to surface property rights, another key question that arises in the mining of coal on private lands is who owns valuable byproducts that may come with coal extraction. The primary resource for which this issue arises is coalbed methane (CBM).

544. CBM is one kind of natural gas. It becomes embedded within or beneath coal reserves and is kept in place by water pressure. Over the past decades, capture of CBM has increased in the United States, though after 2008, its production has

752. *See, e.g.*, Mason v. Peabody Coal Co., 51 N.E.2d 285 (Ill. App. 1943); Paull v. Island Coal Co., 88 N.E. 959 (Ind. App. 1909); Atherton v. Clearview Coal Co., 110 A. 298 (Pa. 1920).

753. *See, e.g.*, Wells v. American Elec. Power Co., 548 N.E.2d 995 (Ohio App. 1988); Porter v. Consolidation Coal Co., No. 88-3637 (W.D. Pa. Aug. 8, 1988), *aff'd*, 870 F.2d 651 (3d Cir. 1989); Karst-Robbins Coal Co. v. Arch of Ky., Inc., 964 S.W.2d 419 (Ky. Ct. App. 1997); Ball v. Island Creek Coal Co., 722 F. Supp. 1370 (W.D. Va. 1989).

754. Martin v. Ky. Oak Mining Co., 429 S.W.2d 395 (Ky. 1968).

755. Akers v. Baldwin, 736 S.W. 294 (Ky. 1987).

756. Ward v. Harding, 860 S.W.2d 280 (Ky. 1993).

757. *See, e.g.*, Stewart v. Chernicky, 266 A.2d 259 (Pa. 1970); Phipps v. Leftwich, 222 S.E.2d 536 (Va. 1976); Belville Min. Co. v. United States, 999 F.2d 989 (6th Cir. 1993).

758. Tennessee's statute was defended against constitutional challenge. *See* Doochin v. Rackley, 610 S.W.2d 715 (Tenn. 1981). Kentucky amended its state constitution after an earlier law was struck down. *See* Akers v. Baldwin, 736 S.W.2d 294 (Ky. 1987).

slightly decreased since its peak in that year.[759] The U.S. Geological Survey has estimated that domestic coal deposits include 700 trillion cubic feet of CBM, and that up to 100 trillion cubic feet of this CBM is recoverable.[760]

545. The United States has not always used CBM. Until the 1980s, CBM was considered a hazardous byproduct of coal mining; operators simply vented it to avoid explosions. As natural gas prices rose, however, and as technology developed, CBM extraction came into vogue. The U.S. government promoted it on federal lands, and private development of this resource began in Alabama's Black Warrior Basin in the 1980s, followed by increasing development in the West, particularly in the San Juan Basin in New Mexico and Colorado and then the Powder River Basin in Wyoming and Montana. Today, CBM production in the Rocky Mountain states, particularly Wyoming and New Mexico, accounts for more than half of the nation's CBM yield, and roughly 8% of domestic natural gas production.[761] Notwithstanding its potential value, much recoverable CBM in the United States is wasted rather than extracted, with about two-thirds of this occurring through intentional venting in the coal mining process.[762] When CBM is extracted, it is recovered by drilling into the coal seams, placing wells, removing groundwater, and then allowing the resource to flow to the surface.

546. In part because CBM presents conflicting interests for different parties in the same piece of property—surface owners, coal mining operators, and, potentially, parties with oil and gas rights in the land—disputes over this resource have produced a large body of law in the United States. The key issue is whether land patents reserving coal rights, or mineral rights leases to extract coal, include CBM within their definition of coal rights.[763] The issue of ownership arises on both federal and private lands.

547. The issue arises on federal lands because in the early 1900s the federal government conveyed surface lands in the West to private owners but reserved for itself rights to the coal. Thus, the question became whether that reservation of coal rights

759. In recent years, production of CBM has slightly decreased from its high of 1,966 billion cubic feet in 2008 to 1,866 billion cubic feet in 2010 and 1,466 billion cubic feet in 2013. U.S. Energy Info. Admin., *U.S. Coalbed Methane Production* (Dec. 4, 2014), *available at* http://www.eia.gov/dnav/ng/hist/rngr52nus_1a.htm.

760. See U.S. Geological Survey, *Coalbed Methane—An Untapped Energy Resource and an Environmental Concern* (Jan. 17, 1997), http://energy.usgs.gov/factsheets/Coalbed/coalmeth.html.

761. *See generally* U.S. Environmental Protection Agency, Coalbed Methane Extraction: Detailed Study Report (Dec. 2010); Aaron M. Flynn, Congressional Research Service, Selected Legal and Policy Issues Related to Coalbed Methane Development (2004). For more on issues involving coalbed methane extraction in the Rocky Mountains, see Gary Bryner, *Coalbed Methane Development in the Intermountain West: Producing Energy and Protecting Water*, 4 Wyo. L. Rev. 541 (2004).

762. L. James Lyman, Comment, *Coalbed Methane: Crafting a Right to Sell from an Obligation to Vent*, 78 U. Colo. L. Rev. 613 (2007).

763. Another issue is priority among oil and gas rights owners and coal rights owners. *See* Flynn, *supra* note 761.

included CBM. In *Amoco Production Co. v. Southern Ute Indian Tribe*,[764] the Supreme Court decided this question in the negative for federal lands, ruling that reservation of the right to extract coalbed methane must be explicit. At issue in *Amoco* was the United States' withdrawal in 1906 of full ownership of 64 million acres of coal-producing land, conveying the surface rights to private patentees.

548. However, in laws passed by Congress in 1909 and 1910, the government reserved the underground coal rights for itself. The Ute Indian Tribe assumed the federal government's ownership of the underground coal rights and, later, argued that ownership of those rights also included ownership of valuable coalbed methane gas found within the underground coal beds. The U.S. Court of Appeals for the Tenth Circuit agreed, and held that there was no need to specifically convey ownership of the coalbed methane gas separate from the underground coal rights, because the mineral rights in the coal itself subsumed ownership of the CBM rights.[765] The United States Supreme Court reversed. It held that more was necessary: an explicit conveyance of CBM, in addition to the coal itself. Thus, applying the Acts of 1909 and 1910, the Court found that because those laws only specifically reserved the underground coal rights and not CBM, the Ute Tribe did not have title to the CBM.[766] This was true, the Court said, even though the right to mine coal implies the right to release CBM incident to coal mining. "There is no reason to think Congress would have been more concerned about the creation of a split coal/CBM gas estate than the creation of a split gas estate."[767] In short, the specific intent of the parties at the time coal and gas rights are reserved or conveyed controls who owns CBM mining rights.

549. Although not binding on state courts, *Amoco* has been influential. Numerous state courts across the country addressing the scope of mineral leases under state law similarly have held that coalbed methane is separate from coal, some of them relying expressly on the reasoning from *Amoco*. In 2002, for instance, the Wyoming Supreme Court ruled in *Newman v. RAG Wyoming Land Co.* that methane was not included in a mineral deed that granted the right to mine "all coal and minerals commingled with that coal."[768] Likewise, relying on the lower court's ruling in *Amoco*, a 1995 decision by the Montana high court held that the coal mining operation had the right to ventilate coal if needed for safety, but that the oil and gas lessee on the property had a "mutual, simultaneous right" to extract CBM from the coal seam.[769] Other courts have followed these same approaches, deeming CBM rights separate from coal rights,[770] or allowing coal operations to ventilate CBM to the

764. Amoco Prod. Co. v. S. Ute Indian Tribe, 526 U.S. 865 (1999).
765. *See* S. Ute Indian Tribe v. Amoco Prod. Co., 151 F.3d 1251 (10th Cir. 1998), *cert. granted*, 525 U.S. 1118 (1999).
766. *Amoco*, 526 U.S. at 865.
767. *Ibid.* at 880.
768. 53 P.3d 540, 549 (Wyo. 2002).
769. Carbon Cnty. v. Union Reserve Coal Co., Inc., 898 P.2d 680, 689 (Mont. 1995).
770. *See, e.g.*, Energy Dev. Corp. v. Moss, 591 S.E.2d 135 (W.Va. 2003); Harrison-Wyatt, LLC v. Ratliff, (Va. 2004).

surface but not capture the gas[771]—although at least a few decisions have gone the other way, finding CBM included within mineral deeds to extract coal.[772] Again, energy lawyers must be aware not only of federal laws but of varying state laws as well.

§2. Coal Mining on Federal Lands

550. While much of the coal extraction in the United States occurs on private lands, mining on public lands is also important. An estimated one-third of the remaining coal reserves in the western United States lie beneath federally owned lands.[773] Companies seeking to mine these coal reserves do not negotiate leases as they would on private lands. Rather, they are bound by standardized federal mineral rights leases, which in turn must comply with federal legislation and regulations governing the extraction of natural resources from federal public lands.[774]

551. Coal mining on federal lands is important, as such coal comprises roughly a quarter of recoverable reserves in the United States. Federal coal mining falls under the jurisdiction of the Department of the Interior. Depending on where the resource is located, the different aspects of coal production and extraction may be controlled by different agencies, including the Bureau of Land Management within the Department of the Interior, and the U.S. Forest Service, which is part of the Department of Agriculture.[775]

552. Specifically, the Mineral Leasing Act,[776] as amended by the Federal Coal Leasing Act Amendments of 1976 (FCLAA),[777] governs coal mining leases on

771. *See, e.g.*, NCNB Texas Nat'l Bank, N.A. v. West, 631 S.2d 212 (Ala. 1993); Carbon Cnty. v. Union Reserve Coal Co., 898 P.2d 680 (Mont. 1995).

772. *See, e.g.*, U.S. Steel Corp. v. Hoge, 468 A.2d 1380, 1382 (Pa. 1983); *see also* Rayburn v. USX Corp., 844 F.2d 796 (11th Cir. 1988) (holding that CBM belonged to person who owned the coal); Vines v. McKenzie Methane, 619 S.2d 1305 (Ala. 1993) (finding drilling for CBM and mining for coal inextricably linked and therefore CBM rights bound up in mineral estate for "all coal" and "minerals"); Cimarron Oil Corp. v. Howard Energy Corp., 909 N.E.2d 1115 (Ind. Ct. App. 2009) (finding that coal lessee, as opposed to oil lessee, had exclusive right to recover all CBM). For more on the topic of coalbed methane ownership, see Wendy B. Davis, *Coalbed Methane: Degasification, Not Ventilation, Should Be Required*, 2 Appalachian J.L. 25 (2003).

773. *A Digital Database of Coal Ownership Status*, U.S. Geological Survey (Feb. 1998), http://pubs.usgs.gov/fs/1998/fs012-98/; *An Assessment of U.S. Federal Coal Royalties: Current Royalty Structure, Effective Royalty Rates, and Reform Options*, Headwater Economics (Jan. 2015), http://headwaterseconomics.org/wphw/wp-content/uploads/Report-Coal-Royalty-Valuation.pdf.

774. Bureau of Land Management, BLM Coal Resources 15–26 (2012).

775. U.S. Dep't of the Interior, *Coal Operations*, http://www.blm.gov/wo/st/en/prog/energy/coal_and_non-energy.html (last visited Feb. 23, 2015); *Coal*, U.S. Forest Serv., http://www.fs.fed.us/geology/energyCoal.html (last visited Feb. 23, 2015).

776. 30 U.S.C. §§ 181–196, 201–209, 211–214, 221–237, 241–242, 261–263, 271–276, 281–287.

777. Federal Coal Leasing Amendments Act of 1976, Pub. L. No. 94-377, 90 Stat. 1083 (codified at 42 U.S.C. § 181 et seq. (1976). The FCLAA were a response to suggestions that parties were sitting on leases, and that the government was not receiving fair value for its leases. Other concerns in the MLA regime were concentration of mining operations, environmental degradation and lack of planning, and lease speculation. *See* H.R. Rep. No. 94-681 (1975).

federal land. These Acts authorize the BLM to lease coal and other subsurface mineral rights for development to private entities. Under this regime, the BLM also administers leases on Forest Service land, but Forest Service consent is required for mining to proceed, and that agency can specify the terms and conditions of leases carried out on Forest Service land.[778]

553. The overarching objective of the MLA/FCLAA regime is to ensure that extraction of coal from federal lands is economically efficient—that the government, and thus the public, receives fair compensation from the resource, and that lessees act in an economically responsible way. Accordingly, these laws require that all federal coal leases be issued through competitive bidding processes.[779] The law further mandates that the federal government shall not accept lease bids that are for less than fair market value.[780] Fair market value of coal, in turn, is determined by the Secretary of the Interior,[781] who has discretion to determine what fair value is. To that end, two mechanisms are used: first, the "comparable sales" approach, which considers bonus bids received for past sales to value the tract in question; and second, the "income" approach, which analyzes information from previous lease sales, and the amount of money companies stand to make from extracting the coal, to determine the price.[782]

554. Under these laws, coal leases are "for a term of twenty years and for so long thereafter as coal is produced annually in commercial quantities from that lease."[783] Federal coal leases thus have both a primary term and a continuing term, the latter of which is contingent on ongoing production of coal from the lease. The amount of the ongoing production must be at "commercial quantities,"[784] which includes a prudence requirement for the lessee. Specifically, the implementing regulations define "commercial quantities" only to include "the coal deposit discovered" that is "of such character and quantity that a prudent person would be justified in further expenditure of his labor and means with a reasonable prospect of success in developing a valuable mine."[785] Thus, federal law seeks to foreclose the possibility that lessees will squat on leases simply by continuing to produce minimal amounts of coal from year to year; the relevant coal production must be economically viable, not just extant.

555. The FCLAA further mandates that each lease require "diligent development" of the coal mines, "except where operations are interrupted by

778. SERVICE MANUAL § 2800, amendment no. 2800-94-2 (1994).
779. *See* 30 U.S.C. § 201(1).
780. *See Ibid.*
781. *Ibid.*
782. U.S. GOVERNMENT ACCOUNTABILITY OFFICE, COAL LEASING: BLM COULD ENHANCE APPRAISAL PROCESS, MORE EXPLICITLY CONSIDER COAL EXPORTS, AND PROVIDE MORE PUBLIC INFORMATION 28 (2013).
783. *See* 30 U.S.C. § 207(a).
784. *Ibid.*
785. *See* 43 C.F.R. § 3430.1–2(a).

strikes, the elements or casualties not attributable to the lessee."[786] Like the law's anti-squatting provision, this prohibition, by requiring immediate efforts to extract coal, is designed to prevent leases from being held for speculation. The aim of the law, like the Mineral Leasing Act more generally, is to stimulate rapid development of coal and to maximize royalties paid to the Department of the Interior.[787] Leases entered into under the FCLAA require payment of a royalty to the federal government in an amount determined by the Secretary of the Interior.[788] Currently, the Secretary of the Interior imposes a royalty rate of 12.5% for above-ground mining, and 8% for underground mining, on coal leases on federal lands.[789]

556. While the FCLAA generally obliges companies that acquire federal leases to engage in extraction and to do so prudently, the law also provides some flexibility, allowing mining companies to respond to market demand. For example, if a lessee is unable to obtain a market because it is not financially feasible at the time, the lessor may be able to retain a lease even if they fail to comply with the covenant to produce.[790] However, this flexibility is not unlimited. To invoke it, the lessor likely needs to show that it expended resources in an effort to make the leased area marketable, and not simply that they waited to act on the lease.[791]

557. Coal leasing under the MLA/FCLAA regime occurs in three stages: planning, regional planning, and leasing itself. The first and last of these steps—planning and procedures and requirements for leasing itself—are mandatory, but the regional planning process can be skipped in two contingencies. The first is if emergency leasing is needed in designated federal coal leasing areas;[792] this is "intended for those limited circumstances where federal coal might be bypassed or where the coal is needed to maintain production levels in the near term."[793] The second contingency is that regional planning need not be used for leases "outside coal production regions."[794] This exception is much broader than it sounds. The government effectively applies it to mean that a region does not exist unless the area has shown a "substantial interest in new leasing."[795] Under that meaning, even the Powder River Basin has been deemed not to be a coal production region, because bids for lease tracts tend to be from the current lessee of the tract.[796]

786. *See* 30 U.S.C. § 207(b).
787. U.S. Department of the Interior, Federal Coal Management Program 3.1.5 (1978).
788. 30 U.S.C. § 207(a), (b); Sally K. Fairfax & Karolyn E. Yale, Federal Lands: A Guide to Planning, Management, and State Revenues, Western Officer, Council of State Governors 83 (1987).
789. U.S. Dep't of the Interior, *Coal Operations*, http://www.blm.gov/wo/st/en/prog/energy/coal_and_non-energy.html (last visited Feb. 23, 2015). The statute allows a lower rate for underground mining. 30 U.S.C. § 207(a).
790. 30 U.S.C. § 188; *see also* Eggleson v. McCasland, 98 F. Supp. 693 (E.D.Okla.1951).
791. *Ibid.*
792. 43. C.F.R. § 3425.1–4.
793. Rasband et al., *supra* note 590, at 1164.
794. 43 C.F.R. § 3425.1–5.
795. Rasband et al., *supra* note 590, at 1164.
796. *Ibid.*

558. The three stages of this leasing regime are designed to ensure coordinated extraction, and to promote competition and thus maximization of revenues from the federal leasing program. The first stage is the planning stage. Under applicable regulations, this planning requires that a proposed lease pass through four screening procedures. The initial screen aims to gauge the extraction potential of the area, including assessing the amount of economically recoverable reserves. The next screen checks whether the land is suitable for coal mining; if it is not, it cannot be leased. After that, the agency assesses multiple-use tradeoffs, such as extraction versus recreation versus aesthetics versus surface uses like grazing. Conflicts at this stage also may cause lands to be removed from consideration. Finally, consent from surface owners is sought. The screening process ultimately culminates in issuance of a final land use plan, often termed a "resource management plan."[797]

559. The next stage of the process is the regional planning stage.[798] Here, a federal-state advisory group known as a Regional Coal Team (RCT), reviews the resource management plan and engages in a market analysis. This analysis is long-term in nature and is used to determine whether to proceed further with the leasing process. If the RCT recommends proceeding, science advisors are engaged and, together with an internal BLM review council, conduct the necessary environmental impact statement under the National Environmental Policy Act.[799] NEPA is a procedural statute that does not foreclose federal actions if they are environmentally detrimental; rather, it simply requires the government to consider the impacts of its actions before proceeding. If, after NEPA review, the agency decides to proceed, it publishes a regional lease sale decision in the Federal Register.[800]

560. The final stage, then, is the leasing itself.[801] This is conducted via an auction aimed to maximize revenues. The BLM seeks public comment on fair market value of the resource in question, along with the appropriate mining methods for the site. The agency then also estimates the value of each tract in the lease itself. Finally, the agency accepts sealed bids. The agency accepts the high bids that also meet its assessment of fair market value.[802] Winning bidders also must "endure an antitrust review by the Department of Justice."[803]

§3. Environmental Regulation of Coal Mining

561. Irrespective of whether coal extraction occurs on private or public land, federal environmental law plays a heavy role in regulating coal mining. Coal exploitation can have major environmental impacts. These include land disturbance

797. 43 C.F.R. § 3420.1–4(e).
798. 43 C.F.R. §§ 3420.3–.5.
799. National Environmental Policy Act of 1969, Pub. L. No. 91-190, 83 Stat. 852 (codified in scattered sections of 42 U.S.C.).
800. 43 C.F.R. §§ 3420.3–.5.
801. *Ibid.* § 3422.
802. *Ibid.*
803. Rasband et al., *supra* note 590, at 1164.

from strip mining and surface mining, as well as potential water pollution, which can be particularly problematic where mountaintop removal is used as the method of coal extraction. Most critical in this respect are the Surface Mining Control and Reclamation Act (SMCRA),[804] and the federal Clean Water Act.

I. Surface Mining Reclamation

562. Since 1977, the federal Surface Mining Control and Reclamation Act has imposed limits on strip and surface mining. The statute's aim is to mitigate the adverse effects of coal mining on the surface of the land.[805] SMCRA establishes a cooperative federal-and-state regulatory program to protect mined land, water, and other resources adjacent to mining sites. It is a comprehensive statute designed to create "a nationwide program to protect society and the environment from the adverse effects of surface coal mining operations."[806] The Office of Surface Mining Reclamation and Enforcement (OSM) within the Department of the Interior is charged with primary responsibility for administering and implementing SMCRA.[807]

563. SMCRA divides the world of coal mining into two categories: abandoned mines and active mines. Thus, two titles of the law are most critical. Title IV of the law establishes a program for remediating abandoned mining sites.[808] It does so by imposing a tax on current coal producers.[809] Those funds are then distributed by the Secretary to states, which can use the funds for remediation efforts.[810] Title V of the law, on the other hand, applies to currently operating mines.[811] Such mines must obtain regulatory permits in order to continue coal extraction.[812] The purpose of the permits is to limit environmental degradation and to ensure remediation following mine closure.

---

804. Surface Mining Control and Reclamation Act of 1977, Pub. L. 95–87, 91 Stat. 445 (codified as amended in scattered sections of 30 U.S.C.).
805. Save Our Cumberland Mountains, Inc. v. Hodel, 651 F. Supp. 1528, 1531 (D.D.C. 1986) ("The Surface Mining Control and Reclamation Act of 1977 . . . is a comprehensive statute designed, *inter alia,* to control and remedy adverse social, economic and environmental effects of surface coal mining operations, to minimize damages and risks affecting productivity of the soil arising from such operations, to protect the health and safety of the public, and to prevent or mitigate adverse effects of present and future surface coal mining operations.").
806. *See* 30 U.S.C. § 1202(a).
807. *See* 30 U.S.C. § 1211(c). SMCRA survived several early constitutional challenges. *See* Hodel v. Va. Surface Mining and Reclamation Ass'n, Inc., 452 U.S. 264 (1981); Hodel v. Indiana, 452 U.S. 314 (1981).
808. *See* 30 U.S.C. §§ 1231–1244; *see also* JAMES M. MCELFISH, JR. & ANN E. BEIER, ENVIRONMENTAL REGULATION OF COAL MINING: SMCRA'S SECOND DECADE 20 (1990).
809. 30 U.S.C. § 1232.
810. 30 U.S.C. § 1232(g).
811. 30 U.S.C. §§ 1251–1279.
812. 30 U.S.C. §§ 1256–1261.

564. The cooperative federalist regime erected by SMCRA was designed to be phased in through two stages: an initial, or interim, regulatory phase,[813] to be followed by a permanent regulatory program.[814] Specifically, SMCRA was designed with the idea that ineffective state regulation needed to be replaced with minimum, nationally uniform reclamation requirements. Thus, under the initial interim phase of SMCRA, the federal government regulated mining directly for all states that did not have existing regulatory programs. For those states that did have such programs, their regulation continued until a full federal program could be developed, as long as the state program's standards complied with federal requirements. The "interim program left any existing state permits in effect, but required all coal mine operators to comply with the federal interim program regulations."[815]

565. Subsequently, minimum federal standards were adopted under SMCRA's "permanent" phase, and today, all mining occurs under this program.[816] Under the permanent phase of SMCRA, states and the federal government can work together. If a state wants to continue regulating mines within its boundaries, it can develop a plan and submit it to the Secretary. Such plans can differ from federal regulations, but they cannot fall below the floor of the federal minimum.[817] If the Secretary determines that a state's program satisfies the federal minimum, the state then may continue to regulate. If, however, the state's program does not comply, or if a state does not submit a plan, then the federal government regulates directly.[818] These rules apply to private lands; all mining of coal on federal lands is under permanent federal government regulation. Today, roughly two-thirds of states have adopted their own regulatory programs under SMCRA.[819]

566. SMCRA's reclamation program operates using four key mechanisms. First, the law mandates that parties must obtain a permit before conducting mining activities. Second, the Act requires the mine to post a bond before commencing operations, and to keep the bond in place while mining remains operational. Third, SMCRA imposes detailed operational requirements for mines. And, fourth, the Act authorizes both the Secretary of the Interior and states to enforce its mandates via inspections and orders to cease operation. SMCRA thus does not foreclose mining, but rather, seeks to ensure that its environmental impacts are limited, and what environmental impacts it does have are remediated.

---

813. *See* 30 U.S.C. § 1251.

814. *See* 30 U.S.C. § 1251.

815. McElfish & Beier, *supra* note 808, at 20.

816. *Ibid.*

817. *See* U.S.C. § 1255(b).

818. Nat'l Wildlife Fed'n v. Lujan, 1990 WL 134826 n.3 (D.D.C. 1990).

819. U.S. Dep't of the Interior, Office of Surface Mining Reclamation and Enforcement, *Budget Justifications and Performance Information, Fiscal Year 2015*, 68, (2015). States that choose not to implement their own regulatory framework are known as "federal program states"; they include Tennessee, Washington, Arizona, California, Idaho, Massachusetts, Michigan, North Carolina, Oregon, Rhode Island, and South Dakota.

567. SMCRA's operating mechanism is its permitting requirement. Before mining can begin, all surface mining operators must obtain a surface mining operating permit.[820] These permits are valid for up to five years, but mining activity must commence within three years of the permit's issuance.[821] To obtain a permit, operators must submit a reclamation plan, an insurance certificate, a blasting plan, and provide for public inspection of the application.[822] In addition, the applicant must pay a review fee and provide notice both to all those with a property interest, including leaseholds, in the proposed extraction tract and to all surface and subsurface property owners of adjacent areas.[823] In addition, the applicant must provide information on: (1) its past mining practices, including whether it has ever had any permits or applications suspended or revoked;[824] (2) its corporate structure and ownership;[825] (3) the starting and termination dates of each phase of the operation;[826] and (4) a description of the method of extraction that will be used, along with the engineering techniques and equipment that will be utilized.[827] Finally, applicants must provide extensive environmental information, including (1) the names of the watershed and location of streams and tributaries into which any drainage will be discharged; (2) a determination of probable hydrologic consequences of the mining and reclamation operations (on- and off-site);[828] (3) maps of the area to be affected, including archaeological sites and any subsurface water;[829] and (4) test bore results.[830]

568. If this permitting requirement is its operational mechanism, SMCRA's bonding requirement is its prophylactic device. Under SMCRA, operators must post a performance bond covering the land that will be mined.[831] The purpose of this bond is to ensure that any damage caused to the land from the mining activity will be remediated. As one court has observed, "Congress enacted SMCRA to provide protection against environmental degradation from coal mining and to clean up areas damaged by past coal mining."[832] The total amount of the bond is determined by looking at the difficulty of reclamation considering factors such as hydrology, revegetation potential, topography, and geology.[833] No bond, however, can be less than USD 10,000.[834] Liability under the bond extends for the period of the permitted mining and also encompasses reclamation efforts, including mandatory

820. 30 U.S.C. § 1257.
821. *See* 30 U.S.C. §§ 1256,1257.
822. *See* 30 U.S.C. §§ 1257(d)(e)(f)(g).
823. *See* 30 U.S.C. §§ 1257(a)(b)(1),(2). The fee cannot exceed the cost of reviewing, enforcing, and administering the permit.
824. *See* 30 U.S.C. §§ 1257(b)(2),(5).
825. *See* 30 U.S.C. §§ 1257(b)(3),(4).
826. *See* 30 U.S.C. §§ 1257(b)(7),(8).
827. *See* 30 U.S.C. §§ 1257(b)(7),(8).
828. *See* 30 U.S.C. §§ 1257(b)(10)–(12).
829. *See* 30 U.S.C. §§ 1257(b)(9),(13),(14).
830. *See* 30 U.S.C. §§ 1257(b)(15),(17).
831. *See* 30 U.S.C. § 1259.
832. Pa. Fed'n of Sportsmen's Clubs, Inc. v. Kempthorne, 497 F.3d 337, 340 (3d Cir. 2007).
833. 30 U.S.C. § 1259(a).
834. *Ibid.*

revegetation success periods (generally five years after the first seeding, or ten years in areas that receive less than 26 inches of annual precipitation).[835] The process for setting bonds is fact-intensive. Bonds range in price, but they can be expensive, with cost ranging between USD 15 and USD 75 million.[836]

569. Perhaps the most critical element of SMCRA, however, is its strict performance standards for mining operations. These dictate, in great detail, "how the operation must be conducted and to what degree reclamation must take place. SMCRA § 515 contains 25 performance standards, many with several sub-parts, for surface mining operations generally, and others for special kinds of surface operations, such as steep-slope mining."[837] SMCRA's standards for underground mining are "almost as numerous" and are provided for in SMCRA § 516.[838]

570. SMCRA's operating standards promote both efficiency and environmental protection. For instance, they require that surface coal mining operations maximize the utilization and conservation of the coal being recovered, so "that reaffecting the land in the future through surface coal mining can be minimized."[839] Crucially, the operating standards also extend past the active life of the mine. Thus, they require coalmine operators to restore the affected land to a condition capable of supporting those uses it could support prior to mining, or for "higher or better uses."[840]

571. More specifically, operators must: (1) restore the land to its original approximate contour by backfilling, grading, and compacting; (2) minimize hydrologic disturbances, particularly by avoiding acid mine drainage and preventing additional sediments to nearby bodies of water from erosion; (3) perform land reclamation activities as soon as practicable following coal extraction, including while the mining operation continues; and (4) build up a permanent vegetative cover in affected areas.[841]

572. To ensure compliance with established performance standards, SMCRA requires ongoing inspections, including at least one complete on-site inspection each quarter and one partial inspection without notice to the operator each month.[842] When inspections reveal violations, SMCRA requires inspectors to take enforcement action, and the law gives inspectors full authority to shut down mining operations if a violation poses an imminent threat to the public or the

835. 30 U.S.C. § 1265(b)(20)(A).
836. U.S. Dep't of Agric.: U.S. Forest Service, Training Guide for Reclamation Bond Estimation and Administration 18 (2004); *see also* U.S. Dep't of the Interior: Office of Surface Mining Reclamation and Enforcement, Handbook for Calculation of Reclamation Bond Amounts (2000).
837. Nat'l Wildlife Fed'n v. Lujan, 1990 WL 134826 n.3 (D.D.C. 1990); *see* 30 U.S.C. § 1265.
838. Nat'l Wildlife Fed'n v. Lujan, 1990 WL 134826 n.3 (D.D.C. 1990); *see* 30 U.S.C. § 1266.
839. *See* 30 U.S.C. § 1265(b)(1).
840. *See* 30 U.S.C. § 1265(b)(2).
841. *Ibid.*
842. *See generally* 30 U.S.C. § 1271.

environment.[843] Where, however, a violation does not risk imminent harm, the inspector must issue a notice of violation.[844] The mining operation then has time to abate the harm in a period established by the notice; but if abatement does not occur, the inspector must then order cessation of operations and compel the operator to remedy the violations.[845] The Office of Surface Mining Reclamation and Enforcement also has the ability to assess monetary penalties.[846] Fines are assessed on a point system and can range from USD 32 for a single point violation to USD 8,500 for a point violation of 70.[847]

573. Finally, SMCRA also establishes lands deemed unsuitable for coal mining.[848] These include lands in National Parks and certain other National Forest lands. The Act also prohibits mining within 300 feet of occupied homes, churches, public buildings and parks, and within 100 feet of cemeteries or public roads. Further, coal mining is prohibited where it will affect a property on the National Register of Historic Sites,[849] and SMCRA grants states with discretion to designate other lands unsuitable for mining.[850]

II. Mountaintop Mining

574. The final set of legal issues[851] related to the exploitation of coal in the United States involves the most controversial type of surface mining: mountaintop mining. In this type of extraction, mine operators use blasting equipment to remove the soil and rock from the top of a mountain, exposing the underlying coal seams. Operators dump that overburden, or "spoil," into the adjacent valleys and commence mining. Once mining is complete, they attempt to replace the spoil onto the mountaintop, re-contouring it as best as possible, though this often does not work

843. *See* 30 U.S.C. § 1271(a)(2).
844. *Ibid.* § (a)(3).
845. *Ibid.*
846. 20 C.F.R. 723.
847. 30 C.F.R. 723.14.
848. *See generally Ibid.* § 1272.
849. *See Ibid.* § 1272(e).
850. 30 U.S.C. § 1272. The only exception to these prohibitions is for valid existing rights (VER). VER seek to protect private property rights from infringement by the state or federal governments if such infringement would otherwise be an unconstitutional taking. In order to claim the protections of VER, a mining operator must be able to show that: (1) they have a legal document vesting them with the right to mine the land at the time it became subject to SMCRA; and (2) either that the landowner at that time had made a good faith effort to obtain all necessary mining permits or that coal was immediately adjacent to and necessary to ensure economic viability of a surface mining operation existing at the time the mine became subject to SMCRA. Parties cannot invoke VER if, for example, a home or park was built after the mining operation already was permitted and approved. *See* 30 C.F.R. § 761.5.
851. As noted, extensive safety regulations also apply to coal mining operations, most notably, the Federal Mine Safety and Health Act of 1977, Pub L. 91-173, 83 Stat. 742 (codified at 20 U.S.C. 801-964). *See supra* note 734.

well. As a result, significant soil erosion and water pollution can occur in the valleys surrounding the mountaintop mining sites, often burying the streams that run through these valleys.[852]

575. Companies that engage in this type of mining assert that it would be uneconomical with more restrictive environmental permitting, and that the steep grades where the coal is extracted make filling the valleys necessary. Those concerned about the practice assert that greater environmental protection is possible, and that if it is not, the practice should not proceed because of the great effect it has on surrounding ecosystems, waterways, and communities. The practice affects approximately 12 million acres of land in the United States, and has buried roughly 1,200 miles of streams since 1992.[853] The practice is used primarily in Appalachia (Kentucky, West Virginia, Virginia, Tennessee, Pennsylvania, and Ohio).

576. Two primary sets of legal issues surround mountaintop mining: one involving SMCRA, and the other over the applicability of the Clean Water Act.[854]

577. The SMCRA question is relatively straightforward. The issue is how restrictive Office of Surface Mining, Reclamation and Enforcement regulations should be in terms of requiring stream and aquatic ecosystem protection under SMCRA. Previously, in 2008, OSM issued a new "revised buffer zone rule" that: (1) designated requirements for mountaintop mines to minimize the amount of spoil they place outside the mined area; (2) required minimization of land disturbance; and (3) required mine design to avoid disturbing streams and lands that are within 100 feet of those streams.[855] However, legal challenges were brought to this rule, and in early 2014, a federal court vacated it for OSM's failure to consult with federal wildlife agencies under the Endangered Species Act.[856] Then, in December 2014, OSM withdrew the rule.[857] The agency now is contemplating modified standards, and has said it will propose a new rule sometime in 2015.[858]

578. The Clean Water Act issues are somewhat thornier. Essentially, coal companies engaging in mountaintop mining might need to obtain two kinds of CWA permits: a National Pollutant Discharge Elimination System, or NPDES, permit (also known as a § 402 permit),[859] and a dredge and fill permit (also known as a § 404 permit).[860] The NPDES permit requirement applies to confined and discrete

---

852. *See* Ohio Valley Envt'l Coal. v. Aracoma Coal Co., 556 F.3d 177, 186 (4th Cir. 2009).
853. Copeland, *supra* note 726, at 1.
854. *See* 30 C.F.R. § 785.14.
855. Excess Spoil, Coal Mine Waste, and Buffers for Perennial and Intermittent Streams, 73 Fed. Reg. 75,814 (Dec. 12, 2008).
856. Nat'l Parks Conservation Ass'n v. Jewell, No. CV 09-00115 (BJR), 2014 WL 5355048 (D.D.C. Feb. 20, 2014); *see also* Nat'l Parks Conservation Ass'n v. Salazar, 660 F. Supp. 2d 3 (D.D.C. 2009).
857. Excess Spoil, Coal Mine Waste, Diversions, and Buffer Zones for Perennial and Intermittent Streams, 79 Fed. Reg. 76,227 (Dec. 22, 2014).
858. *See* Copeland, *supra* note 726, at 10.
859. 33 U.S.C. § 1342.
860. 33 U.S.C. § 1344.

manmade conveyances that discharge pollutants into the waters of the United States, so most mining operations have these permits for wastewater they put off through pipes or ditches or the like. The harder question for mountaintop mining operations is whether they also need a § 404 permit, which covers placing fill into jurisdictional waters. Two kinds of § 404 permits are available—nationwide permits, which are supposed to be used when there are minimal environmental effects from the filling, or individualized permits, which apply to a specific site and tend to impose more stringent environmental limits on the activity.

579. For years, the mountaintop mining industry in the United States has operated under a nationwide permit, Nationwide Permit 21 (NWP 21). However, when the Obama administration came into office, it quickly took several steps aimed at limiting mountaintop mining's environmental effects. One of those steps was to modify NWP 21. This process began in 2009,[861] was followed by the administration suspending NWP 21 in 2010,[862] and culminated in the reissuance of NWP 21 in 2012.[863] The modified permit has certain limitations: It can only be used for operations that have impacts on one-half an acre or stream beds of 300-linear feet or less. "Further, the reissued permit strictly prohibits use of NWP 21 to authorize discharges of dredged or fill material into U.S. waters to construct valley fills associated with surface coal mining."[864] Thus, new mountaintop mining operations that cannot function within the modified NWP 21 now must obtain individualized § 404 permits.[865]

580. Also relevant to CWA implementation for mountaintop mining operations is whether the EPA may veto the issuance of § 404 permits after the fact. The CWA is jointly administered by two agencies: the EPA and the U.S. Army Corps of Engineers. EPA issues § 402 permits and promulgates environmental guidance applicable to § 404 permits,[866] but the Corps is the entity that actually issues § 404 permits. However, the Act also includes a provision that allows EPA to veto § 404 permits under certain conditions.[867]

---

861. Proposed Suspension and Modification of Nationwide Permit 21, 74 Fed, Reg. 34311-01.
862. Suspension of Nationwide Permit 21, 75 Fed. Reg. 34,711 (June 18, 2010).
863. Reissuance of Nationwide Permits, 77 Fed. Reg. 10,184 (Feb. 21, 2012).
864. COPELAND, *supra* note 726, at 11.
865. Recently, the U.S. Court of Appeals for the Eleventh Circuit remanded the Corps' issuance of NWP 21 on the grounds that additional assessment of whether it will have minimum environmental impacts is warranted. *See* Black Warrior Riverkeeper, Inc. v. U.S. Army Corps of Engineers, 781 F.3d 1271 (11th Cir. 2015).
In addition to NWP 21, EPA, in April 2010, also issued interim guidance, which it finalized in July 2011, applicable to both § 404 and § 402 permits for coal mining operations in Appalachia. *See* U.S. Envt'l Protection Agency, *Final Memorandum: Improving EPA Review of Appalachian Surface Coal Mining Operations Under the Clean Water Act, National Environmental Policy Act, and the Environmental Justice Executive Order*, July 21, 2011. This guidance also seeks to increase the amount of environmental protection ensured in conjunction with mountaintop mining. Industry groups challenged this guidance, initially successfully, *see* Nat'l Mining Ass'n v. Jackson, 856 F. Supp. 2d 150 (D.D.C. 2012), but the government prevailed on appeal. *See* Nat'l Min. Ass'n v. McCarthy, 758 F.3d 243 (D.C. Cir. 2014).
866. 33 U.S.C. § 1344(b).
867. *Ibid.* § 1344(c).

581. EPA rarely uses its veto power,[868] but the issue arose in 2011 when the agency decided to veto a permit for the Spruce No. 1 mine in Logan County, West Virginia—a mountaintop mining operation—based on wildlife and fishery impacts.[869] Perhaps unsurprisingly, the mining operation challenged EPA's action, and was initially successful in district court, which agreed that EPA lacks the authority under the CWA to retroactively veto a permit.[870] The case did not end there, however, and on appeal, the government prevailed. Specifically, the U.S. Court of Appeals for the D.C. Circuit, which is typically regarded as the best authority on administrative law in the United States (save the Supreme Court), held: "The unambiguous language of subsection 404(c) manifests the Congress's intent to confer on EPA a broad veto power extending beyond the permit issuance."[871]

582. Thus, the recent back-and-forth over mountaintop coal mining under the Obama administration reveals two truths, one general and one specific to coal in the United States: First, today as always, the details of how any administrative regime is implemented in the United States depends heavily on who is in office and what their policy priorities are, in large part because U.S. administrative law is designed to give agencies flexibility. Second, the Obama administration's efforts to ameliorate environmental impacts from mountaintop mining play to a certain segment of the populace, but also might be put in a broader context regarding U.S. exploitation of coal. While the United States remains rich in this resource, and it remains heavily relied upon for electricity production in this country, there is generally a broad effort to move away from it[872]—and how that will play out over time remains to be seen.

868. "EPA has used this veto authority . . . only 12 times since 1972. None of the previous vetoes involved a surface coal mining or mountaintop mining project." COPELAND, *supra* note 726, at 14.

869. Final Determination of the Assistant Administrator for Water Pursuant to Section 404(c) of the Clean Water Act Concerning the Spruce No. 1 Mine, Logan County, WV, 76 Fed. Reg. 3126 (Jan. 19, 2011).

870. Mingo Logan Coal Co. v. EPA, 850 F. Supp. 2d 133 (D.D.C. 2012).

871. Mingo Logan Coal Co. v. EPA, 714 F.3d 608, 613 (D.C. Cir. 2013).

872. Thomas L. Casey, David W. Mitchell, *Has EPA's "War on Coal" Stalled?*, 27 ABA NAT. RESOURCES & ENV'T 3, 20 (Winter 2013); Patrick McGinley, *Collateral Damage: Turning A Blind Eye to Environmental and Social Injustice in the Coalfields*, 19 J. ENVTL. & SUSTAINABILITY L. 305 (2013). *See generally* TOMAIN, *supra* note 21.

Chapter 3. Transportation and Distribution

583. Coal is a relatively heavy fuel and therefore expensive to ship. Moreover, transport prices for coal heavily affect the price paid for the resource by the end consumer, so the relative efficiency and price of transport are significant factors for coal users. On average, shipping costs account for 20% of the price of coal in the United States, though this can of course vary; the cost of Power River Basin coal can be comprised of almost 60 % shipping costs.[873]

584. As noted, coal is used primarily for electricity production in the United States. Upwards of 90% of U.S. coal consumption goes to this purpose, so the vast majority of coal customers—and thus the ultimate consumers of transport services used to move coal—are electric utilities or merchant power producers who operate coal-fired power plants. Moreover, because coal is not easy to transport, and because electricity plants need such large quantities of it, the function and regulation of coal transportation and distribution in the United States are heavily intertwined, and we thus treat them together here. Indeed, for many large coal consumers in the United States, transportation and distribution of coal are effectively the same act. This is because of the way coal is moved.

585. Roughly 75% of all coal delivered to coal-fired power plants in the United States arrives by rail.[874] "In the United States, coal trains carried 816 million tons of coal in 2011 from Eastern mines and the Powder River Basin in Wyoming."[875] Much of this coal moves long distances—over 1,000 miles—and loads can be qu ite large—coal-bearing trains can weigh in at over 18,000 tons. Thus, these three industries—coal, railroads, and electricity—heavily depend on each other, and the contemporary story of coal transport regulation in the United States is really the story of railroad regulation.

586. Federal railroad regulation began in the late nineteenth century with passage of the Interstate Commerce Act of 1887.[876] Under that Act, the Interstate Commerce Commission received authority to regulate railroad transport extensively, in a way that should sound familiar to anyone knowledgeable about modern electricity

873. Richard J. Campbell et al., Congressional Research Service, Prospects for Coal in Electric Power and Industry (2013); Ass'n of American R.R.s, Railroads and Coal (June 2012), *available at* https://www.aar.org/keyissues/Documents/Background-Papers/Railroads-and-Coal.pdf. Eastern coal, which is typically mined underground, often is shipped by barge and does not travel long distances, making the transportation of eastern coal less expensive than western coal. Western coal, however, while cheaper to mine at the surface, must be transported greater distances, normally by rail, making transportation costs higher.

874. Other transportation options for coal include barge or ship, truck, and coal slurry pipelines. Coal is generally transported by truck over short distances, and rail or ships are used to cover longer distances. World Coal Ass'n, *Coal Market and Transportation*, http://www.worldcoal.org/coal/market-amp-transportation/ (last visited Apr. 29, 2015).

875. Davies et al., *supra* note 29, at 613.

876. Interstate Commerce Act of 1887, 24 Stat. 379 (codified as amended at 49 U.S.C. § 501–508, 521–526).

and natural gas regulation under the Federal Power Act and Natural Gas Act. Specifically, the Interstate Commerce Act charged the ICC with: (1) regulating railroad rates to ensure their justness and reasonableness; (2) preventing discrimination in rates and services; (3) precluding higher rates for shorter distances than longer routes in the same direction; and (4) ensuring that railroads publicly file their rates. Thus, the ICC became the blueprint later relied upon for general public service regulation in the United States. Congress lifted its basic architecture in later adopting the Federal Power Act, and that law then became the template for the Natural Gas Act.

587. Since the ICA's enactment in 1887, Congress subsequently and repeatedly revised how railroads would be regulated, eventually arriving at roughly the same status natural gas and electricity regulation have today—a regulatory system premised heavily on markets. These reforms included: (1) the Elkins Act in 1903,[877] which foreclosed railroad rebates and gave the ICC civil and criminal penalty authority for intentional discrimination and filed tariff violations; (2) the Hepburn Act three years later,[878] which gave the ICC jurisdiction to set maximum rates and to regulate both fuel pipelines and sleeping-car, express, and steamship companies; (3) the Mann-Elkins Act of 1910,[879] which further augmented the agency's rate procedures and gave it the power to suspend tariffs on its own initiative pending hearing proceedings; and (4) the Esch-Cummings Act, or Transportation Act, of 1920,[880] which compelled the ICC to draft a plan to consolidate parallel rail companies and gave the agency jurisdiction over minimum rates, market entry and exit, and intercorporate relationships.[881]

588. Then, in the 1970s and 1980s, Congress adopted the most significant changes to railroad regulation yet. This was the time in the nation's history when there was the greatest push for deregulation and industry restructuring, and efforts to reform the railroad industry were only one movement in a symphony of reforms that also swept up the telecommunications, airline, and eventually natural gas and electricity industries as well. Thus, Congress in 1973 adopted the 3R Act,[882] in 1974 the 4R Act,[883] and in 1980 the Staggers Rail Act, which, among other things, helped begin regulatory reform in the railroad industry, and then limited the ICC's authority over railroad ratemaking. These laws largely had their intended effects. After

877. Elkins Act of 1903, 32 Stat. 847 (codified in scattered sections of 49 U.S.C.).
878. Hepburn Act of 1906, 34 Stat. 584 (codified in scattered sections of 49 U.S.C.).
879. Mann-Elkins Act of 1910, 36 Stat. 539 (enacted at 49 U.S.C. §§ 501 et al.).
880. Transportation Act of 1920 (Esch-Cummings Act), 41 Stat. 456 (codified as amended at 40 U.S.C. § 316).
881. Paul Stephen Dempsey, *The Rise and Fall of the Interstate Commerce Commission: The Tortuous Path from Regulation to Deregulation of America's Infrastructure*, 95 Marq. L. Rev. 1151 (2012).
882. Regional Rail Reorganization Act of 1973, Pub. L. 93-236, 87 Stat. 985 (codified at 45 U.S.C. §§ 701-97).
883. Railroad Revitalization and Regulatory Reform Act of 1976 (4-R Act), Pub. L. 94-210, 90 Stat. 31 (codified at 45 U.S.C. §§ 801 et al.).

passage of the Staggers Act, for instance, coal-shipping prices by rail fell over 40% between 1984 and 2001.[884]

589. Nevertheless, Congress was not done reforming how railroads are regulated. In 1995, Congress passed the ICC Termination Act,[885] which "sunset the Interstate Commerce Commission itself, deregulated and amended certain of the agency's functions, and transferred jurisdiction over rail, motor, bus, broker, freight forwarder, and pipeline services to the newly created Surface Transportation Board (STB) within the DOT."[886] Specifically, the ICC Termination Act instituted three key reforms: First, it eliminated the requirement that common carrier rail rates be filed. Second, it kept within the STB the ability to directly regulate railroad tariffs, but it limited that authority to those carriers with market power. And, third, the Act continued the practice of rail carriers entering directly into contracts negotiated with customers (called "shippers" in industry parlance), though it gave the STB limited authority to regulate those contracts. The STB may review such contracts, as the ICC previously could, under a narrow set of circumstances, but unless the agency holds a contract unlawful, the transportation proceeds under the contract and is not subject to agency oversight.[887]

590. Thus, today, regulation of coal transportation falls largely to the STB.[888] The result is that many coal shipment rates are not really regulated at all, but rather, are determined by the market through arms-length negotiations between the railroads and customers.[889] This system has largely worked, in part because competition for terms and rates is possible for long-haul transportation of coal over different rail lines. Where disputes have continued to arise, however, is in the delivery of coal directly to power plants that rely on the fuel—what are sometimes called "captive shippers," captive because they cannot turn to an alternate rail transport provider to obtain the service they need. Numerous cases have arisen in this context.

591. Two key facets of law address the issue. First, under the Termination Act, any rail carrier determined to be a dominant service provider must offer service on a common carrier basis, that is, to all who seek service and on the same terms to all comers.[890] The Board does so by imposing "common carrier" obligations on the rail carriers controlling bottlenecks. The Surface Transportation Board has ruled that

---

884. U.S. Energy Info. Admin., *Coal Transportation Issues*, *available at* http://205.254.135.7/oiaf/aeo/otheranalysis/cti.html.
885. ICC Termination Act of 1995, Pub. L. 104-88, 109 Stat. 803 (codified in scattered sections of 42 U.S.C.).
886. Dempsey, *supra* note 881, at 1185.
887. Joseph D. Kearney, *Will the FCC Go the Way of the ICC?*, 71 U. Colo. L. Rev. 1153, 1163 (2000).
888. The Department of Transportation retains authority to regulate railroad safety; under its auspices, the Federal Railroad Administration promulgates regulations to maintain safety in the industry. These regulations are extensive and include rules on track gage, roadbed construction, track geometry, track structure, the regular inspection of completed lines, and operation of those lines. *See generally* U.S. Dep't of Transp., Fed. R.R Admin., *Regulations and Rulemaking*, https://www.fra.dot.gov/Page/P0084 (last accessed Apr. 29, 2015).
889. 49 U.S.C. § 10701(c).
890. 49 U.S.C. § 10701(d).

"bottleneck carriers satisfy their common carrier duties and thus comply with the Act by providing origin-to-destination service that includes the bottleneck, or by providing joint or proportional service with other carriers that includes transportation over the bottleneck."[891] This ruling and others have been given deference by federal courts, which recognize the Board's considerable expertise in the economic underpinnings of the rail industry, and the need to balance competing objectives within the Act.[892]

592. Second, the STB has instigated what it refers to as standalone cost (SAC) rates for railroad spurs that serve such captive shippers.[893] Under these rules, where a railroad is deemed a dominant carrier, a SAC rate allows the railroad to charge the cost it would take to build a rail spur to the customer (a "standalone railroad") rather than merely the marginal cost of moving the coal from the mainline of the railroad to its ultimate destination. This, the STB has asserted, balances its twin obligations to protect customers but also to ensure railroad revenue adequacy. Perhaps predictably, electricity companies have challenged both the general SAC test as well as its particular application, but it has been upheld in a number of contexts.[894] As one court has observed, "Nothing in the Act explicitly requires carriers to provide separate local rates for bottleneck service. Furthermore, requiring carriers to provide separately challengeable rates on bottlenecks would prevent them from exploiting bottlenecks and charging rates up to SAC for complete origin-to-destination service. In the Board's view, this would impede the industry's efforts to achieve revenue adequacy, which is necessary for long-term capital investment and, ultimately, for a safe and efficient rail system."[895]

593. While deregulation of the rail industry initially led to decreases in the price of shipping coal, the beginning of the twenty-first century has seen a rise in coal transport costs. The cost of transporting coal by rail for electricity production increased, on average, from USD 13/short ton to USD 16/short ton from 2001 to 2010. Such averages, of course, mask differences among regions, which can be substantial. Indeed, over recent years, Powder River Basin shipping rates have increased only 14%, while costs for shipping from Southern Appalachia have gone up 149%. At the same time, of course, the quantities transported from the Powder River Basin dwarf those from Southern Appalachia.[896]

594. Some electricity producers, then, have looked for alternatives to shipping coal by rail. One such option is for railroads to build their own spurs, from the mainline railroads. MidAmerican Energy, for instance, has done this, owning a six-mile rail spur from its Council Bluffs, Iowa power plant to a neighboring main rail

891. *See* Cent. Power & Light Co. v. S. Pac. Transp. Co., No. 41242, 1997 STB LEXIS 91, at *28 (Surface Transp. Bd. Apr. 28, 1997) (*Bottleneck II*).

892. *See generally* MidAmerican Energy Co. v. Surface Transp. Bd., 169 F.3d 1099 (8th Cir. 1999).

893. Coal Rate Guidelines, Nationwide, 1 I.C.C.2d 520, 542–46 (I.C.C. Aug. 8, 1985).

894. Metro. Edison Co., 37931S, 1989 WL 238264, at *1 (DCSS Mar. 10, 1989); Major Issues in Rail Rate Cases W. Texas Utils. Co., 41191, 2006 WL 513502, at *1 (Feb. 27, 2006).

895. *MidAmerican Energy*, 169 F.3d at 1109.

896. DAVIES ET AL., *supra* note 29, at 620.

line.[897] Another option is to utilize slurry pipelines instead of rail. Coal slurry pipelines move crushed coal that has been purified and mixed with water. When this slurry mixture reaches its destination, it is dewatered for use. Coal slurry pipelines face a number of obstacles, including acquiring both land and water rights and, sometimes, resistance from railroads if a proposed pipeline needs to cross rail property or rights of way.[898] Nonetheless, some have been successful. As of 2010, about 7% of coal in the United States moved via slurry pipeline.[899] Finally, some coal facilities have switched fuel, typically to natural gas. This, however, like any energy problem, presents tradeoffs, including that many older coal-fired power plants have already recovered their capital costs, that natural gas costs historically have been more volatile than coal prices, and that the cost of getting the gas to the facility also can be high.[900]

897. Josh Funk, *Berkshire Hathaway Selling 2 Short Line Railroads*, AP (Oct. 2, 2012), http://bigstory.ap.org/article/berkshire-hathaway-selling-2-short-line-railroads; *see also* Burlington N. R. Co. v. Surface Transp. Bd., 114 F.3d 206, 211 (D.C. Cir. 1997) (noting that, in a rate battle with the Burlington Northern Santa Fe railroad, Western Texas Utilities threatened to build its own USD 70 million rail spur); Donald G. Avery, *Do-it-Yourself "Electric" Coal Trains*, 127 No. 12 FORTNIGHTLY 27, 28 (1991).

898. *See, e.g.*, MONT. CODE ANN. § 85-2-104 (1979); Energy Transp. Sys., Inc. v. Union Pac. R.R. Co., 606 F.2d 934 (10th Cir. 1979); Sporhase v. Nebraska, 458 U.S. 941 (1982); Energy Transp. Sys., Inc. Pipeline Project v. Missouri, 484 U.S. 495 (1988).

899. U.S. Energy Info. Admin., *Coal Transportation Rates to the Electric Power Sector* (Nov. 6, 2012), *available at* http://www.eia.gov/coal/transportationrates/print_version.cfm.

900. *See, e.g.*, Brian Reinhart et al., *Paper of the Year: A Case Study on Coal to Natural Gas Fuel Switch* (Dec. 21, 2012), http://bv.com/Home/news/solutions/energy/paper-of-the-year-a-case-study-on-coal-to-natural-gas-fuel-switch; Rebecca Smith, *Turning Away From Coal: Utilities Are Increasingly Looking to Natural Gas to Generate Electricity*, WALL ST. J. (Sept. 13, 2010).

Chapter 4. Consumption

595. The legal issues that flow from coal consumption hinge heavily on the uses that coal is put to in the United States. As noted, coal is used for two primary functions in the United States: dominantly, for electricity production, and secondarily, for industrial purposes (particularly anthracite coal). For each of these uses, coal is burned, or combusted.

596. Coal combustion produces an array of harmful pollutants, particularly air pollutants. Coal combustion produces a fine dust called particulate matter that is a source of respiratory problems, heart and lung disease, and haze. The particulate matter produced from coal combustion can contain varying levels of mercury, sulfur dioxide, nitrogen oxides, and carbon dioxide. These pollutants are responsible for food and water pollution, acid rain, smog, and increased levels of greenhouse gases in the earth's atmosphere. As the Congressional Research Service has summarized, "Coal-fired power plants are among the largest sources of air pollution in the United States. In 2005, they accounted for 10.2 million tons of sulfur dioxide emissions (70% of the U.S. total), 52 tons of mercury emissions (46% of the U.S. total) and 3.6 million tons of nitrogen oxides (19% of the U.S. total). . . . In addition, they account for about 40% of U.S. anthropogenic emissions of the greenhouse gas carbon dioxide."[901] Thus, it should not be surprising that a 2011 study placed coal's cost on U.S. society at approximately one-third to over a half a trillion dollars per year.[902]

597. The combustion of coal, and the environmental externalities associated with it, raise three chief legal issues. First, any coal-fired power plant will need to obtain an emissions permit under the federal Clean Air Act.[903] That law, like many of the United States' environmental laws, operates under a cooperative federalist regime. The federal government, via the EPA, establishes national air quality standards, referred to as national ambient air quality standards, or NAAQS. States then have the charge to meet these standards, and to create state implementation plans (SIPs) to do so, which in turn require approval from the EPA. State-issued air emissions permits inevitably are a key part of any state's SIP, and thus, its compliance process with the CAA. How emissions are limited from coal-fired power plants—both in how EPA sets NAAQS and in how states issue emissions permits under the Act—are key factors that affect coal consumption in a given area.

598. Second, in 1990, Congress amended the Clean Air Act to deal with a particular kind of pollution caused by coal combustion.[904] Specifically, Congress placed new limits on sulfur dioxide emissions from coal-fired power plants, which

901. *See* James E. McCarthy, Congressional Research Service, Clean Air Issues in the 113th Congress: An Overview 3 (2013).
902. *See* Paul R. Epstein et al., *Full Cost Accounting for the Life Cycle of Coal*, 1219 Ann. NY. Acad. Sci. 73 (2011).
903. 42 U.S.C.A. § 7411.
904. *See generally,* 42 U.S.C.A. §§ 7401-7671.

had been linked to causing acid rain. The 1990 Act set up an emissions trading program that aimed both to limit these emissions and to do so efficiently; polluters were assigned allowances, and if one plant could reduce its emissions more cheaply than another facility, they could trade their allowances, thus driving down the overall cost of compliance. This program had a rapid effect on coal consumption. Initially, power plants and utility companies responded by purchasing higher quantities of western coal from Wyoming's Powder River Basin, which is lower in sulfur content than eastern coal. Electricity producers reasoned that this solution—switching fuel types—was more cost efficient than installing expensive pollution control equipment.[905] Their decision, in turn, shifted coal consumption and production patterns, and helped reduce sulfur dioxide emissions. The recent trend, however, has been for power plants to install new technology to comply with the Clean Air Act emission restrictions and return to using higher sulfur- and energy-content Midwestern and eastern coal.[906]

599. Finally, coal consumption increasingly may be impacted in the United States by climate regulation. While the United States has not passed federal legislation (either a carbon tax or a cap-and-trade system) to deal with carbon dioxide emissions, many states have begun to take action on their own.[907] Moreover, the EPA is in the process of implementing rules that may significantly impact—that is, reduce—coal consumption for electricity production.

600. In June 2014, the EPA released its proposed Clean Power Plan, which it finalized in August 2015.[908] This rule will require the electricity sector to reduce carbon dioxide emissions by roughly one-third from 2005 levels by 2030, with states required to implement plans to achieve this goal. Different states will have different goals, but the steps that can be taken to meet them—called "building blocks" under the proposed rule—are the same. These include increasing plant efficiency, increased reliance on low-carbon sources, and, importantly, fuel switching from sources like coal to sources like natural gas. Thus, while the means of obtaining the required results will vary from state to state, compliance in at least some states will likely be achieved through retiring old power plants and using more natural gas and

---

905. Shi-Ling Hsu, *Reducing Emissions from the Electricity Generation Industry: Can We Finally Do It?*, 14 TUL. ENVTL. L.J. 427 n.46 (2001).

906. In 2005, EPA instituted another kind of tradable permit system, this time to reduce mercury emissions from coal-burning power plants. In 2012, however, EPA abandoned this approach in favor of the Mercury Air Toxics Standards (MATS), which imposes even stricter requirements for reducing mercury and other toxic metal emissions. 40 C.F.R. §§ 60, 63; *see also U.S. Business Group Challenges EPA Mercury Rule*, 33 WESTLAW J. ENVTL. 4 (2012).

907. Neil Keenan, *Global Warming Due to Greenhouse Gas Emissions: The Success of State Solutions As A Model for A Federal Solution*, 34 J. LEGIS. 168 (2008); Carolynne C. White, Kevin Holst, *Colorado Local Government Regulation of Land Use for Climate Change*, 37 COLO. LAW. 12, (Dec. 2008), at 29; Robert I. McMurry & Max B. Talpalar, *California's Push Against Global Warming: Through AB 32 and SB 375, California Combines Air Quality and Land Use Legislation to Fight the Effects of Climate Change*, SR004 ALI-ABA 431 (Aug. 2009).

908. Carbon Pollution Emission Guidelines for Existing Stationary Sources: Electric Utility Generating Units, 79 Fed. Reg. 34,830 (June 18, 2014).

renewable energy sources.[909] If this rule survives legal challenge, then the Clean Power Plan may have a tremendous effect on coal consumption in the United States, driving it down.[910] The Clean Power Plan is discussed in more detail in Part V, *Interaction between Energy Law and Environmental Law*.

909. *See* McCarthy, *supra* note 901.
910. *See* U.S. Envt'l Protection Agency, *Clean Power Plan Proposed Rule*, http://www2.epa.gov/carbon-pollution-standards/clean-power-plan-proposed-rule (last visited Apr. 29, 2015).

Part V. Interaction between Energy Law and Environmental Law

Chapter 1. Overview of Energy and Environmental Regulation in the United States

601. Throughout the fuel cycle, environmental consequences follow: from exploration and extraction to production and distribution and, ultimately, consumption and disposal of the natural resources used to produce the energy in the first place. Most prominently, these environmental consequences are the result of burning fossil fuels that emit conventional and greenhouse gas pollutants into the atmosphere. Additionally, energy activities contribute to land, water, and air degradation all along the fuel cycle. Irrespective of the energy source, there is always a tradeoff. If the full life cycle is taken into account, no energy source is free of environmental consequences.

602. For example, coal mining results in surface and subsurface land degradation, including polluting streams, aesthetically scarring mountaintops, and yielding human fatalities in mining disasters. Nuclear power is largely favorable from a climate change perspective, but the recent disaster at Fukushima and the prior disasters at Three Mile Island and Chernobyl remind of the risks of radiation releases that come with this resource—not to mention the dilemma of nuclear waste storage or the environmental consequences that can arise from uranium mining. Similarly, offshore oil and gas exploration presents environmental risks, such as those that occurred in the BP *Deepwater Horizon* explosion in 2010. That explosion resulted in water pollution, ecological damage, and damage to wildlife as well as 11 human fatalities. And, even comparatively benign energy resources, such as solar, wind, and hydropower, present environmental challenges, including the large amounts of land required for utility-scale plants, as well as dangers to the surrounding ecology and wildlife.

603. The regulation of energy and the environment should coincide and complement each other. As a result of historic anomaly, however, the legal and regulatory interface between the disciplines of energy and the environment in the United States is more disjointed than coordinated. The discipline of environmental law began in full with the passage of the National Environmental Policy Act of 1969,[911] together

911. National Environmental Policy Act, 42 U.S.C. §§ 4321, 4331–4335, 4341–4347, 4361–4370h.

with associated substantive environmental laws, such as the 1970 amendments to the Clean Air Act,[912] the amendments to the Clean Water Act in 1972,[913] the Coastal Zone Management Act in 1972,[914] and the Endangered Species Act of 1973,[915] among other legislation.[916]

604. Prior to the 1970s, environmental legislation addressing water and air had been passed, but the legislation of the 1970s was more detailed, comprehensive, and aggressive.[917] Additionally, 1970s environmental legislation embodied an environmental ethic based upon such books as Rachel Carson's *Silent Spring* (1962), which addressed the dangers of pesticides and the risks of increased in DDT in the human body, and, earlier, Aldo Leopold's *Sand County Almanac* (1949), which promoted the idea of a "land ethic" urging environmental protection. The United States, moreover, had a history of conservation and natural resource protection dating back to the late nineteenth century, with the establishment of national parks and the protection of wilderness areas. This new round of legislation was consistent with those earlier laws as well as a growing concern for environmental protection within the nation—a modern environmental ethic.

605. This new environmental movement was driven by a desire to ameliorate the effects of the modern world, particularly health effects felt by cities from dirty air, polluted waters, and broad-scale impacts of a growing reliance on modern chemicals. The nation's new environmental laws matched this ethic. It concerned itself with humans' relationship to the natural environment, and aimed to provide legal protections to help negotiate that relationship. Thus, new national environmental organizations became influential, such as the Natural Resources Defense Council (NRDC) and the Environmental Defense Fund, which added themselves to the long-standing stalwarts, such as the Sierra Club.

606. The environmental movement, its rhetoric, its tactics—and, as a result, the legislation it helped produce—emphasized conservation, species protection, ecological care, protection of public lands, and caution regarding residential, commercial, and industrial growth and development. Environmentalism was cautious about unrestrained economic growth and, in some instances, expressed concerns about the limits to natural resources and the environmental consequences of their unchecked exploitation. This caution was exemplified and popularized in *The Limits to Growth:*

---

912. Clean Air Act Amendments of 1970, 42 U.S.C. §§ 7401–7671.
913. Clean Water Act, Pub. L. No. 92-500, 86 Stat. 816 (codified as amended in scattered sections of 33 U.S.C.) (Originally known as the Water Pollution Control Act of 1972).
914. Coastal Zone Management Act of 1972, 16 U.S.C. §§ 1451–1465.
915. Endangered Species Act of 1973, 16 U.S.C. § 1531–1544.
916. *See, e.g.*, Resource Conservation and Recovery Act of 1976, Pub. L. No. 94-469, 90 Stat. 2003 (codified as amended in scattered sections of 42 U.S.C.; Toxic Substances Control Act of 1976, Pub. L. No. 94-469, 90 Stat. 2020 (codified as amended in scattered sections of 15 U.S.C.); Comprehensive Environmental Response, Compensation, and Liability Act of 1980, Pub. L. No. 96-510, 94 Stat. 2726 (codified as amended in scattered sections of 42 U.S.C.).; Surface Mining Control and Reclamations Act, 30 U.S.C §§ 1201–1328.
917. *See generally* RICHARD J. LAZARUS, THE MAKING OF ENVIRONMENTAL LAW (2004).

A Report for the Club of Rome's Project on the Predicament of Mankind (1974). *Limits to Growth* attracted a widespread critical response. Its argument was that there are inherent limits in the natural world, including with resource consumption, and that there are significant environmental risks that accompany resource consumption. Critics of this report argued that we were "not running out of" natural resources, and that limits to economic growth are unwise.

607. At bottom, critics of the environmental movement were largely concerned about economic growth and the increased cost of doing business due to environmental regulations. Similarly, the United States' energy sector developed largely on the core assumption that increased energy production and consumption were necessary for continued economic expansion. This, then, set up the clash at the core of environmental regulation in the United States: protection versus production and consumption. That focus in the environmental discourse, moreover, highlighted environmental law's deep division from energy law.

608. Energy law, as noted in Part I, grew as a discipline out of several strands of extant law in the United States: public utility law, oil and gas law, and natural resources law. Nevertheless, as a result of the energy crises experienced in the middle of the 1970s, President Carter introduced, and Congress passed, a welter of legislation that helped give rise to the discipline we now refer to as energy law.

609. In part, then, it was a historical coincidence that both environmental and energy law developed during the same decade. Particularly noteworthy for how the two fields function in the United States, however, is that they grew up separately and largely remain separate today. Different agencies implement the rules and regulations that constitute energy law and environmental law. Different statutes govern their substance. And they use different tools and have different aims—environmental law focuses on pollution reduction and ecological protection, and energy law largely centers on governing markets that lead to plentiful and inexpensive energy production. In short, these two fields of law attend to different concerns; rely on different concepts, metrics, and vocabularies; and push in divergent directions. Where environmental law and policy zeroes in on ecological protection, energy law is more attuned to the economic consequences of the energy fuel cycle than its environmental costs.

610. Indeed, energy advocates address concerns different from environmentalists. Energy advocates promote energy production and consumption and job creation, and view these as linked to economic growth. Additionally, energy advocates argue that technological advances can both keep the cost of energy reasonable as well as address pollution concerns. Limits, they say, are imposed only by the extent of human innovation. Energy advocates either downplay or tend to avoid the negative externalities of pollution, tend to ignore the tragedy of the commons, and continue to rely on the idea that there is a direct correlation between energy production and consumption and economic growth. This belief in energy and the economy ignores the fact that the United States has increased its energy intensity. In other words, for four decades, consumers have been paying less for the same amount of energy received.[918]

611. Energy law and policy and environmental law and policy do share some common features. Each examines its field in silos. Electricity, natural gas, hydroelectricity, and nuclear power, for example, are all regulated by separate statutory regimes and by separate administrative agencies or departments. Similarly, the environment is regulated by statutory regimes that separately address public lands, air, water, hazardous wastes, and the like. In other words, energy regulation and environmental regulation proceed not only independently of each other but also in an uncoordinated, disaggregated way within each field.

612. As a result, both perspectives pay inadequate attention to the reality of the energy fuel cycle. From the environmental side, too little attention is paid to the role of energy as an input into the economy, to the possibilities of lifestyle change, to past investments, and to the costs of transition from one energy paradigm to another. From the energy side, too little attention is paid to the social, as well as the economic, costs of pollution, including disturbing or depleting the commons.

613. Of course, the separation between energy law and environmental law is not complete. These are not entirely mutually exclusive regulatory systems. The modern energy lawyer in the United States needs to know both energy law and environmental law. Coal-fired power plants need Clean Air Act permits for air emissions. Coalmines require permits under the Clean Water Act and the Surface Mine Control and Reclamation Act to operate and to discharge waste into streams. Similarly, hydroelectric dams are required by law to pay attention to their consequences on fish, wildlife, and recreation, as well as electricity production. Nevertheless, these points of intersection do not constitute a significant merger between energy and the environment. Nor do these points of intersection mean that energy and the environment are treated either comprehensively or in coordination with each other.

614. Increasingly, some aspects of traditional energy law and environmental law are becoming more coordinated. CAFE standards, for instance, discussed in Part III *supra*, combine both energy and environmental goals: They maximize energy supplies by increasing fuel efficiency—energy law's goal of energy security—and they reduce air emissions from vehicles—environmental law's goal of pollution reduction. Likewise, state renewable portfolio standards use a traditional energy mechanism, regulation of utilities' generation portfolios, to advance the environmental objective of cleaner energy and the energy goals of fuel diversity and enhanced technological innovation. These are two examples; others continue to emerge.[919]

615. A merger of environmental and energy law and policy has much to commend it, not the least of which is the need to address the full fuel cycle integral to

918. U.S. Department of Energy, *Monthly Energy Review* 14–18 (Feb. 2015).
919. *See generally, e.g.*, Wildermuth, *supra* note 23, at 524, 528; *The Next Step*, *supra* note 23, at 388.

both systems of law. More notably, the United States has developed a policy consensus in favor of promoting clean energy—a consensus that only continues to grow.[920] Indeed, the cost of generating electricity from renewable resources, such as wind and solar power, are decreasing significantly such that those resources contributed more new power generation than any other resource 2014.[921] In other words, the United States, together with other parts of the world, is experiencing an energy transition through the greater adoption of renewable resources and energy efficiency.

616. A clean energy future, then, may be valuable in and of itself as new sources of energy increase the diversity of resources available, introduce new technologies and new markets, increase competition, and increase consumer choice. Nevertheless, a clean energy future is also consistent with addressing the environmental damage that occurs from fossil fuel use. The transition away from fossil fuels to clean energy highlights the need for a closer merger between these two disciplines, a need that has become even more acute as a result of heightened concerns about climate change.

617. As it now stands, however, energy law and environmental law in the United States remain largely separate. While energy law consists of statutes and doctrines that date back to the New Deal and earlier, environmental law is more modern, beginning in earnest in the 1970s. That new set of statutes undeniably plays an important role in how energy systems function in this country.

618. The laws that comprise the field of environmental law might be divided into four key types.[922] First, there are laws that seek to limit, or mitigate, the amount of pollution added to society. These include the Clean Air Act;[923] the Clean Water Act;[924] the Federal Insecticide, Fungicide, and Rodenticide Act (FIFRA);[925] and the Toxic Substances Control Act (TSCA).[926] Second, there are a set of environmental laws focused on pollution remediation and cleanup. These include the Resource Conservation and Recovery Act[927] and the Comprehensive Environmental

920. *See generally* Tomain, *supra* note 21.
921. U.S. Energy Info. Admin., *Electric Power Monthly with Data for December 2014*, tbl. ES1.A (Feb. 2015).
922. *Cf.* Aagaard, *supra* note 62; Todd S. Aagaard, *Environmental Law Outside the Canon*, 89 Ind. L.J. 1239 (2014).
923. Clean Air Act (Air Pollution Control Act), Pub. L. 84-159, 69 Stat. 322 (codified as amended at 42 U.S.C. § 7401–7431).
924. Federal Water Pollution Control Act (Clean Water Act), Pub. L. 80-845, 62 Stat. 1155 (codified as amended in scattered sections of 33 U.S.C.).
925. Federal Insecticide, Fungicide, and Rodenticide Act, Pub. L. 92-516, 61 Stat. 163 (codified as amended at 7 U.S.C. § 135–136y).
926. Toxic Substances Control Act, Pub. L. 94-469, 90 Stat. 2003 (codified as amended in scattered sections of 15 U.S.C.).
927. Resource Conservation and Recovery Act of 1976, Pub. L. 94-580, 90 Stat. 2795 (codified in scattered sections of 42 U.S.C.).

Response, Compensation, and Liability Act (CERCLA, or Superfund).[928] Third, there is a set of laws designed to manage natural resource use, typically with the idea in mind that multiple values and uses must be balanced. These include the National Forest Management Act (NFMA)[929] and the Federal Land Policy and Management Act (FLPMA).[930] Finally, there are process-oriented statutes, which tend not to impose substantive requirements, but rather, seek to ensure that environmental and other considerations are taken into account in the decision-making process. These include the National Environmental Policy Act[931] the National Historic Preservation Act,[932] and the Emergency Planning and Community Right-To-Know Act (EPCRA).[933]

619. Throughout this encyclopedia, we have sought to address specific environmental laws where they play a key role in the different parts of a given energy resource's fuel cycle. Below, we now address the key environmental statute, NEPA, which cuts across use of all energy resources in the United States. Finally, we overview the EPA's Clean Power Plan, which has the potential to merge energy and environmental regulation in the United States more than ever before.

928. Comprehensive Environmental Response, Compensation, and Liability Act of 1980, Pub. L. 96-510, 94 Stat. 2767 (codified as amended in scattered sections of 42 U.S.C.).
929. National Forest Management Act, Pub. L. 94-588, 90 Stat. 2949 (codified as amended in scattered sections of 16 U.S.C.).
930. Federal Land Policy and Management Act of 1976, Pub. L. 94-579, 90 Stat. 2743 (codified as amended in scattered sections of 43 U.S.C.).
931. National Environmental Policy Act of 1969, Pub. L. 91-190, 83 Stat. 852 (codified as amended in scattered sections of 42 U.S.C.).
932. National Historic Preservation Act, Pub. L. 89-665, 80 Stat. 915 (codified as amended in scattered sections of 16 U.S.C. § 470(a)).
933. Emergency Planning and Community Right-To-Know Act of 1986, Pub. L. 99-499, 100 Stat. 1728, (codified in scattered sections of 42 U.S.C.).

Chapter 2. National Environmental Policy Act

620. NEPA can be considered both the first modern environmental law as well as the linchpin of all other environmental laws in the United States. NEPA and its associated legislation had predecessors. NEPA, however, stands out from other environmental law because it is different in nature. Prior environmental laws, together with the environmental laws passed contemporaneously with and subsequent to NEPA, adopted substantive goals and imposed measurable targets. Air, water, coastal zones, species, and hazardous waste, as well as rivers and other waterways, were all addressed by specific legislation that either set standards or provided protection for specifically designated areas. NEPA, by contrast, is procedural in nature. It is directed at federal agencies rather than regulated industries, and it requires those agencies to consider both the environmental impacts of their proposed actions as well alternatives to them.

621. NEPA "declare[s] a national policy which will encourage productive and enjoyable harmony between man and his environment; . . . promote[s] efforts which will prevent or eliminate damage to the environment and biosphere . . .," and "enrich[es] the understanding of the ecological systems and natural resources important to the Nation."[934] In so doing, the law recognizes the "profound impact of man's activities on the interrelations of all components of the natural environment."[935] Included among those impacts are population growth, urbanization, industrial expansion, resource exploitation, and expanding technological advances. NEPA thus aims to use all "practicable means and measures . . . calculated to foster and promote the general welfare, create and maintain conditions under which man and nature can exist in productive harmony."[936] These policies are to be carried out (1) so that each generation can serve as trustee of the environment as well as for succeeding generations; (2) to assure healthful, productive, and aesthetically and culturally pleasing surroundings; (3) to obtain the widest beneficial uses of the environment without degradation; and (4) to preserve important historical, cultural, and natural aspects of our heritage among other ends.[937]

622. NEPA established the Council on Environmental Quality (CEQ) within the executive office of the President. The CEQ has no direct enforcement power but is responsible for coordinating federal environmental efforts, and for issuing implementing regulations and guidance for NEPA. CEQ also works closely with agencies to develop environmental policies and initiatives. Thus, in addition to setting policy, CEQ can serve as referee when federal agencies disagree on the adequacy of the environmental assessment.

623. The central mechanism for executing NEPA is, as noted, not substantive but procedural. Section 102 of the statute imposes on all federal agencies the obligation

934. 42 U.S.C. § 4321.
935. *Ibid.* at § 4331(a).
936. *Ibid.* at § 4331(a).
937. *Ibid.* at § 4331(b).

to utilize a systematic, interdisciplinary review of the environmental impacts of each of the agency's major decisions that affect the human environment. More precisely, NEPA mandates that "every recommendation or report or proposals for legislation and other major Federal action significantly affecting the quality of the human environment" must contain an environmental impact statement.[938]

624. The EIS must identify environmental harms and describe the adverse environmental effects that cannot be avoided should a proposal be implemented. Additionally, the EIS must identify and discuss alternatives to the proposed action that can mitigate the adverse consequences. Decisions that are determined likely to harm the environment, however, are not foreclosed by NEPA. Rather, the statute merely requires that the agency assess the likely impacts of its decision through the EIS.

625. The trigger for the preparation of an EIS is that the action must be (1) "federal," (2) "major," and (3) "significantly" affect the environment.

626. The first requirement of a "federal" action is relatively easy to satisfy. It is met where the federal government is the acting agency. Thus, CEQ's implementing regulations for NEPA state, "Federal actions tend to fall within one of the following categories: (1) [a]doption of official policy . . . "; "(2) [a]doption of formal plans, such as official documents prepared or approved by federal agencies which guide or prescribe alternative uses of Federal resources, upon which future agency actions will be based"; "(3) [a]doption of programs, such as a group of concerted actions to implement a specific policy or plan;" and "(4) [a]pproval of specific projects, such as construction or management activities located in a defined geographic area. Projects include actions approved by permit or other regulatory decision as well as federal and federally assisted activities."[939]

627. Often, federal actions include concrete decisions or proposals an agency is considering, such as building a dam, leasing federal lands for coal mining, or siting a natural gas pipeline. However, some federal actions can be more diffuse, if more pervasive. For such actions, agencies often conduct "programmatic" EISs rather than EISs specific to a given project.[940] Programmatic reviews examine such broad initiatives as rulemakings or policies. By way of example, the Department of the Interior, through the Bureau of Land Management, has prepared a programmatic EIS evaluating issues associated with wind energy development on public lands.[941]

628. The second requirement—that an agency's action be "major"—also demands some definition. CEQ regulations define "major" as including "actions with effects that may be major and which are potentially subject to Federal control and responsibility. Major reinforces but does not have a meaning independent of

938. *Ibid.* at § 4332(c).
939. 40 C.F.R. § 1508.18.
940. Exec. Office of the President, Council on Envt'l Quality, *Memorandum for Heads of Federal Departments and Agencies: Effective Use of Programmatic NEPA Reviews* (Dec. 18, 2014).
941. Dep't of Interior, Bureau of Land Mgmt., *Wind Energy Final Programmatic Environmental Impact Statement* (June 2005), *available at* http://www.windeis.anl.gov/documents/fpeis/index.cfm.

significantly."[942] Because CEQ regulations intertwine the question of whether an action is major with its impacts, court decisions have played an important role in further defining this requirement.

629. Frequently, the assessment of whether an action is "major" poses little problem. A decision by the National Institutes of Health, for example, amending its regulations to allow for approval of releases of DNA was determined to be a major action that required an EIS.[943] Similarly, the decision to build dams along navigable rivers satisfies the major requirement of NEPA.[944] If a project involves a substantial commitment of financial resources that the federal government directs,[945] then it is "major" for NEPA purposes.

630. In energy, major federal actions often arise in the context of permit issuances. A decision to build a nuclear reactor, for example, is a major federal action because the Nuclear Regulatory Commission issues power plant licenses, and the courts have ruled that NEPA applies to such licenses, even if issued by an independent agency like the NRC.[946] Similarly, construction projects by the Army Corps of Engineers also constitute federal actions, as did the State Department's decision recommending to the President the construction of the TransCanada Keystone Pipeline. All of these federal actions require an agency to undertake an EIS analysis.

631. By contrast, the third requirement—that the action have a significant impact on the environment—tends to generate more controversy. Under CEQ regulations, three outcomes are possible when NEPA is invoked: the agency may conduct a full-blown EIS; it may complete only an environmental assessment (EA); or it may determine that its action will not significantly affect the environment, thus obviating the need for an EIS at all. Accordingly, before undertaking a full-blown EIS, the agency prepares an "environmental assessment." An EA is like a mini-EIS. The EA considers alternatives to the proposed action even if the environmental impacts are not "significant." Further, through the EA process, an agency may avoid the necessity of preparing an EIS if it can show that it can mitigate the environmental impacts of the proposed action. If an agency decides not to prepare an EIS, it must make a "finding of no significant impact"—or FONSI—and it must then make that determination available to the public. If, however, the EA reveals that the action will have significant environmental impacts, the agency must proceed with an EIS.

632. Because a FONSI removes an agency's action entirely from the NEPA process, the issue of whether an action will have significant impacts on the

942. 40 C.F.R. § 1508.18.

943. Found. on Econ. Trends v. Heckler, 756 F.2d 143, 153 (D.C. Cir. 1985).

944. March v. Oregon, 490 U.S. 360 (1989).

945. 40 C.F.R. § 1508.18 ("Actions do not include funding assistance solely in the form of general revenue sharing funds, distributed under the State and Local Fiscal Assistance Act of 1972, . . . with no Federal agency control over the subsequent use of such funds. . . .").

946. *See, e.g.*, Calvert Cliffs' Coordinating Comm'n, Inc. v. Atomic Energy Comm'n, 449 F.2d 1109 (D.C. Cir. 1971); Vermont Yankee Nuclear Power Corp. v. Natural Resources Defense Council, 435 U.S. 519 (1978).

environment has generated substantial litigation and a significant number of judicial decisions. For example, courts have determined that the EIS requirements apply to a wide variety of activities, including both quite specific actions (such as constructing roads through national parks and granting permits and licenses for public and private projects) and more general agency actions (such as approving rules, regulations, and other policies pertaining to their jurisdiction or adopting formal plans or programs). NEPA applies to all federal actions with the exception of those that are specifically exempt from its requirements, including some actions in parts of the Clean Air Act and the Clean Water Act.

633. Generally, courts have viewed the human environment expansively under NEPA, including aesthetic and other impacts. In *Hanley v. Mitchell*, for example, the court held that the construction of a new jail in Manhattan fell under NEPA because of its impacts on the human environment, including the possibility of increased crime, and because of the aesthetic impact of the building.[947] There are, though, limits to what must be addressed in an EIS. In *Metropolitan Edison Co. v. People Against Nuclear Energy*, for instance, the U.S. Supreme Court held that major federal actions, such as resuming operation of reactors after the Three Mile Island nuclear incident, did not require that an EIS address psychological disturbance caused by the mere risk of another nuclear incident, unless those psychological effects had a reasonably close causal relation to changes in the physical environment.[948]

634. One recent example of judicial disputes about NEPA's scope involve climate change, and whether EISs must evaluate the impact of federal actions in this light. In *Center for Biological Diversity v. National Highway Traffic Safety Administration*,[949] for instance, the agency had set fuel economy standards for light trucks. Eleven states and four public interest organizations challenged the rule for not paying adequate attention to its likely impact on global warming. The court ruled that the agency's NEPA analysis was insufficient because it failed to monetize benefits of greenhouse gas emission reductions.

635. The EIS process is intended to be educative and inform agencies about environmental consequences and alternatives to mitigating environmental harms. Thus, prior to making "any detailed statement," the responsible federal official must consult with and obtain comments from any other federal agency that has jurisdiction "by law or special expertise" over any environmental issue in question. The agency also must make copies of the statement, together with comments and views of appropriate federal, state, and local agencies, available to the President and to the Council on Environmental Quality. Additionally, the statement must accompany the

947. Hanley v. Mitchell, 460 F2d 640 (2d Cir. 1973).

948. Metropolitan Edison Co. v. People Against Nuclear Energy, 460 U.S. 766 (1983). *Compare* New Jersey Dep't of Envtl. Protection v. NRC, 561 F.3d 132 (3d Cir. 2009), *with*, San Luis Obispo Mothers for Peace v. NRC, 449 F.3d 1016 (9th Cir. 2006).

949. Center for Biological Diversity v. Nat'l Highway Traffic Safety Admin., 538 F.3d 1172 (9th Cir. 2008).

proposal throughout all agency review processes.[950] Thus, the EIS requirement is intended to be comprehensive, as well as collaborative, with the specific purpose of providing an explicit and detailed explanation of the environmental consequences of federal actions.

636. In reality, however, the EIS process tends to be quite technical and detailed. EISs rely heavily on natural and biophysical sciences, and a cottage industry of environmental consultants has grown up since NEPA's passage to draft these statements. Any given EIS can be hundreds, and sometimes thousands, of pages long, including supporting documentation and analyses.[951]

637. In considering the environmental consequences of a proposed action, the EIS is required to discuss not only the environmental effects of the proposal and its alternatives but also energy requirements, the potential for resource conservation, possible depletion of natural resources, and such things as urban quality, historic and cultural resources, and the design of the built environment.[952] An agency must prepare an EIS if substantial questions, after considering both the context and intensity of the proposal,[953] are raised as to whether the project may cause significant degradation of some human environmental factor. Those who challenge the EIS need not show that the action will necessarily impact the environment significantly, but rather, only that substantial questions have been raised about whether a project may have such significant impacts.[954] The more difficult calculation is to figure out the magnitude of those effects; often, these are "judgment calls" made by courts, with a tendency to favor the filing of an EIS.

638. The adequacy of EISs is often resolved in court through judicial review. Generally, the idea behind the creation of administrative agencies is that they are objective, non-political, expert bodies that have the authority to review and make technical decisions. Following such a scheme, courts often grant a significant degree of deference to agency decision-making. For EISs, however, the U.S. Supreme Court has directed lower courts not to automatically defer to federal agencies without carefully reviewing the record and satisfying themselves that the agency has made a reasoned decision. This standard of review is sometimes referred to as the "hard look" test, which effectuates the statutory standard that an agency's decision is not arbitrary or capricious.[955]

639. The preparation of an EIS involves several steps. The agency responsible for the EIS must first seek the participation of other agencies, as well as the public,

950. 42 U.S.C. § 4332(c).

951. Given the burden of putting together an EIS, at times agencies have been reluctant to engage the process, and some, including FERC, have categorically determined that many of their actions are exempt from NEPA because they do not materially affect the human environment.

952. 40 C.F.R. § 1508.16.

953. Nat'l Parks & Conserv. Ass'n, 241 F.3d 722 (9th Cir. 2001).

954. Idaho Sporting Cong. v. Thomas, 137 F.3d 1146 (9th Cir. 1998).

955. March v. Oregon, 490 U.S. 360 (1989).

in planning the scope of the document. During this phase, the agency determines the significant environmental issues that should be addressed in the EIS. After the EIS is outlined, a draft is prepared (Draft EIS), comments are invited, and a final EIS is published. Once a final decision is made, the agency then prepares a record of the decision (ROD), which summarizes its actions and explains why it may have rejected environmentally preferable alternatives and other mitigation efforts.

640. Two key issues often arise in the preparation of EISs. One is whether NEPA applies at all, that is, whether the agency has proposed an actual course of action or is merely weighing possible avenues it may or may not eventually proceed down. This matters not just for triggering NEPA but also for ensuring that the agency actually evaluates the full impacts of its action. Another issue is assessing the appropriate scope of alternatives that the agency must weigh in its EIS. This dictates the scope and content of the EIS itself.

641. Determining exactly when an agency has "proposed" an action has presented courts with some difficulties.[956] The word "proposal" is not free from controversy. Agencies need leeway to conduct day-to-day business. Agency heads, as well as staff, routinely make recommendations regarding future courses of conduct. Similarly, the daily work of agencies, including holding hearings, issuing analyses and reports, and other such activities, make it difficult to determine whether or not a NEPA-triggering "proposal" has occurred. Thus, during the course of developing a federal program, there are often dozens or even hundreds of memoranda prepared by various agencies suggesting possible courses of action, and often evaluating them. Some reports may even signal an agency's direction; some may not. Accordingly, courts have observed that "no bright line divides those memoranda that are merely evaluations of possible courses of action from those that constitute recommendations or reports on proposals for action."[957]

642. Whether an agency has proposed an action is relevant not just to trigger NEPA. It also influences the scope of the review undertaken. Courts have been reluctant to allow agencies to proceed piecemeal under NEPA. An agency cannot segment its activity, thus masking the extent of its action's impacts on the environment. An agency cannot, for instance, "propose" to build one portion of a road at a time, so that, once segmented, the road would appear to have little environmental impact. Instead, courts require that such actions be assessed as a whole. At the same time, courts have been careful not to expand what constitutes a proposal and, therefore, the necessary scope of the EIS. In *Kleppe v. Sierra Club*,[958] for instance, the U.S. Supreme Court ruled that an EIS was not required to address the regional impacts of coal leasing in the Northern Great Plains, because no such regional proposal had been made. Instead, the Department of the Interior had issued leases that had been local in scope.

956. *See, e.g.*, Aberdeen & Rockfish R.R. Co. v. Students Challenging Regulatory Agency Procedures, 422 U.S. 289 (1975); Weinberger v. Catholic Action of Hawaii, 454 U.S. 139 (1981).
957. Daniel A. Farber, Environmental Law in a Nutshell 44 (2014).
958. 427 U.S. 390 (1976).

643. The CEQ has attempted to more clearly define "proposal," saying that a proposal "exists at that stage in the development of an action when an agency subject to the Act has a goal and is actively preparing to make a decision on one or more alternative means of accomplishing that goal and the effects can be meaningfully evaluated . . . . A proposal may exist in fact as well as by agency declaration that one exists."[959] Nevertheless, determining whether or not an actionable "proposal" has been made will continue to be the source of litigation.

644. Determining what alternatives must be explored in an EIS also presents some difficulty. The EIS must discuss alternatives, and the scope of those alternatives can often challenge agency decision makers. Courts have ruled that an agency must consider alternatives not only within its own jurisdiction but also within the jurisdiction of any part of the federal government. Further, agencies must consider international actions when there are impacts within the United States.[960]

645. In determining the scope of these alternatives, courts have applied a "rule of reason" test. The idea behind this test is to assess the likelihood and reasonableness of particular alternatives, including taking no action whatsoever. Implausible alternatives need not be discussed. Rather, the issue is whether a reasonable person would think that an alternative is sufficiently significant to warrant discussion.[961] In other words, the agency need not consider every alternative, nor even unlikely ones. It must not, however, ignore important or likely alternatives.[962]

646. Again, this puts courts in the position of making judgment calls. In one case, for example, the U.S. Court of Appeals for the D.C. Circuit court ruled that a Class Nine nuclear accident was too speculative to be included in an EIS. After that ruling, the NRC staff did in fact classify the accident at Three Mile Island as a Class Nine incident, thus clearly indicating the "likelihood" of such an alternative.[963] On the other hand, other courts have ruled that an EIS must consider a "worst case scenario" under CEQ's regulations. The worst-case scenario requirement, however, was rejected by the U.S. Supreme Court in favor of the CEQ's requirement that the EIS include a summary of existing credible scientific evidence and present an evaluation of such impacts based upon generally accepted theoretical approaches or research methods.[964] In short, the range of alternatives evaluated within an EIS need not be exhaustive, but it must be reasonably based on credible scientific standards.

647. Finally, it must be emphasized that NEPA is strictly procedural. Thus, even in the face of a completely inadequate EIS, NEPA does not empower courts to engage in a substantive review of proposed projects. In reality, however, by closely

---

959. 40 C.F.R. § 1508.23.

960. Nat'l Org. for the Reform of Marijuana Laws v. U.S. Dep't of State, 452 F. Supp. 1226 (D.D.C. 1978).

961. FARBER, *supra* note 957, at 46–51; NRDC v. Morton, 458 U.S. F.2d 827 (D.C. Cir. 1972); Vt. Yankee Nuclear Power Corp. v. NRDC, 435 U.S. 519 (1978).

962. California v. Block, 690 F.2d 753 (9th Cir. 1982).

963. Carolina Environmental Study Group v. United States, 510 F.2d 796 (D.C. Cir. 1975).

964. Robertson v. Methow Valley Citizens Council, 490 U.S. 332 (1989).

scrutinizing the content, scope, and alternatives presented in an EIS, courts often remand EISs they have found inadequate to agencies in a way that effectively forestalls government action. Further, even though the statute is procedural, "NEPA has played an important role in opening agency decision-making to the public. NEPA has provided constant pressure on agencies to broaden their missions to consider and adopt environmental values. And it has spurred agencies to modify proposals and mitigate adverse impacts."[965]

---

965. James Salzman & Baron H. Thompson, Jr., Environmental Law and Policy 335 (3d ed. 2010).

Chapter 3. Clean Power Plan

648. Many have observed that the best way to begin integrating energy law and environmental law is through climate change regulation. In the United States, efforts to adopt a national policy on climate change have remained largely stalled, even as state and regional efforts have progressed. Indeed, sub-national activities on this score include renewable portfolio standards; regional arrangements to curb carbon emissions; and individual state efforts, such as those in California to regulate greenhouse gas emissions, or those of New York to restructure the electricity sector. Nevertheless, federal action to link energy and environment more closely is beginning to occur, despite continuing gridlock in Congress. Instead, the Obama administration has chosen to take action administratively.

649. On June 2, 2014, the EPA issued a proposed rule known as Clean Power Plan,[966] which the EPA then issued as a final rule on August 3, 2015.[967] The Clean Power Plan has ambitious goals. It seeks to use federal-state cooperation to develop state-based plans that will significantly reduce greenhouse gas emissions from fossil fuel electric generating plants. These plants are responsible for nearly 40% of total U.S. CO_2 emissions—the nation's largest source of this pollutant.

650. Specifically, the CPP seeks to reduce U.S. CO_2 emissions by 26% (below 2005 levels) by 2020, and by 32% by 2030. The EPA estimates that the total combined climate and health benefits from adopting the CPP range from USD 3.5 billion to USD 8.1 billion in 2020 and from USD 34 billion to USD 48 billion in 2030.[968] These benefits include health risks from heat stroke, heat-related deaths, and reduced particulate pollution, as well as the costs incident to a decrease in the intensity of extreme weather events. In the CPP, EPA also envisions emissions reductions of sulfur dioxide, nitrogen oxides, and fine particular matters.[969] The agency also estimates that the health and climate benefits will outweigh the estimated annual cost of meeting the standards, which are projected to run from USD 5.1 billion to USD 8.4 billion in 2030.

651. The EPA intends the CPP to be flexible. Consistent with the overall design of the Clean Air Act (the legal source of EPA's authority to regulate in this arena), EPA will establish overall goals for each state, and then states themselves will have leeway to craft compliance plans. This makes sense, EPA notes, because different states have different energy resource mixes, and different states are already engaged in a variety of emission reduction activities.

---

966. Carbon Pollution Emission Guidelines for Existing Stationary Sources: Electric Utility Generating Units, 79 Fed. Reg. 34,830 (June 18, 2014) (proposed rule).
967. Carbon Pollution Emission Guidelines for Existing Stationary Sources: Electric Generating Units [EPA-HQ-OAR-2013-0602; FRL-XXXX-XX-OAR] RIN 2060-AR33 (August 4, 2015) (hereinafter **Final Rule**).
968. Final Rule at 92 and 91-99.
969. *Ibid.* at 34,839.

652. According to the proposed rule, each state must first meet an interim goal from 2022 to 2029 that achieves a certain annual average amount of pollution reduction during that period. To meet that goal, states can adopt one of two approaches. First, a state can reduce its carbon intensity (*i.e.*, the amount of carbon per unit of power generation), or, second, the state can meet a mass-based goal of pollution reduction (*i.e.*, the number of tons of carbon emitted). After this interim period, states must then meet EPA's 2030 target for emission reductions, and continue to do so from that point onward.

653. State goals will be tethered to EPA's calculation of the "best system of emissions reduction" (BSER) available to the state from existing generating units. In other words, EPA seeks to establish state-by-state emissions reduction rate targets, which states can decide how to meet under the CPP's flexible framework.

654. The BSER will be calculated based upon the mix of the power sources in each state and the application of three "building blocks" to achieve an individual state's emissions reduction goals. The building blocks are: (1) increasing the efficiency of fossil fuel-fired power plants through heat-rate improvements, including retiring older, less efficient coal plants; (2) using lower emitting energy resources such as natural gas; and (3) utilizing more zero or low-carbon energy sources such as renewable energy. The proposed rule included a fourth building block allowing states to use electricity more efficiently via, for example, demand-side management programs and more efficient transmission systems. For legal reasons, the fourth building block was eliminated; however, the final rule notes that while this is not a formal building block under the Plan, states may still achieve targets through energy efficiency Each building block merits further description.

655. Building Block 1 seeks heat rate improvements by EGUs. By increasing the heat rate—that is, by adopting improvements that increase the efficiency with which an EGU converts fuel to electricity—carbon intensity can be reduced. In short, increasing the heat rate reduces the amount of fuel needed to produce the same amount of electricity. These improvements can be realized through a variety of measures, such as equipment upgrades and further optimizing operational practices.

656. Building Block 2 aims to modify fleet composition and dispatch practices. The intent is to encourage EGUs that are carbon-intensive and the utilities that own them to switch to less carbon-intensive generation. In effect, the CPP is encouraging reduction in carbon intensity, rather than perpetuating the existing practice of dispatching least-cost, carbon-heavy resources first for electricity production. Much of this change may come from increased reliance on natural gas-fired units, and particularly natural gas combined cycle units. EGUs may also choose to dispatch larger amounts of solar and wind power. To facilitate this re-dispatching, the EPA expects affected EGUs to conduct their dispatch analysis by using the Integrated Planning Model (IPM). The IPM is a multi-regional, dynamic programming model that has been used by the electric power sector for over two decades to evaluate the economic and emission impacts of prospective environmental policies. In the context of the CPP, the IPM also assesses the costs of requiring increasing levels of re-dispatch from higher to lower emitting EGUs.

657. Building Block 3 is similar to Building Block 2 in that it encourages greater use of lower-carbon generation. In particular, Building Block 3 contemplates greater use of renewable generating capacity. Renewable electricity, including conventional hydropower, currently constitutes 12% of total U.S. electricity generation. The CPP's suggested expansion of renewable resources should be aided by the fact that more than half of the states have developed renewable portfolio standards, which require utilities to sell a certain percentage of electricity from renewable resources.

658. Although proposed Building Block 4 has been eliminated, states can still achieve targets by increasing energy efficiency or by reducing demand for generation at the affected EGUs, if emissions reductions will occur. Most states—forty-seven—have established a demand-side energy efficiency policy and, therefore, the CPP can contribute to encouraging their use. The proposed rule recognizes that significant gains have been made in states such as California and Minnesota, which have reduced demand for electricity by 12.5% and 13.1%, respectively, through such programming.

659. The CPP envisions that states will adopt a wide variety of measures to achieve their respective BSERs. Indeed, in addition to the examples listed above, states may choose to expand use of natural gas combined cycle units, install carbon capture and sequestration facilities, increase use of nuclear power, improve energy storage, adopt energy conservation programs, or use other zero or low-carbon strategies to generate electricity. Additionally, state plans will be designed to balance CO_2 reductions with other policy goals, including: (1) utilizing a diversity of energy resources; (2) maintaining electric reliability; (3) providing affordable electricity; (4) recognizing investments already made; and (5) meeting states' specific energy, environmental, and economic needs. States can proceed individually or in collaboration via multi-state or regional plans.

660. States are required to submit their final plans by September 6, 2016 unless granted an extension until September 6, 2018. State compliance will be accomplished through two general approaches: (1) direct limits on emissions for EGUs and (2) composing a portfolio of building blocks to attain an individual state's goal. In other words, the CPP is directed not only at existing EGUs but also at states' provision of electricity more broadly.

661. The CPP was promulgated under the Clean Air Act, which requires EPA to regulate air pollutants that endanger public health and welfare.[970] The Act's reach to greenhouse gas emissions has been affirmed by the U.S. Supreme Court.[971] Nevertheless, there have been, and will continue to be, legal challenges to the exercise

970. Clean Air Act, § 111(d); 42 U.S.C. § 7411(d).

971. Massachusetts v. EPA, 549 U.S. 497 (2007); American Elec. Power Co. v. Connecticut, 131 S. Ct. 2527 (2011).

of that authority as it applies to power plants. Recently, for example, the U.S. Supreme Court has been critical of EPA's broad reading of the Clean Air Act, and has denied the EPA a certain degree of deference when it sought to use the Act to broadly regulate "a significant portion of the American Economy."[972] Instead, the Court noted that it expects Congress to speak clearly if it wishes to assign an agency power to make decisions that have vast "economic and political significance," including when it intends to "regulate millions of small sources."[973] Certainly, the CPP can reach into both these areas.

662. In the end, the CPP is a significant attempt by the federal government to address climate change as part of President Obama's Climate Action Plan,[974] and to do so by recognizing the interdependence between energy and the environment. The CPP sets ambitious goals, encourages state flexibility and participation, and is an attempt by the federal government to take a leadership role in this area. Over the next few years, the CPP will be subject to numerous legal challenges, policy disagreements, and political debates. Regardless, this step is a significant one, acknowledging that energy and the environment are of a whole and should be regulated as such.

---

972. Utility Air Regulatory Group v. EPA, 134 S. Ct. 2427, 2443 (2014) (quoting FDA v. Brown & Williamson Tobacco Corp., 529 U.S. 120, 159 (2000).
973. *See, e.g.*, Utility Air Regulation Group v. EPA, 134 S. Ct. 2427, 2444, 2446 (2014).
974. Executive Office of the President, *The President's Climate Action Plan* (June 2013).

Part VI. Interaction between Energy Law and Tax Law

Chapter 1. Introduction

663. Energy firms are taxed for two predominant reasons. First, firms are taxed for revenue collection. Second, and more importantly, taxation is also used to create markets and to provide firms with incentives to participate in them. These two objectives often act at cross-purposes. To the extent that "tax breaks" are given to stimulate or create markets, revenue collection is reduced. Tax breaks are sometimes the result of political influence to lobbying; other times, tax breaks are part of broader policy goals and objectives. Consequently, governments often resort to tax policies to advance their energy policies.

664. Most recently, current U.S. energy policy has been described as focusing on three major goals: (1) assuring a secure supply of energy; (2) keeping energy costs low; and (3) protecting the environment.[975] To this list, of course, must be added economic stability and growth, as in many ways energy policy is fundamentally industrial policy. Thus, because U.S. energy policy aims are diverse, tax policies and incentives used to support them also are diverse.

665. Governmental spending policy is used to support energy objectives in two ways: by reducing the tax obligations of firms or individuals that government wants to promote ("tax expenditures"), and by spending tax revenues to support desirable energy activities, such as research, demonstration, and development ("direct spending"). Both of these receive attention in the literature, and are often criticized as the government unfairly putting a thumb on market outcomes.[976] Our focus here is on the tax system itself, and thus we do not address direct spending.[977]

975. Brent D. Yacobucci, Congressional Research Service, Energy Policy: 114th Congress (2015).

976. Davies, *supra* note 75; Lisa Larrimore Ouellette, *Addressing the Green Patent Global Deadlock Through Bayh-Dole Reform*, 119 Yale L.J. 1727 (2010); Michael Shellenberger et al., *Fast, Clean, & Cheap: Cutting Global Warming's Gordian Knot*, 2 Harv. L. & Pol'y Rev. 93 (2008).

977. For more on direct spending and energy, *see* U.S. Energy Info. Admin., *Direct Federal Financial Interventions and Subsidies in Energy in Fiscal Year 2013* (Mar. 2015), *available at* http://www.eia.gov/analysis/requests/subsidy/pdf/subsidy.pdf; Congressional Budget Officer, *Energy Security in the United States* (May 2012), *available at* http://www.cbo.gov/sites/default/files/05-09-EnergySecurity.pdf.

666. Indeed, energy policy in the United States must be understood at least in part as tax policy, or a policy of "tax expenditures"—that is, "expending" or *using* (such as through tax credits and deductions) funds that otherwise would be captured as tax revenues for a specific policy purpose. More technically, tax expenditures can be broadly defined as "federal revenue losses (reduced government receipts) associated with tax allowing for special exemptions, deductions, credits, income deferrals, or reduced rates."[978] Tax expenditures are made to accomplish specific economic purposes, such as promoting the exploration and development of oil and gas reserves or stimulating renewable resource markets in solar or wind.

667. Energy firms, depending upon their corporate structure, are susceptible to being taxed by both individual states and the federal government. This Part examines tax provisions that are specifically addressed to the energy sector. In addition to these more specific provisions, energy firms are also subject to general tax laws. In the United States today, the general goals of the energy policy stated above apply to individual states as well as to the federal government. Individual states, however, have particular interests, often related to their available natural resources or desire to encourage or attract business, which they advance through their energy tax policies.

978. Molly F. Sherlock, Congressional Research Service, Energy Tax Policy: Historical Perspectives on and Current Tax Status of Energy Tax Expenditures 1 (2011).

Chapter 2. State Taxation

668. While most states impose income taxes, the more significant taxes for energy firms are ad valorem taxes, severance taxes, and sales taxes. The key difference between income taxes and ad valorem and severance taxes is the profitability of the business. An energy firm will pay an income tax if they are profitable. Ad valorem and severance taxes, however, do not depend upon the firm's profitability.

669. An ad valorem tax is a tax on the property owner. Typically, a taxing authority will determine an identified property's market value, and a percentage of that value will be paid as an ad valorem or property tax. It is not unusual for property owners to contest the property valuation. Nevertheless, once the valuation is determined, the amount of the tax can be computed.

670. Severance taxes work differently. A severance tax is imposed on the amount of a natural resource that is "severed" from the land. The idea behind the severance tax is that as a natural resource becomes depleted, a state's wealth is diminished. Several states impose a severance tax based on a specific percentage of the gross value of all the natural resources on a given property. Thus, a severance tax is a "tax on the wealth that is severed and removed from its natural state by production; it is not a tax based upon elements of value added after the act of severance."[979]

671. With fugacious resources such as oil and gas, the amount of these taxes can be contested because the extent and value of the oil and gas reserves are essentially unknown. Consequently, the property valuation is often based upon the price that an unrelated third party would pay for the resource (*i.e.*, the oil or natural gas) in an arm's length transaction.

672. Thus, state ad valorem and severance taxes serve primarily as revenue generation mechanisms for their jurisdictions, designed in part to compensate the sovereign for the utilization of its resources but also, fundamentally, to raise state funds. Of course, the level at which a state sets its tax rates matters. If a state wishes to attract industry that might otherwise go to another state (if, for instance, neighboring states both have desirable oil and gas or coal reserves), it might use its tax structure to compete with other states for that business.[980]

673. While state natural resources taxation, particularly through severance taxes, have a long history, more recently states also have begun to provide tax incentives for encouraging investment in certain types of energy endeavors. Often, these are

979. J. Clifford Rogillio, *State Taxation of Oil and Gas*, *in* 2 ENERGY LAW AND TRANSACTIONS, *supra* note 81, at § 41.02[1][a].

980. DaimlerChrysler Corp. v. Cuno, 547 U.S. 332 (2006); Gen. Motors Corp. v. Tracy, 519 U.S. 278 (1997).

renewable resource and efficiency projects.[981] These incentives generally are applied either through tax credits for a reduction in sales taxes.

674. Utah, for example, encourages renewables installations by offering an investment tax credit. For residential installations, that credit is worth 25% of the eligible system's cost or USD 2000, whichever is less, and for commercial systems, 10% of eligible system cost or USD 50,000, whichever is less. Eligible technologies include solar photovoltaic, solar thermal, wind, geothermal, hydropower, and biomass energy systems. Likewise, Colorado offers investment tax credits and production tax credits to corporations for renewable energy investments.[982] And California offers 202 different incentive programs for renewable and clean energy investments, including a variety of tax credits. Additionally, among the 45 states that levy a state sales tax, 24 grant partial or full exemptions for PV and other solar systems.[983]

---

981. DSIRE, *Database of State Incentives for Renewables & Efficiency*, *available at* http://www.dsireusa.org/ (last viewed Apr. 1, 2015).

982. *See Ibid.*

983. Kate Daniel et al., N.C. Clean Energy Technology Center, Understanding Sales Tax Incentives For Solar Energy Systems: Fact Sheet For Customers, Industry, and Local Governments (Mar. 17, 2015).

Chapter 3. Federal Taxation

675. Although the U.S. income tax began in 1916, tax expenditures have been used to affect the behavior of fossil fuel firms since 1913, with accelerated depreciation as part of an excise tax deduction.[984] Beginning in 1918, however, energy tax expenditures were more formally contained in the Internal Revenue Code. Thus, since the early twentieth century, U.S. tax policy has favored fossil fuel industries, and especially coal, oil, and natural gas. Toward the end of that century, however, as concerns about U.S. energy independence increased in the oil crises of the 1970s, tax policy began to shift and also provided incentives for renewable resources and other non-fossil fuel energy sources.

676. In the early years, federal tax policy promoted the development of oil and gas through various mechanisms. Initially, two tax provisions favored the exploration and initial production of these resources. First, intangible drilling costs, such as wages and machinery used for drilling, as well as dry hole costs, were afforded favorable tax treatment. This began as early as 1917,[985] by allowing these costs to be fully deducted in the first year that they were incurred rather than having them capitalized and then depreciated over a period of time like most other business assets.[986]

677. Second, tax policy afforded a "percentage depletion allowance." This provision allowed for a deduction on a fixed percentage of gross receipts of the natural resources extracted, rather than on the resources' actual value.[987] This provision was introduced in the tax code for oil and natural gas in 1926 and extended to coal in 1932. Over the years, the depletion allowance has been the subject of political controversy, as several presidents have attempted to eliminate it. Although it has been adjusted in part, it remains part of the tax code. Together, these provisions infused capital into energy firms specifically to encourage exploration, and both "remain in the tax code in limited form today."[988]

678. Thus, from 1916 into the 1970s, the accelerated depreciation and the depletion allowance provisions afforded fossil fuel firms with favorable tax treatment. In the 1970s, energy tax policy changed dramatically. The Arab Oil Embargo of 1973 was a wake-up call for the United States. The price of a barrel of oil quadrupled, and the country experienced gas shortages, double-digit inflation, and concerns about national security. Policymakers, then, set about creating a change. The change was driven by the twin concerns of trying to wean the nation from dependence on foreign oil and, concomitantly, to increase domestic exploration and production for fossil fuel resources. To these ends, Congress redesigned energy tax policy to

984. Tariff Act of 1913, 38 Stat. 114 (1913).

985. *See* Mona Hymel, *Environmental Tax Policy in the United States: A "Bit" of History*, 3 ARIZ. J. ENVTL. L. & SOC. 157, 169–74 (2013).

986. 26 U.S.C. § 263(c) (intangible drilling and development costs); 26 U.S.C. § 616(b) (development costs of minerals and ores other than oil and gas).

987. 26 U.S.C. §§ 611, 612, 613 and 613A.

988. SHERLOCK, *supra* note 977.

promote domestic exploration and production of fossil fuels, and to create new energy markets in non-fossil fuel energy and energy efficiency.

679. To accomplish these goals, the tax code made three shifts. First, the decades-old tax expenditures for intangible drilling costs and the depletion allowance were reduced. Second, excise taxes penalizing fossil fuel use took effect. Third, tax incentives were introduced to promote energy efficiency, stimulate alcohol fuel production, and promote unconventional fuels. Congress implemented many of these goals in the Energy Tax Act of 1978,[989] which constituted one of the major portions of President Carter's National Energy Act of that year.

680. The Energy Tax Act contained several provisions that aimed to promote renewable and unconventional energy sources, promote efficiency, and discourage the use of fossil fuels. On the energy efficiency side, businesses and residences were allowed to claim tax credits for energy efficiency investments, and a "gas guzzler" tax was imposed on automobiles that failed to meet specified fuel economy standards. The gas guzzling tax, then, both sought to increase efficiency as well as reduce demand for oil.

681. Regarding renewable energy, residences and businesses became eligible to receive tax credits for the installation of solar, wind, and geothermal energy applications. Nonconventional fuels, such as gasoline blended with ethanol, became eligible for a partial excise tax exemption. Additionally, geothermal projects were made eligible for percentage depletion allowances.

682. In 1980, also part of President Carter's energy program, Congress enacted the Crude Oil Windfall Profit Tax Act (WPT).[990] The problem this law sought to address was that the world price of oil had escalated noticeably. Consequently, the price of domestic oil reserves would rise to the world level without any increased production investments by producers. Domestic oil firms, then, would stand to gain significant economic rents, called windfall profits, without making additional capital investments. The idea behind the tax, then, was to tax those "unearned" gains and use the revenue for the development of other energy resources.

683. The Windfall Profit Tax thus embodied several goals. First, the tax sought to act as an excise tax on oil produced domestically. Second, by making tax credits available for certain domestic energy sources, the law aimed to promote domestic energy production from "unconventional fuel sources." Those "unconventional sources" included oil from tar sands and oil shale, shale gas and gas from other tight formations, and gas produced from biomass, coal, or methane. Investments in these resources were eligible to receive production tax credits. Additionally, the act provided for tax-exempt interest bonds on state and local electric utilities to develop facilities that produced fuel for electricity generation from solid wastes.

989. Pub. L. No. 95-618.
990. Pub. L. No. 956-223.

684. In 1980, while campaigning for the presidency, Ronald Reagan promised to repeal the WPT, and he finally did so in 1988. The WPT bore a problematic history. First, the tax turned out to collect significantly less revenue than was anticipated. Originally, it was estimated that the WPT would bring in gross revenues of over USD 390 billion and actual revenues (after deductions) of USD 175 billion. In actuality, it is estimated that the Treasury Department recovered gross revenues of about USD 80 billion and actual revenues of USD 38 billion.[991] Additionally, the tax was imposed on domestic oil, and not imported oil, and thus served as a disincentive for domestic production when the cost of oil (together with the tax) rose higher than the world price. In other words, at that price level, the WPT became an incentive for oil importation—exacerbating precisely the problem it sought to combat.

685. Further, the tax was very difficult to apply and administer. Among the difficulties encountered by the Internal Revenue Service were: (1) identifying properties subject to examination; (2) ensuring that the tax was assessed on eligible sales of oil; and (3) examining tax refund claims to ensure that the tax was only levied on barrels of oil that yielded a profit. Given the difficulties surrounding such a complex tax structure, oil producers both contested and evaded the tax in significant numbers, which partially contributed to the unexpectedly low level of revenue production from the WPT.[992]

686. In addition to repealing the WPT, the Reagan administration also attempted to conform energy tax policy with the administration's free market approach to economic regulation. The administration thus allowed most business energy tax credits to expire in 1982, as well as residential energy tax credits in 1985, both as previously scheduled. However, tax credits for specifically defined solar, geothermal, ocean geothermal, and biomass energy "properties" were extended during that administration.

687. After the Reagan administration, energy tax policy became more aggressive and interventionist. Congress enacted legislation aimed at promoting energy production from various specific resources, including alcohol fuels and unconventional fuels. Additionally, the oil and gas industry continued to enjoy favorable tax provisions. In the Omnibus Budget Reconciliation Act of 1990, Congress both increased the gasoline tax by five cents per gallon and doubled the gas-guzzler tax. The law also introduced a 10% tax credit for enhanced oil recovery,[993] liberalized restrictions on the percentage depletion allowance, and added production tax credits for unconventional fuels and small ethanol producers.[994]

---

991. Salvatore Lazzari, Congressional Research Service, The Crude Oil Windfall Profit Tax of the 1980s: Implications for Current Energy Policy (2006).

992. U.S. Comptroller General, *Report to the Chairman, Subcommittee on Commerce, Consumer and Monetary Affairs, House Committee on Government Operations: IRS Administration of the Crude Oil Windfall Profit Tax Act of 1980* (June 18, 1984).

993. 26 U.S.C. § 43.

994. Omnibus Budget Reconciliation Act of 1990, Pub. L. No. 101-508; Lazzari, *supra* note 990, at 5.

688. Significant energy legislation was also passed in 1992. The Energy Policy Act of 1992 included within it several energy tax provisions. The law established tax credits for energy produced from renewable resources, including for electricity generated using wind power and biomass systems. Additionally, the act provided for income tax deductions for the cost of clean fuel vehicles, more liberalized alcohol fuels tax credits, expanded unconventional fuel production credits, and further favorable tax treatment for oil and gas producers.[995]

689. Production tax credits, particularly for solar and wind power, became both popular and successful. The PTC reduced federal income taxes on a dollar-for-dollar basis for qualifying taxpayers based upon the electrical output of grid-connected renewable energy facilities. The PTC thus encouraged investments in renewable energy by lowering production costs. The idea, in short, was to bridge the cost gap between the production of traditional fossil fuels and wind power and, later, other renewables. The PTC has had staying power. Although it has expanded today to encompass additional resources, the renewable energy industry often asserts that the PTC is necessary to level the playing field, and renewable energy investments have been observed to cyclically track the presence or dormancy of the PTC.[996]

690. Investment tax credits (ITC)[997] have a similar effect. The ITC reduced federal income taxes for qualifying taxpayers based upon capital investments made in renewable energy projects. The ITC is earned once equipment is put in place. To qualify for an ITC, the system must be in service before December 31, 2016. The ITC provides: (1) a credit equal to 30% of expenditures for solar installations that generate electricity, heat, or cooling; (2) a 30% credit of expenditures for fuel cells of certain capacity and efficiency; (3) a 30% credit for expenditures for small wind turbines; (4) a 10% credit for geothermal systems; (5) a 10% credit for micro-turbines; and (6) a 10% credit of expenditures for eligible combined heat and power systems. There has been some debate whether PTCs or ITCs are more effective at incentivizing investment in energy systems, but what is clear is that both have had important impacts.[998]

691. President Obama has sought extension of both the PTC and the ITC, though extensions appear unlikely at this time. The PTC has been instrumental in the growth of wind power by 2700% from 2000 to 2013. And the solar ITC has been instrumental in the growth of solar power by 1600% from 2006 to 2013.

995. Energy Policy Act of 1992, Pub. L. No. 102-486.
996. Eric Lantz et al., Nat'l Renewable Energy Laboratory, *Implications of a PTC Extension on U.S. Wind Deployment* (Apr. 2014), *available at* http://www.nrel.gov/docs/fy14osti/61663.pdf.
997. 26 U.S.C. § 48.
998. Mark Bolinger et al., *PTC, ITC, or Cash Grant? An Analysis of the Choice Facing Renewable Power Projects in the United States*, National Renewable Energy Laboratory & Berkeley National Laboratory, Mar. 2009, *available at* http://www.nrel.gov/docs/fy09osti/45359.pdf; Mormann, *Beyond Tax Credits*, *supra* note 77, at 303.

692. At the turn of the twenty-first century, oil prices were increasing, raising concerns about energy security. This put energy at the forefront again of the policy conversation. Congress thus retroactively extended several tax subsidies that had previously expired. These included (1) production tax credits for renewable energy; (2) liberalized oil and gas depreciation allowances; (3) electric vehicle tax credits; and (4) clean fuel vehicle deductions.[999] In addition, Congress passed the Energy Policy Act of 2005.[1000]

693. EPAct 2005 was a comprehensive energy bill that pushed in many directions, but clearly sought to address increased energy prices and oil imports. The Act contained a variety of tax provisions directed to the (1) electricity infrastructure, (2) domestic fossil fuel production; (3) energy efficiency; and (4) cleaner vehicles and fuels. The law also substantially increased energy tax subsidies across a wide spectrum of resources.[1001]

694. Regarding electricity infrastructure, EPAct 2005: (1) extended modified tax credits for renewable energy production; (2) established tax credit bonds to finance investments in clean renewable energy projects; (3) made available special capital gains treatment for the sale of certain electricity transmission facilities for the purpose of industry restructuring; (4) provided production tax credits for advanced nuclear power facilities; and (5) instituted investment tax credits for clean coal projects.

695. Regarding fossil fuels, the Act: (1) extended production tax credits for facilities producing fuel from coke or coke gas; (2) provided for modified tax credits for producing fuel from nonconventional sources; (3) permitted refineries to expense qualified refinery properties without limitation; (4) gave natural gas pipelines favorable depreciation treatment; and (5) provided favorable amortization schedules for certain expenses for domestic oil and gas exploration.

696. Regarding energy efficiency, the Act made available: (1) tax deductions for expenditures on energy-efficient heating, cooling, lighting, and ventilation, and for the building envelopes in commercial building; (2) tax credits for more energy-efficient home construction and efficiency improvements in existing homes, such as investments in efficient furnaces, boilers, windows, insulation, and roofs; (3) credits for solar power and solar heating purchases; (4) credits for the production of energy efficient dishwashers, clothes washers, and refrigerators; and (5) business tax credits for investments in solar energy properties and for the installation of qualified fuel cells or micro-turbines.

---

999. Working Families Tax Relief Act of 2004, Pub. L. No.108-311; American Jobs Creation Act of 2004, Pub. L. No. 108-357.

1000. Pub. L. No. 109-58.

1001. The tax provisions of EPAct 2005 were extensive and are explained, with citations to the tax code and Treasury regulations. U.S. Congress, Joint Committee on Taxation, *General Explanation of the Tax Legislation Enacted in the 109th Congress* 12–82 (Jan. 17, 2007).

697. Regarding clean vehicles and fuels, the Act: (1) provided individual tax credits for qualified motor vehicles and hybrid vehicles; (2) gave tax credits for the installation of alternative fueling stations; (3) reduced fuel excise taxes for certain mixtures of diesel fuels; (4) extended tax credits for biodiesel fuels; and (5) gave additional biodiesel fuel credits for small agricultural producers.

698. EPAct 2005, then, can be seen as a microcosm of the developing U.S. energy policy in the twenty-first century. On the one hand, traditional fossil fuels continue to receive tax expenditures, and are seen as desirable, particularly for energy security purposes. On the other hand, reduced fuel consumption and cleaner energy resources also receive governmental support, and are promoted both for economic growth purposes but also to help protect the environment. Thus, U.S. energy policy, including as expressed in tax policy, embraces many goals, including energy efficiency, security, and independence.

699. After the passage of EPAct 2005, further legislation reduced tax subsidies to oil and gas firms,[1002] extended about-to-expire renewable energy provisions,[1003] and promoted cellulosic ethanol production.[1004] Additional legislation extended tax breaks for renewable energy production, energy efficiency, alcohol fuels and cleaner vehicles, while reducing existing tax breaks for the oil and gas industry.[1005] In 2009, amidst growing concerns about the Great Recession, Congress passed the American Recovery and Reinvestment Act of 2009,[1006] which furthered the energy tax policies set out in previous recent legislation.

700. The ARRA both modified and extended tax incentives for renewables, energy efficiency, and clean vehicles. Given that the country was in a credit crunch, ARRA § 1603 converted tax credits for renewables into direct cash payments. At the end of 2010, a number of tax provisions were set to expire, including favorable treatment for alcohol fuels and energy efficiency improvements in existing homes. Under the ARRA and subsequent legislation,[1007] those credits were extended.

701. In addition to its tax provisions, the ARRA also expanded the amount of spending on energy initiatives, such as improving the electric grid, increased funding for clean energy research and development, and a wide variety of other energy projects, altogether estimated at approximately USD 90 billion. The ARRA was the last major energy legislation to come from Congress.

702. Today, then, U.S. energy tax policy favors both fossil fuels and non-fossil fuels. There is a significant difference in tax treatment, however, that must be recognized. For most of its history, energy tax policy regarding fossil fuels has remained largely stable, although it certainly has been modified from time to time.

1002. Tax Increase Prevention and Reconciliation Act, Pub. L. No. 109-222.
1003. Tax Relief And Health Care Act of 2006, Pub. L. No. 109-432.
1004. Food, Conservation, And Energy Act of 2008, Pub. L. No. 110-234.
1005. Emergency Economic Stabilization Act of 2008, Pub. L. No. 110-343.
1006. American Recovery and Reinvestment Act of 2009, Pub. L. No.111-5.
1007. Tax Relief, Unemployment Reauthorization, And Job Creation Act of 2010, Pub. L. No. 111-312.

Tax policy regarding renewable resources and energy efficiency, however, has been quite sporadic. Congress, for example, has provided tax credits and other tax expenditures for renewables on many occasions, but also has let those provisions lapse—or has threatened to let them lapse.[1008] Thus, these programs may be seen as less stable, which can erode investor confidence, thereby dampening their desired economic effect. Although these programs have been extended from time to time, extension is not automatic, making investment planning in these energy resources more risky.

703. Indeed, over the last four decades, tax expenditures for renewable resources and energy efficiency consistently have been short-term and temporary. By way of example, for businesses, a 15% investment tax credit was permitted for investments in solar property made during 1986. If an investment was made during 1987, the credit was reduced to 12%—and 10% thereafter. The Energy Policy Act of 1992 extended a 10% credit for businesses for qualified investments in specifically defined solar properties. Consequently, energy firms and individuals who invested in renewable resource projects were required to pay close attention to the availability and timing of production tax credits and investment tax credits.[1009]

704. To be sure, the intermittent nature of these tax schemes is clear, and the number of expired tax provisions substantial. By the end of 2014, for example, 52 temporary tax provisions expired.[1010] The current Congress may choose to extend those provisions beyond their current lives,[1011] as they have in the past. However, there is no guarantee that energy tax provisions will be extended.[1012]

705. Some important current tax expenditures include: Solar investment tax credits of 30% for federal corporate income taxes (though this decreases to 10% starting in 2017);[1013] homeowner solar PV investment tax credits of 30%;[1014] and 30% credits for geothermal heat pumps, small wind turbines for residential use, and solar energy systems in fuel cells through 2016.

706. Another important debate currently taking place in U.S. energy tax policy is whether to impose a carbon tax. Unlike the European Union and other jurisdictions, the United States to date has not taken national action to address climate change, although, at the sub-national level, many U.S. states have implemented their own programs either within their boundaries or through regional cooperation. Part

1008. Alan J. Wilensky, *Federal Taxation of Energy*, *in* 2 ENERGY LAW AND TRANSACTIONS, *supra* note 81, at § 40.04.

1009. For more on environment project finance *see*, ANDREA KRAMER & PETER FUSARO, ENERGY & ENVIRONMENTAL PROJECT: FINANCE LAW & TAXATION (2013).

1010. Molly F. Sherlock, *Tax Provisions that Expired in 2014* ("tax extenders") (Feb. 6, 2015).

1011. The Tax Increase Prevention Act of 2014, Pub. L. No. 113-295, signed into law on December 19, 2014, for example, made tax provisions set to expire at the end of 2013 available to taxpayers for the 2014 tax year.

1012. MOLLY F. SHERLOCK & JEFFREY M. STUPAK, CONGRESSIONAL RESEARCH SERVICE, ENERGY TAX INCENTIVES: MEASURING VALUE ACROSS DIFFERENT TYPES OF ENERGY RESOURCES (2015).

1013. 26 U.S.C. § 48.

1014. 26 U.S.C. § 25D.

of the debate in the United States, then, is whether to take action on carbon emissions at all, and part of the debate is, if action is taken, whether to do so via a cap-and-trade mechanism or through a carbon tax. The closest the country has gotten to implementing a national greenhouse gas limit was in 2009, when the House of Representatives passed the American Clean Energy and Security Act of 2009 (ACES), also known as the Waxman-Markey Act,[1015] which would have instituted a cap-and-trade regime. The bill died in the Senate, however, and never became law. Now, as the prospect that the Obama administration will institute emissions rules under its Clean Power Plan grows near,[1016] the idea of a carbon tax has been resurrected, perhaps as a way to more cost-effectively achieve the same goals the Clean Power Plan seeks. An early 2015 survey revealed that there is broad support among the American public for such a tax, at 61%, and even more support if the tax sends rebate money to households, at 67%.[1017] Whether Congress ultimately adopts such a bill, however, is far more questionable, though it obviously could have a significant impact on the overall shape of U.S. energy tax expenditures.[1018]

707. Through history, U.S. tax expenditures for energy have been substantial. In 2013, for example, tax related support for energy was estimated at USD 23.3 billion. Of that amount, USD 4.8 billion was attributed to tax incentives for fossil fuels, while tax support for renewables was estimated at USD 13.4 billion. This snapshot indicates a shift in tax expenditures away from fossil fuels and toward renewable resources and energy efficiency. Again, however, that snapshot must be taken in historical context—a context where, increasingly, many resources are promoted, but historically, fossil fuels received the bulk of government favor.

1015. American Clean Energy and Security Act, H.R. 2454, 111th Cong. (2009).

1016. For more on this, *see* Part I, *Electricity*.

1017. *Global Warming National Poll: Resources for the Future*, New York Times & Stanford University, 48–49 (Jan. 2015), *available at* http://rff.org/Documents/RFF-NYTimes-Stanford-global-warming-poll-Jan-2015-topline-part-3.pdf.

1018. For more on the United States policy on energy taxation *see*, GILBERT E. METCALF, U.S. ENERGY TAX POLICY (2014).

Part VII. Interaction between Energy Law and Competition Law

Chapter 1. Introduction

708. Energy law in the United States is intrinsically intertwined with competition law and regulation. This is for at least two reasons. First, energy law seeks to address economic failures in markets, and many of those problems, such as natural monopoly power, relate to competition. Second, in a significant sense, both the U.S. energy industry and its regulation developed in tandem with U.S. competition law. Indeed, many of the most famous decisions and statutes in U.S. competition law involved or targeted energy firms.

709. Competition law, referred to in the United States generally as antitrust law, arose in response to the increasing consolidation of large businesses in the final decades of the nineteenth century. In response to this trend, Congress in 1890 passed the Sherman Antitrust Act with an extraordinary amount of support: It passed the House of Representatives without opposition, and breezed through the Senate by a 52 to 1 vote. Notably, the Act did not establish a particular administrative agency for its enforcement. Instead, the Act anticipated that federal prosecutors and private litigants would carry out antitrust enforcement. The theory was that those most affected would invoke the law to reduce monopolistic practices.

710. As its name implies, the Sherman Act made its target trusts, or legal agreements that companies had devised to work around state laws which prevented corporations from holding stock in other companies. Perhaps the most famous of these trusts was John D. Rockefeller's. Beginning in the latter half of the nineteenth century, Rockefeller quickly amassed an industrial juggernaut. Rockefeller's Standard Oil Company was a giant in the oil industry. It became famous for its aggressive, non-competitive practices,[1019] and it exerted market power over railroads, pipelines, and shipping. It was the largest industrial combination in the country and had a reputation for exercising its influence in Congress as well.

1019. YERGIN, *supra* note 42, at 19–55; *see also* Benjamin Klein, *The "Hub-and-Spoke" Conspiracy That Created the Standard Oil Monopoly*, 85 S. CAL. L. REV. 459 (2012).

711. Because of this, Standard Oil soon gained much negative attention, and trusts became a notable social and political concern. Progressive journalist Ida Tarbell set her particular sites on Rockefeller and his Standard Oil Company. Writing for *McClure's Magazine*, she published a series of articles critical of the oil giant. In 1904, these articles culminated in the book, *The History of the Standard Oil Company*, which not only captured a wide public audience but also the attention of President Theodore Roosevelt. Roosevelt then took on the task of "breaking up" the nation's trusts. Indeed, it was Roosevelt's focus on addressing the accumulation of corporate power that earned him the nickname "Trust Buster"—and to him, Standard Oil was the mother of all trusts.

712. Under Roosevelt's direction (and later), the federal government pursued Standard Oil through numerous investigations and lawsuits. In 1905, for example, after an investigation, President Roosevelt reported that Standard Oil had taken enormous profits from secret rail rates through its exercise of intimidating market power. By 1907, the federal government had suits pending against seven of Rockefeller's corporations and subsidiaries. The colorful Judge Kenesaw Mountain Landis, later the first commissioner of Major League Baseball, fined Standard Oil over USD 29 million for violating the Elkins Act, which prohibited railroads from offering rebates to shippers.[1020] Standard Oil routinely violated those prohibitions in a concerted effort to squeeze out competitors. Standard Oil's response to such a steep fine was indicative of the company's ruthlessness in competition. The story is that Rockefeller was playing golf when a messenger delivered an envelope with the judge's decision imposing the fine. Rockefeller opened the envelope, read it, and put it back into his pocket. He then resumed his golf game. When one of his playing partners asked him about the amount of fine, he responded, "Judge Landis will be dead a long time before this fine is paid."[1021] Rockefeller was correct; he never paid the USD 29 million fine because it was overturned later in court.

713. Roosevelt's pursuit of Standard Oil became a hallmark of his presidency. In 1906, the Roosevelt administration brought suit against Standard Oil under the Sherman Act, alleging that it conspired to restrain trade. The suit was under investigation for more than two years and culminated in 1911 in two arguments before the United States Supreme Court. The Court ruled against Standard Oil, ordering dissolution of the trust. Reading the opinion in open court, Chief Justice Edward White explained that Standard Oil intentionally and purposely, and thus illegally, sought to exclude others from the market. As a result of this decision, Standard Oil was divided into multiple entities.[1022] Though initially upset, Rockefeller and

---

1020. *See, e.g.*, JAMES MACGREGOR BURNS, THE WORKSHOP OF DEMOCRACY: FROM THE EMANCIPATION PROCLAMATION TO THE NEW DEAL 349–50 (1985); *see also* DORIS KERNS GOODWIN, THE BULLY PULPIT: THEODORE ROOSEVELT, WILLIAM HOWARD TAFT, AND THE GOLDEN AGE OF JOURNALISM (2014).

1021. YERGIN, *supra* note 42, at 108–09. The story continues that Rockefeller had the best golf game of his life.

1022. The new entities included Standard Oil of New Jersey, which eventually became Exxon; Standard Oil of New York, which became Mobil; Standard Oil (California), which became Chevron; Standard Oil of Ohio, which became Sohio and later the American arm of British Petroleum. There

Standard's other shareholders were soon assuaged. Within a year of dissolution, the value of the shares in the successor companies nearly doubled, and Rockefeller's personal wealth increased to nearly USD 1 billion.[1023]

714. Thus from the beginning, energy and competition law in the United States were intrinsically linked. In Standard Oil, the risks of consolidation and market power became prominent. And in the government's use of the Sherman Act to break up Standard Oil, the power of law to correct market problems was plain.

715. Antitrust law, however, is not the only tool used to address competition questions in the U.S. energy industry. Other types of regulation also grapple with these questions. The nature of the energy industry makes this necessary. While U.S. law long has generally left the function of oil markets to competition and competitors, other portions of the energy industry in the United States have received different treatment. For example, the transmission and distribution segments of electricity and natural gas industries effectively function as natural monopolies, and the law has treated them as such. Likewise, in part because of past practices and in part because of the significant capital investments needed to build them, oil pipelines have been treated as common carriers. It is a curiosity of our regulatory system that concerns about market power have been addressed by giving electric and natural gas utilities exclusive control of (and obligations within) specific service territories. The idea, however, is that by limiting competition, regulatory law can act as a surrogate for the result that competition would produce. In other words, in order to avoid the exercise of market power, U.S. law segmented markets in natural gas and electricity by giving those utilities exclusive territories where others could not compete, but in turn requiring these utilities to provide service for all and controlling the price at which they did so. In this way, U.S. law treated the risk of monopoly power in different markets in different ways, relying primarily on antitrust law to quell the concern in oil markets but public utility regulation to rein it in for electricity and natural gas. The existence of governmentally protected monopolies, then, adds a new twist to competition law. One question is how antitrust law functions in the energy context. Now, another question is whether, and to what extent, antitrust law applies to regulated industries. Thus, since the beginning of the twentieth century, the relationship between energy regulation and competition law is not only critical but somewhat intricate as well.

---

was also Standard Oil of Indiana, which became Amoco; Continental Oil, which became Conoco; and Atlantic, which became a part of ARCO and eventually of Sun. *Ibid.* at 94.

1023. *Ibid.* at ch. 5.

Chapter 2. General Antitrust Principles

716. In the United States, there are three primary laws that govern competition: the Sherman Act, the Clayton Act, and the Federal Trade Commission Act. Each act has been amended, supplemented, and modified over the years. Initially, these statutes were enacted to address gaps left by common law rules and principles that were deemed insufficient to curb market power and anticompetitive practices in an increasingly complex and concentrated industrial economy.

717. Prior to these laws, the United States began to consider and regulate modern monopolistic practices, first incrementally via judicial decisions, and then, through regulation. Thus, in 1887, following the Supreme Court's decision in *Munn v. Illinois*,[1024] Congress created the Interstate Commerce Commission.[1025] The ICC was charged with regulating railroad rates in the public interest: ensuring that they would be just and reasonable. Otherwise, railroads could exercise both monopoly and monopsony power over both shippers and consumers. As described extensively in the Parts on Natural Gas and Electricity, the ICC model was later adopted in the Federal Power Act and subsequently the Natural Gas Act and is still used today to ensure that rates in these industries are "just and reasonable."

718. Many motives and goals have been ascribed to the Sherman Antitrust Act of 1890.[1026] Regardless, competition clearly is at the heart of the law, which enunciates the value of competition as a national economic policy.[1027] The Act is general in nature. Thus, rather than specifically identifying prohibited conduct, Congress essentially directed federal courts to create a federal common law of antitrust. The two key sections of the act are:

> §1: Every contract, combination in the form of trust or otherwise, or conspiracy, in restraint of trade or commerce among the several States, or with foreign nations, is hereby declared to be illegal . . . and [violators] shall be deemed guilty of a felony.
> §2: Every person who shall monopolize, or attempt to monopolize, or combine or conspire with any other person or persons, to monopolize any part of the trade or commerce among the several States, or with foreign nations, shall be deemed guilty of a felony.[1028]

719. Congress designed these two provisions to complement each other. Section 1 is directed at conspiracies or agreements among a number of persons or

1024. Munn v. Illinois, 94 U.S. 113 (1876).
1025. Interstate Commerce Act of 1887, 49 U.S.C. § 1, 24 Stat. 379 repealed in part in 1978 with energy reorganization in Pub. L. No. 95-473. Oil pipeline regulation, though, continues and can be found in appendix to 49 App. U.S.C. § 1.
1026. Sherman Antitrust Act of 1890, 15 U.S.C. §§ 1–7.
1027. *See, e.g.*, Ernest Gellhorn et al., Antitrust Law and Economics in a Nutshell 23–25 (2004).
1028. 15 U.S.C. § 1–2.

organizations. Section 2 applies to unilateral conduct, such as the creation and maintenance of a monopoly. Both sections, however, are directed at anticompetitive behavior and make trade restraints and the creation of anticompetitive structural conditions, such as monopolies, not only illegal but also criminal. As in the case of Standard Oil, the Sherman Act also empowered courts to dismantle monopolies, once found.

720. Courts, then, received significant interpretive leeway in the implementation of the Sherman Act and, over the course of history, they have expanded and contracted the scope of its two core provisions. Nevertheless, the U.S. Supreme Court's first significant, and historically important, foray into the application of the Sherman Act was the Standard Oil antitrust case of 1911.[1029] After the Court found that Standard Oil constituted a monopoly, the oil trust was dissolved into 33 companies, even though the defendants claimed that such dissolution would cause economic disaster. The Court ordered both horizontal and vertical divestiture: Horizontal divesture was ordered of potentially competing oil refining and sales business units that had been combined into a holding company. Vertical divesture was ordered between pipeline operations and refining operations.

721. *Standard Oil* remains a significant case for the interpretation of the Sherman Act. It highlights several important aspects of that Act, including: (1) monopoly and other agreements in restraint of trade are prohibited when they restrict output and raise prices; (2) competitive markets, in the absence of barriers to entry, can counteract monopoly power; (3) monopoly power can be inferred from the presence of dominant market share as well as barriers to entry; (4) to prove monopolization, anticompetitive conduct that goes beyond usual competition must be shown; and (5) the prohibition of contracts, combinations, and conspiracies that restrain trade is intended to protect competitive markets from the evils of monopoly while applying a rule of reason.[1030]

722. The Court's opinion attracted significant criticism. Chief Justice White ruled that not all restraints of trade were illegal *per se* but only those that were unreasonably anticompetitive, thus establishing the "rule of reason." By way of example, the Chief Justice wrote that determining whether or not a restraint is unreasonable can be based on whether it tends to "bring about the evils such as enhancement of prices, which were considered to be against public policy."[1031] In other words, the Chief Justice was referencing prior decisional law as well as prior common law to define and interpret the Sherman Act. Critics were concerned that the rule of reason would gut the Act—that the standard was too loose and that many restraints on trade would not be able to satisfy that test. The rule of reason, however, has withstood the test of time and remains a key method for analyzing antitrust claims today.

1029. Standard Oil Co. v. United States, 221 U.S. 1 (1911).

1030. *See* Robert P. Edwards, Jr., *Introduction to Federal Antitrust Law*, *in* 5 ENERGY LAW AND TRANSACTIONS, *supra* note 81, at § 100.01[1].

1031. *Standard Oil Co.*, 221 U.S. at 58.

723. Following the Court's decision in *Standard Oil*, Congress passed the Clayton Act in 1914.[1032] The Clayton Act was a response to criticisms of *Standard Oil*, including the tendency of courts to apply the rule of reason rather broadly, limiting the circumstances in which an illegal monopoly or restraint of trade could be shown. The Clayton Act expanded the activities that were deemed illegal under U.S. antitrust law, though it did not make these prohibitions criminal. Such covered acts included: (1) price discrimination, *i.e.*, selling the same product at different prices to similarly situated purchasers; (2) tying arrangements, *i.e.*, conditioning the purchase of one product or service with the purchase of another product or service; (3) exclusive dealing contracts, *e.g.*, requiring that one buyer stop dealing with other buyers (usually the seller's competitors); (4) the acquisition of competing companies through corporate mergers; and (5) interlocking directorates, *i.e.*, creating corporate boards consisting of members from one company sitting on the boards of their competitors. Thus, the Clayton Act serves as a supplement to the Sherman Act. First, the Clayton Act includes express authority for injunctive relief from threatened loss or damage. Additionally, the law authorizes private actions that may recover treble damages.

724. The Clayton Act subsequently was supplemented and modified by several other laws. The Robinson-Patman Act in 1936 addressed price discrimination.[1033] It was spurred to adoption by an increasing spread of large chain stores. The concern was that competition would be reduced as chain stores supplanted independent merchants. The Clayton Act was also supplemented by the Celler-Kefauver Antimerger Act of 1950,[1034] which extended the Clayton Act's scope to asset transactions as well as to stock acquisitions. Finally, the Hart-Scott-Rodino Antitrust Improvements Act of 1976 provided for premerger review of major mergers and corporate acquisitions.[1035] Together, these statutes both expanded as well as clarified general antitrust provisions and practices.

725. In 1914, Congress also passed the Federal Trade Commission Act.[1036] That law seeks to foreclose unfair methods of competition and unfair or deceptive trade practices. The Act charges the Federal Trade Commission with its enforcement. The FTC is an "independent" agency. Rather than functioning with a single head who serves at the pleasure of the president, it is comprised of five members appointed by the president and confirmed by the Senate. No more than three members may be of the same political party. While the Department of Justice, as well as the Federal Trade Commission, share enforcement of the Clayton Act, only the FTC enforces the Federal Trade Commission Act. The idea behind the statute was to create an administrative agency that furthers competition policy in the United States. The FTC does so through a professional staff drawn from a variety of disciplines, including accounting, economics, and law.

1032. Clayton Act of 1914, 15 U.S.C. §§ 12–27.
1033. Robinson-Patman Act of 1936, 15 U.S.C. § 13.
1034. Celler-Kefauver Antimerger Act of 1950, 64 Stat. 1225.
1035. Hart-Scott-Rodino Antitrust Improvements Act of 1976, Pub. L. No. 94-435.
1036. Federal Trade Commission Act of 1914, 15 U.S.C. §§ 41–58 (amended in 1938 1975).

726. It should be noted that, as a general matter, these antitrust laws limit the role of the federal judiciary to identifying conduct that the statutes deem to be illegal. In other words, the judiciary does not have a broader authority to regulate economic and commercial relationships. While competition policy may be desirable, that policy must emanate from Congress or from other authorized administrative agencies. Courts, then, must protect competition from statutorily identified actions, as opposed to organically creating common law doctrines over time.

Chapter 3. Antitrust Laws and Energy Regulation

727. A first principle in U.S. antitrust law is that market power must be controlled. So too for antitrust law's intersection with the energy industry: The core aim is to prevent the exercise of monopoly behavior or other forms of anticompetitive activities.

728. At the same time, one of the most central functions of energy regulation in the United States is to ensure that energy firms do not exercise market power and provide products at unfair prices or on discriminatory terms. This, of course, is itself a form of competition regulation. Importantly, however, the mechanism typically used to stamp out anticompetitive behavior in much of U.S. energy law, and in particular for electricity and natural gas, is public utility regulation: limiting the competition that firms encounter by giving them an exclusive service territory, and then heavily regulating their behavior within that area. In this way, the law, rather than protecting competition, attempts to mimic the outcomes that competition would achieve. Moreover, because energy law and competition law occupy some of the same space, conflicts are inevitable, and so the way accommodations are made between the two fields is critical.

729. A core point about the intersection of energy regulation and competition law is that even though public utilities are regulated entities, there is no broad immunity from antitrust principles. In *California v. FPC*,[1037] for example, the U.S. Supreme Court ruled that even though the FPC had authority to approve an asset acquisition by natural gas companies, such approval did not bar an antitrust challenge. There, the federal government had begun an antitrust action, but the acquiring company subsequently applied to the FPC under Section 7 of the Natural Gas Act[1038] for authority to merge the two companies' assets. The FPC authorized the merger while the antitrust action was pending. The Supreme Court found this immaterial for antitrust purposes. It held that the FPC should not have proceeded to a decision on the merits of the merger while the antitrust litigation was pending, and that regulatory approvals do not immunize parties from antitrust enforcement. Rather, for an antitrust exemption to apply, the exemption must be explicitly stated.[1039]

730. Thus, as a general matter, regulated industries are not exempt from antitrust laws. The fact that an agency has jurisdiction over the behavior of a regulated industry does not, without an express exemption, protect the regulated entity's activities from court challenges under antitrust laws. *Otter Tail Power Co. v. United*

1037. California v. FPC, 369 U.S. 482 (1962).

1038. 15 U.S.C. § 717(f)(c).

1039. By contrast, the Natural Gas Policy Act of 1978 provides an example of a statute where an exemption is provided. That statute provides a defense to any state or federal antitrust action taken pursuant to any order of the President under his emergency allocation authority under that act. *See* 15 U.S.C. § 3364(e). Note, however, that such provisions tend to be the exception, not the rule.

States is the leading case.[1040] There, Otter Tail sold retail electric power in three states. Otter Tail also sold retail power in municipalities in its service territory, and those municipalities lacked the financial ability to construct distribution lines, making Otter Tail a natural monopoly. A competing municipal system sought to use Otter Tail's distribution network to purchase wholesale power, but Otter Tail refused. The U.S. Supreme Court ruled that Otter Tail's refusal constituted a form of non-price predation, which had the effect of driving competition out of the market and deterring new entrants. Thus, even though the FPC could order interconnections between producers and consumers in the public interest, that authority did not immunize a regulated utility from antitrust regulation. This, then, established the default rule: Merely because an entity is regulated does not exempt them from antitrust restrictions.[1041] Further, regulatory agencies, without specific authorization, do not have jurisdiction to determine violations of antitrust laws.[1042]

731. However, the fact that regulated entities generally are not immune from antitrust enforcement does not mean that they are always subject to it. Rather, courts have developed the doctrine of implied immunity from antitrust in certain circumstances. The idea is analogous to the doctrine of field preemption. If a court finds that Congress has devised a comprehensive and pervasive regulatory scheme, then that scheme may evince a Congressional intent to preempt antitrust laws. Such implied immunities, however, are not broadly available. Instead, they are disfavored and not easily invoked.[1043] As the U.S. Supreme Court has held, "Congress, exercising the full extent of its constitutional power, sought to establish a regime of competition [as] the fundamental principle governing commerce in this country."[1044]

732. City of Lafayette v. Louisiana Power & Light Co. is indicative.[1045] In that case, cities that owned and operated electric utility systems brought a lawsuit alleging antitrust violations against an investor-owned electric utility (IOU) with which they competed. The IOU counterclaimed that the cities had violated antitrust laws. The cities moved to dismiss the counterclaim on basis that the "state action" doctrine[1046] immunized them from federal antitrust laws. Their theory was that because, as governmental entities they were operating pursuant to state law, they were exempt from antitrust limits.[1047] The Court ruled that the cities did not enjoy an implied immunity from antitrust laws. Instead, they were compelled to comply with

1040. Otter Tail Power Co. v. United States, 410 U.S. 366 (1973).

1041. Similarly, while FERC has jurisdiction over natural gas pipelines, that jurisdiction is no defense against the claim that a pipeline may have exploited its monopoly power to gain a competitive advantage in the market. *See* Illinois v. Panhandle E. Pipeline Co., 603 F. Supp. 786 (C. D. Ill. 1985).

1042. Northern Natural Gas Co. v. FPC, 399 F.2d 953 (D.C. Cir. 1968).

1043. California v. FPC, 369 U.S. 482 (1962).

1044. City of Lafayette v. La. Power & Light Co. 435 U.S. 389, 398 (1978) (footnotes omitted).

1045. *Ibid.*

1046. Parker v. Brown, 317 U.S. 341 (1943).

1047. Indeed, the state action doctrine is not broadly applied. The Supreme Court had previously ruled that Sherman Act and Clayton Act jurisdiction extends to "person" or "persons," and that those definitions apply to cities and states. "Congress, exercising the full extent of its constitutional

the substantive standards established by federal law; their suggestion that they were operating in the public interest and not accumulating profits was insufficient to shield them from antitrust scrutiny. The Court reasoned that even though the municipal utilities were serving their citizens, that service did not mean that their behavior could not be anticompetitive. Indeed, one city had agreed to provide gas and water service only on the condition that those customers purchase electricity from that municipality, and not from the local IOU. In other words, the cities had engaged in a tying arrangement, which is one of the practices that antitrust law precludes. The Court reasoned that such a tie-in may in fact be an illegal restraint of trade.

733. The Supreme Court also denied an invocation of implied immunity in *Cantor v. Detroit Edison Co.*[1048] There, Detroit Edison had distributed light bulbs to its customers in exchange for used bulbs. The utility was sued for having engaged in an illegal anticompetitive practice to the economic detriment of other sellers. The utility claimed that since its tariff allowed them to engage in the light bulb exchange program, with the approval of the Michigan Public Service Commission, it had antitrust immunity. The Court rejected that assertion, noting that even if the exchange program was terminated, the exercise of antitrust authority did not impair the state's ability to regulate the distribution and sale of electricity. Thus, the Court found implied immunity unavailable to Detroit Edison as an antitrust defense.

734. Another way that arguments for implied antitrust immunity can arise is under the Constitution. The First Amendment of the U.S. Constitution protects the right of free speech, including the rights of citizens to petition the government. Thus, parties sometimes claim immunity from antitrust protection on the basis of First Amendment rights, and the Supreme Court has recognized this defense. Under the *Noerr-Pennington* doctrine,[1049] legitimate attempts to influence a branch of government are allowed even if they result in anticompetitive effects. However, determining legitimacy is not easy. Consequently, "federal courts have attempted to draw the line between political activities with commercial impacts and commercial impacts that incidentally trigger political responses."[1050] In the former case, political activities are protected; in the latter, they are susceptible to antitrust challenges.

735. Specifically, immunity for a particular transaction must satisfy three requirements: (1) government, not private, action must be involved; (2) the private parties must intend to achieve an anticompetitive result through the governmental process; and (3) the anticompetitive result must stem from the government's actions,

---

power, sought to establish a regime of competition the fundamental principle governing commerce in this country." City of Lafayette, 435 U.S. at 398.

1048. Cantor v. Detroit Edison Co. 428 U.S. 579 (1976).

1049. United Mine Workers v. Pennington, 381 U.S. 657 (1965); E.R.R. Presidents Conference v. Motor Freight Inc., 365 U.S. 127 (1961).

1050. J.A. Bouknight, Jr., *Relationship Between Federal Antitrust Laws and Federal and State Regulation of Energy Industries*, *in* 5 ENERGY LAW AND TRANSACTIONS, *supra* note 81, at § 101.02[1][a].

not from private activities.[1051] If private actors caused the anticompetitive harm, then there is no immunity. Further, petitioning must be made before a formal agency, branch, or department of the government. In other words, general political activity is not protected. Instead, filing a complaint, seeking a license, or pursuing such other formal legal avenues must be utilized to protect an activity from antitrust action in favor of exercising one's legal and political rights.

736. In *Modesto Irrigation District v. Pacific Gas & Electric*,[1052] for example, a power marketer complained that PG&E's refusal to wheel power for it to a substation was an illegal anticompetitive practice. PG&E responded to the demand for wheeling, in part, by filing a petition with FERC to determine whether it was obligated to provide interconnection and transmission service. The power marketers argued that the FERC petition was an anticompetitive practice that should be enjoined. The court, however, held that even though the refusal to wheel power came prior to filing the FERC petition, PG&E enjoyed antitrust immunity on that issue because it was legitimately petitioning a formal branch of government for a determination concerning their obligation to wheel or not.

737. Notwithstanding the ruling in *Modesto Irrigation District*, the *Noerr-Pennington* exception cannot be read broadly. Otherwise, one competitor need merely file a petition or complain against another competitor in court in order to obtain immunity from their own anticompetitive behavior. In other words, the action of petitioning the government must be done in good faith and cannot constitute either "sham" or "vexatious litigation."

738. The case of *In re Burlington Northern, Inc.* illustrates this doctrine.[1053] There, a railroad encouraged and supported third parties bringing lawsuits against a coal-slurry pipeline. The railroad did this because it lacked standing to sue but wanted to prevent competition from the pipeline. The court found the *Noerr-Pennington* exception unavailable to the railroad. If the railroad wanted to petition a court, it could do so. But it could not advance its "anti-competitive goals through funding, encouraging, and assisting the lawsuits of others."[1054]

739. While implied immunity cannot be easily invoked, the Supreme Court also has said that because First Amendment rights are so fundamental in the United States, the sham exception to the *Noerr-Pennington* doctrine cannot be lightly applied.[1055]

1051. *Ibid.*
1052. Modesto Irrigation Dist. v. Pacific Gas & Elec., 61 F. Supp. 2d 1058 (N.D. Cal. 1999).
1053. In re Burlington Northern, Inc., 822 F.2d 518 (5th Cir. 1987).
1054. *In re* Burlington, 822 F.2d at 532. It was also relevant that the railroad "continued to litigate each case long after their arguments had been repeatedly rejected by the courts." *Ibid. But see* South Dakota v. Kan. City S. Indus., Inc., 880 F.2d 40 (8th Cir. 1989) (finding, under similar facts, the same series of lawsuits to be protected activity), *cert. denied*, 493 U.S. 1023 (1990).
1055. City Of Columbus v. Omni Outdoor Adver., 499 U.S. 365 (1991).

740. A good example is *Greenwood Util. Comm'n v. Miss. Power Co.*[1056] In that case, a municipally owned electric utility, Greenwood, brought suit against the Southeastern Power Administration (SEPA), a division of the Department of Energy, and an electric utility, Mississippi Power. SEPA and Mississippi Power had entered into a contract under which the utility would transmit surplus electric power from a federal dam project on behalf of SEPA. The municipality argued that the SEPA contract was a product of monopoly power exercised by Mississippi Power and its affiliates over the transmission of electricity. The court ruled that SEPA's decision on transmitting power, and the contract between SEPA and Mississippi Power incorporating that decision, did not violate antitrust laws. The court reasoned that the contract was lawfully created through Mississippi Power's efforts to persuade SEPA to adopt the proposal for marketing federal hydropower. SEPA did not have its own transmission facilities and was ultimately persuaded to use those of Mississippi Power and its affiliates even though anticompetitive consequences may have followed. Basically, the utility lobbied Congress and SEPA to favor its marketing and transmission plan rather than have Congress allocate funds for SEPA's transmission facilities, and that lobbying was protected—that is, not a sham under *Noerr-Pennington.*

741. One key to understanding the scope of *Noerr-Pennington* is that the doctrine protects governmental activity, not private activity. Thus, in *Cantor*, the local utility did not have immunity from its own anticompetitive practices regardless of the fact the government had approved its tariff. Similarly, in *City of Mishawaka v. Am. Elec. Power Co.*,[1057] the court determined that the *Noerr-Pennington* doctrine did not protect the utility that filed for numerous rate increases with anticompetitive consequences. By contrast, in *Greenwood*, the government petitioning in question resulted in a contract precisely for the services that the municipality was complaining about. This stood in contrast to the activities in *City of Mishawaka* and *Cantor*, where the utilities were petitioning for a general tariff and not for the behavior that resulted in the anticompetitive consequences.

742. In other words, there must be a causal nexus between what is being petitioned for and the government action in order for *Noerr-Pennington* to apply. Purely private actions, such as the settlements of disputes, do not immunize the parties to those settlements from antitrust claims. In *New Mexico Natural Gas Antitrust Litigation*,[1058] for example, the court held that a settlement between producers and suppliers of natural gas was not protected from subsequent antitrust attack, precisely because the parties in question were private entities. Government action was not present. When, however, a government agency, such as FERC, is involved in crafting the settlement in order to further its regulatory agenda, then *Noerr-Pennington* protection will be extended.[1059]

1056. Greenwood Util. Comm'n v. Miss. Power Co., 751 F.2d 1484 (5th Cir. 1985).
1057. City of Mishawaka v. Am. Elec. Power Co., 616 F.2d 976 (7th Cir. 1980), *cert. denied*, 449 U.S. 1096 (1981).
1058. New Mexico Natural Gas Antitrust Litigation, 1982-1 Trade Cas. ¶ 64,685 (D.N.M. 1882).
1059. Tenn. Gas Pipeline Co., 20 F.E.R.C. ¶ 61,069 (1982).

743. In addition to *Noerr-Pennington*, the so-called "state action" doctrine is important as an exception to antitrust prohibitions. Energy firms that are subject to state regulatory requirements may be afforded some protection from federal antitrust laws this doctrine, as articulated in *Parker v. Brown*.[1060] The idea is to enable states to exercise their legitimate governmental authority, which plainly includes regulatory authority over different aspects of U.S. energy systems. The key question here, then, tends to be whether or not a state is acting in its governmental capacity or, instead, in its proprietary capacity. In *New Mexico v. American Petrofina Inc.*,[1061] for instance, oil companies and other asphalt suppliers sued the state of New Mexico on antitrust grounds, alleging that the state and some of its political subdivisions,[1062] as consumers, conspired to fix prices to eliminate competition among themselves. The court, however, ruled that those governmental entities were immune from suit on the basis of the state action doctrine. Relying on the Supreme Court's decision in *Parker*, the *Petrofina* court wrote: "'The Sherman Act makes no mention of the state as such, and gives no hint that it was intended to restrain state action or official action directed by a state.' . . . [Given that, and] . . . the absence of any mention of state action in either the Act or the legislative history, . . . [we] conclude that sections 1 and 2 of the Sherman Act do not apply to the activities of a state."[1063] The more difficult question for a court to decide is "whether the anticompetitive result actually is a goal of the state entitled to the state's immunity rather than a private group masquerading under the banner of 'state action.'"[1064]

744. Energy firms, such as local utilities, can also avail themselves of state action immunity. In the case of *Lease Lights Inc. v. Public Service Co. of Oklahoma*,[1065] for example, outdoor lighting companies sued the utility, Public Service, arguing that its rate structure attempted to monopolize the outdoor lighting market. Public Service also sold and leased outdoor lights in the area, and the Oklahoma Corporation Commission oversaw the rates charged for leasing outdoor lights. In 1975, the Commission ordered Public Service to reduce its lighting rates and restructure its rate schedule. Public Service reduced its rates while increasing rates charged to contractors for electricity used in offering competing outdoor lighting services. Competing electrical contractors then alleged that Public Service was engaging in monopoly practices. In court, however, Public Service successfully argued that the rates were approved by and closely supervised by the Commission, and thus, that those activities were protected by the state action doctrine.

745. Other utilities have successfully invoked the state action doctrine. In *Rural Electric Co. v. Cheyenne Light, Fuel & Power Co.*,[1066] the state action doctrine

1060. Parker v. Brown, 317 U.S. 341 (1943).
1061. New Mexico v. American Petrofina Inc., 501 F.2d 363 (9th Cir. 1974).
1062. City of Lafayette v. La. Power & Light Co., 435 U.S. 389 (1978).
1063. *New Mexico*, 501 F.2d at 366–67 (9th Cir. 1974).
1064. *Ibid.* at 369.
1065. Note, however, that the state action doctrine does not automatically extend to municipalities and other state subdivisions unless there is a clearly articulated state policy being pursued by those subdivisions. Lease Lights Inc. v. Pub. Serv. Co. of Okla., 849 F.2d 1330 (1991).
1066. Rural Electric Co. v. Cheyenne Light, Fuel & Power Co., 762 F.2d 847 (10th Cir. 1985).

protected a utility against the charge of conspiring to monopolize a market with the local municipality. In that case, the unsuccessful bidder for a franchise to supply electricity to an industrial plant sued the successful bidder and the city for conspiracy under the Sherman Act. The court, however, dismissed the claims, holding that the award of the nonexclusive franchise to the winning bidder fell within the state action doctrine.[1067]

746. The holding in *Lease Lights* may appear inconsistent with the earlier case of *Cantor v. Detroit Edison Co.*[1068] In the former, the state action doctrine protected the utility from antitrust liability; in the latter, it did not. These two cases thus evince the evolution of the state action doctrine. As the doctrine has evolved, a private actor, such as a utility, "may claim the state action exemption only if (1) anticompetitive activities are conducted pursuant to a 'clearly expressed state policy' and (2) the policy is 'actively supervised' by the state itself."[1069]

747. The first requirement of a clearly expressed state policy tends to be easily met by public utilities, at least in cases where a utility is acting to comply with a state's rate-setting authority. The second element is more difficulty to satisfy. It is met when the private actor's behavior is closely supervised and limits the discretion of that actor, such as when the PUC supervises a utility's promotional practices.[1070] In other words, if a private actor is not compelled by state policy and can exercise discretion, then state action immunity does not apply.[1071] The distinguishing factor between *Cantor* and *Lease Lights*, then, is the degree of state supervision over electric lighting (in *Lease Lights*) and the state's basic granting of a tariff (in *Cantor*).

748. This comports with the overall energy regulatory landscape. One argument for extending the state action exemption to public utilities is that they are regulated as natural monopolies. That regulation, in turn, is premised on the idea that it is more efficient for one provider, rather than several, to deliver energy service, and thus governments bestow on utilities exclusive service territories, free from competition.[1072] In exchange for such a territory, the utility assumes an obligation to serve and consequently is not free to deny service. Thus, when acting under the thumb of government regulation to fulfill its side of this regulatory compact, the state action exemption may protect the utility from antitrust inquiry. But, when

---

1067. Trigen Okla. City Energy v. Okla. Gas & Elec. Co., 244 F.3d 1220 (10th Cir. 2001).

1068. Cantor v. Detroit Edison Co. 428 U.S. 579 (1976).

1069. Bouknight, *supra* note 1049, at § 101.02[1][b].

1070. Gas Light Co. v. Georgia Power Co., 440 F.2d 1135 (5th Cir. 1971), *cert. denied*, 404 U.S. 1062 (1972); Yeager's Fuel Inc. v. Pa. Power & Light Co., 804 F. Supp. 700 (E.D. Pa. 1992).

1071. *See, e.g.*, Fed. Trade Comm'n v. Ticor Title Ins. Co., 504 US 621 (1992).

1072. *See* North Star Steel Co. v. MidAmerican Energy Holdings, 184 F.3d 732 (8th Cir. 1999); Columbia River Peoples Util. Dist. v. Portland Gen. Elec., 40 F. Supp. 2d 1152 (D. OR. 1999); Praxair, Inc. v. Fla. Power & Light Co., 64 F.3d 609 (11th Cir. 1995); Mun. Utils. Board of Albertville v. Ala. Power Co., 21 F.3d 384 (11th Cir. 1994); North Star Steel Tex., Inc. v. Entergy Gulf States, Inc., 33 F. Supp. 557 (S.D. Tex. 1998).

the utility acts on its own and then simply attempts to invoke the background presence of regulation as a way to bootstrap the state action exemption, it remains susceptible to antitrust investigation.[1073]

1073. Columbia Steel Casting Co., Inc. v. Portland General Electric Co., 111 F.3d 1427 (9th Cir. 996).

Chapter 4. Antitrust, Energy Markets, and Market-Based Rates

749. A central issue in antitrust litigation involves the definition of the applicable market. The market can either be a product market, such as particular type of light bulb or service, or a geographic market, meaning the area where substitutable products compete against each other. Once the market is defined, the inquiry focuses on whether the conduct has anticompetitive effects within that market. Historically, electricity and natural gas sales are deemed to be in the public interest and, as noted, are regulated as natural monopolies. Both of these energy services are also relatively price inelastic, which, in turn, means that energy providers can exercise market power if left unchecked.

750. Historically, regulators solved the risk of anticompetitive effects in natural gas and electricity markets through direct control: price regulations that attempted to approximate the price level a competitive market would reach, plus other legal mandates, such as limits on discrimination and imposition of duties to serve. This model of regulation served the public well, particularly in times of economic growth and industrial expansion, but miscues occurred once expansion started slowing and regulators began a general move away from top-down regulation and toward competition in the United States utility industry. Today, then, both the electricity and natural gas industries are experiencing different degrees of restructuring,[1074] predominantly to make their markets more competitive.

751. To this end, both industries have been experiencing "unbundling," in which traditionally vertically integrated functions are split apart and are provided as separate services. In the electric industry, IOUs have been encouraged (and sometimes compelled) to divest generation assets, and FERC has mandated that they reform how they provide transmission service. In the natural gas industry, pipelines have been ordered to unbundle their merchant and transportation functions. In both instances, the policy is to open access to the bottleneck portions of those industries in an effort to promote competition in other segments. From a competition policy perspective, the restructuring of these industries has had two key effects. First, as these industries and markets become more competitive, antitrust issues emerge. Second, in determining how to regulate these restructured markets, policymakers have drawn heavily on economic theory and, in turn, antitrust principles.

752. Even though utilities have exclusive franchise territories, competition still exists among fuels. Large consumers, such as industrial plants, have greater fuel-switching capacity than smaller customers such as homeowners. Now, with industry restructuring, competition is available from suppliers as well as from different fuels. As federal regulators attempt to further open access to both the electricity and natural gas transportation systems, they have engaged in a significant degree of legislative and regulatory reforms. Among those reforms is the idea that market-based rates may send better price signals, lower costs, and increase competition.

1074. *See* Part I, *Electricity* and Part II, *Natural Gas* above.

753. The idea behind market-based rates is simple. Instead of formal ratemaking, where government regulators attempt to approximate the price that would be paid in a hypothetical competitive market, prices will be set based on the contracts that utilities voluntarily enter. This, the idea goes, should lead to more efficient outcomes because parties can use their competitive advantages to negotiate mutually favorable terms, without the administrative cost of government oversight.

754. A threshold question is which producers should be allowed to sell energy at market-based rates. For this determination, regulators have relied heavily on antitrust theory. As a general principle, FERC allows parties to sell energy at market-based rates where the party can show that it lacks the ability to unduly influence market prices—that is, that it lacks market power. In the wholesale electricity market, for example, an electricity generator that wishes to price its electricity using market-based rates must demonstrate a lack of market power, at which time FERC grants it "market-based rate authority": the authorization to sell wholesale electricity without direct cost-of-service regulation.[1075] Similarly, natural gas pipelines can offer gas at market-based rates when they show that they lack market power.[1076]

755. Thus, in reviewing a request to sell at market-based rates, FERC seeks to assess whether the applicant can unduly influence the price of energy in the identified market. If so, then market power is present and market-based rate authority will be denied. FERC has long had the authority to monitor acquisitions and mergers in the electric industry,[1077] and with restructuring, the agency has drawn on its tests analyzing those activities to determine if there will be market power in a given product market.[1078] Specifically, in deciding whether an applicant should be allowed to use market-based rates, FERC tests for three types of market power: horizontal, vertical, and affiliate abuse. FERC's analysis can be understood using its tests for electricity.

756. First, FERC tests for *horizontal market power* by determining whether a utility can affect price by withholding electricity from the market. To determine that, FERC applies both a "market share analysis" and a "pivotal supplier" test. If

---

1075. *See, e.g.*, DAVIES ET AL., *supra* note 29, 412–26; Robert P. Edwards, Jr., *Application of Federal Antitrust Laws to Energy Industries*, *in* 5 ENERGY LAW AND TRANSACTIONS, *supra* note 81, at § 102.01[2].

1076. Alternatives to Traditional Cost-of-Service Ratemaking for Natural Gas Pipelines and Regulation of Negotiated Transportation Services of Natural Gas Pipelines, 61 Fed. Reg. 4633, 4633 (Feb. 7, 1996); *see also* Elizabethtown Gas Co. v. FERC, 10 F.3d 866 (D.C. Cir. 1993).

1077. 16 U.S.C. § 824b; 18 C.F.R. §§ 33.1 et seq.

1078. FERC's market-based rate tests trace directly to antitrust law. Those tests rely on FERC's analyses for mergers. Its merger policy relies heavily on "merger guidelines" jointly developed by the Federal Trade Commission and the Department of Justice. And the DOJ/FTC merger guidelines are those agencies' implementation of antitrust policy. The common theme running throughout all these, then, is the promotion of competition using economic theory. *See generally Horizontal Merger Guidelines*, *supra* note 41.

a firm exercises market power across the relevant product market (here, wholesale electricity generation), then it exercises horizontal market power. Specifically, FERC calculates available resources, committed resources, and transmission constraints, to determine how much supply is available in a market during different periods of the year. If an applicant for market-based rates has a market share of greater than 20%, FERC deems it to have market power, because it can use that share to unduly influence prices. If, however, its market share is less than 20%, then the utility passes that initial screen for horizontal market power. Additionally, the utility must not be a "pivotal supplier," which means that its uncommitted generation capacity at peak is less than the total capacity available in the market on the peak day. The idea here is also simple: If the applicant's available capacity is less than the total of possibly competing capacity, then supplies are available to overwhelm any effect the applicant's withholding of power from the market may have. That is, both the market share test and the pivotal supplier test aim to measure whether a party controls so much of, or so important a part of, the market that it can increase prices above competitive levels.

757. Second, FERC tests for *vertical market power* by examining how much indirect control a firm can exercise over market prices. The term "vertical" is used in contrast to "horizontal." Whereas horizontal market power might be exercised *across* a single product market (*e.g.*, wholesale power sales), vertical market power refers to using *upstream* inputs into a market to unduly influence prices (*e.*g., fuel inputs for electricity generation). Does, for example, a utility have the ability to control an upstream market, such as natural gas pipelines, in a way that allows it to constrain downstream electricity sales? FERC has issued critical policies to limit vertical market power in both the electricity (Order No. 888) and natural gas (Order No. 636) industries, so the agency generally presumes that compliance with these rules removes any vertical market power concerns for market-based rate applicants. Nevertheless, such applicants still must disclose ownership of upstream resources that might create vertical market power concerns, and if a party expresses such concerns, FERC will assess them as part of its market power analysis.

758. Finally, FERC tests for *affiliate abuse or preference*. If wholesale electricity markets are to work competitively, then affiliates cannot be preferred over competitors. Given restructuring in the electric industry, vertically-integrated IOUs that have been divesting assets often also engaged in corporate restructuring, creating marketing arms that compete as brokers, buyers, and sellers of electric power. FERC did not want such corporate family members receiving preferential treatment from the existing IOU part of the corporate family, whether the IOU continued to run generation and transmission or became a "wires only" company. Thus, FERC's market-based rate test regarding affiliate abuse compels parties in such situations to separate out their functions and, when dealing with each other, to engage in arms-length bargaining while avoiding sweetheart deals. FERC has adopted extensive regulations on this score, and compliance with them is a prerequisite for obtaining market-based rate authority.[1079] It is another way that the agency aims to reduce parties' ability to exercise market power.[1080]

1079. Order No. 717, Affiliate Restrictions (Formerly the Codes of Conduct), F.E.R.C. Stats. & Regs. ¶ 31,280 (2008) (Standards of Conduct for Transmission Providers).

1080. In addition to violating FERC regulation, such tying arrangements also can violate antitrust law. *See, e.g.*, Illinois v. Panhandle E. Pipeline Co., 603 F. Supp. 786 (C. D. Ill. 1985).

Chapter 5. Energy Mergers and Acquisitions

759. Because they consolidate assets, mergers and acquisitions in energy industries naturally raise competition problems. Thus, they also raise competition law questions. Historically, energy mergers and acquisitions have been heavily regulated. This occurs in a number of ways. First, generally applicable antitrust laws apply, triggering review of mergers by the Federal Trade Commission or the Department of Justice.[1081] Second, electric utility mergers are reviewed by FERC under Section 203 of the Federal Power Act.[1082] Third, those mergers also receive parallel scrutiny at the state level, typically by state public service commissions. And, finally, laws specific to some portions of the energy industry, such as the Public Utility Holding Company Act, have played a role.[1083]

760. That there are overlapping levels of antitrust review for mergers is critical. As detailed below, electric utilities that wish to merge are subject to review both by FERC and DOJ/FTC, and in many cases by state agencies. Other energy firms, such as merchant generators, coal companies, and oil producers must obtain DOJ/FTC approval for their mergers but are not subject to FERC review. Consequently, it is necessary for the energy lawyer to understand the different approvals that are required, as well as the different methodologies employed to assess merger proposals.

§1. General Antitrust Review

761. Generally applicable federal antitrust laws include those already discussed, namely, the Sherman Act and the Clayton Act. In addition, the Federal Trade Commission Act prohibits unfair methods of competition and empowers the FTC to bring enforcement actions against practices that violate either the Federal Trade Commission Act or the Clayton Act. The FTC can act on its own or can act on private complaints. Competitor standing, however, is closely scrutinized, to avoid the misuse of the laws in a way that inhibits, rather than promotes, competition. Historically, the Department of Justice and the Federal Trade Commission conduct merger reviews under this general antitrust jurisdiction.

762. The standards for finding violations of the Sherman Act or the Clayton Act are detailed above. While both laws originally targeted railroads and other early industries, they remain applicable today, including to modern energy companies. Again, important principles include: (1) The existence of price and service regulation does not immunize energy firms from antitrust oversight. (2) Firms whose operations are generally complementary may nonetheless create antitrust problems, for example, by merging in a way that might create vertical integration in an

1081. *Horizontal Merger Guidelines, supra* note 41.
1082. 16 U.S.C. § 824b.
1083. Public Utility Holding Company Act of 2005, Pub. L. No. 109-58, 119 Stat. 972 (codified at 42 U.S.C. § 15801, 16451–463).

industry and thus generate anticompetitive effects. (3) All firms operate in several economic markets, so the merger of entities might create antitrust problems in one market (geographic or product) but not in others. (4) Producers' ability to shift their production from one market to another must be considered, as market power might be exercised that way.[1084]

763. The Federal Trade Commission Act differs from the Sherman Act and the Clayton Acts. The FTC Act makes it unlawful to engage in unfair methods of competition, or to engage in unfair or deceptive acts or practices affecting competition. The FTC is authorized to prevent persons from using such unfair methods of trade.[1085] If a violation is found, multiple types of relief are available, including: (1) injunctive or other equitable relief; (2) orders restraining or requiring divestitures; (3) state attorney general action on behalf of the public of their state; (4) private party injunctive actions; and (5) actions for damages by persons suffering economic losses. While private parties may have standing to enforce such laws, they must demonstrate actual or threatened injury of the type the antitrust laws were designed to prevent.

764. Mergers with energy firms can occur horizontally by combining competitors in one organization, such as two electricity utilities that both own generation resources. Mergers can also occur vertically, where a firm acquires critical upstream or downstream functions, such as if an electric utility that owns gas-fired generation acquires a natural gas company. Both types of mergers must undergo pre-merger review to assess their potential competitive effects.

765. The Department of Justice and the Federal Trade Commission conduct these reviews. In their reviews, both the Clayton Act and the Sherman Act are relevant. The key provision of the Clayton Act is § 7:

> No person engaged in commerce or in any activity affecting commerce shall acquire, directly or in directly, the whole or any part of the stock or . . . assets of another person engaged also in commerce or in any activity affecting commerce, where in any line of commerce in any section of the country, the effect of such acquisition may be substantially to lessen competition, or to tend to create a monopoly.[1086]

766. The key portion of the Sherman Act is § 1:

> Every contract, combination in the form of trust or otherwise, or conspiracy, in restraint of trade or commerce among the several states, or with foreign nations, is declared to be illegal. . . .

1084. Robert P. Edwards, Jr., *Mergers, Acquisitions and Joint Ventures*, *in* 5 ENERGY LAW AND TRANSACTIONS, *supra* note 81, at § 103.02[5].
1085. Federal Trade Commission Act, 15 U.S.C. § 45.
1086. 15 U.S.C. § 18.

The object of these provisions, then, is to preclude the acquisition of market power, so that firms may not increase prices above, or reduce output below, competitive levels. That is, both the Clayton Act and the Sherman Act seek to ensure that markets properly function and that economically efficient outcomes are achieved.[1087]

767. As noted, to ascertain whether an antitrust problem exists, both the relevant product market and the geographic market must be defined. For this purpose, agencies—and the parties the agencies oversee—utilize economists to assess relevant evidence. Economics play a key role, because how the market functions on the ground is relevant to a merger's likely effect, and thus, whether a statutory violation can be found. If, for example, substitute goods are available to consumers, then that may lessen the effect of the merger on competition. It also may speak to what the relevant product market is. Similarly, within a particular geographic area, whether or not consumers have options to switch to alternate providers may be indicative of the presence or absence of market power.

768. One of the most important merger review decisions is *United States v. General Dynamics*,[1088] a case, notably, that involved a merger in the energy industry. At the time of the case, the coal industry was experiencing a rapid decline in the number of producers. The merger at issue involved major coal producers: If it proceeded, it would have resulted in the combination of two top coal firms, which together would then control about half of all coal sales.

769. Even though the merger would create industry consolidation, the lower court ruled that the merger could proceed.[1089] The court rejected coal production as the product market because consumers, including electric utilities, had forms of energy other than coal that they could purchase. Additionally, the court did not limit the geographic area involved to the Midwest. Instead, it looked to multiple markets based upon transportation costs, and made a finding that the merging companies did not operate in the same geographic market. Therefore, the court held, they could not exercise market power and the merger could proceed.[1090]

770. The U.S. Supreme Court upheld the district court.[1091] The Court agreed that the merger did not represent an exercise of market power, in large part because the acquired firm's coal reserves were under contract to existing customers. Therefore, the Court reasoned, those coal supplies could not be withheld from the market in an effort to restrain prices. "The mere possibility that United Electric, in common with all other companies with the inclination and the corporate treasury to do so, could someday expand into an essentially new line of business does not depreciate the

1087. 15 U.S.C. § 1.

1088. United States v. Gen. Dynamics, 415 U.S. 486 (1974).

1089. United States v. Gen. Dynamics Corp., 341 F. Supp. 534 (N.D. Ill. 1972).

1090. *Ibid.* at 559; *see also* Edwards, *supra* note 1083, at § 103.01[1][f].

1091. United States v. General Dynamics Corp., 415 U.S. 486 (1974).

validity of the conclusion that United Electric at the time of the trial did not have the power to compete on a significant scale for the procurement of future long-term contracts . . . ."[1092]

771. According to one commentator, the "significance of *General Dynamics* may lie in its affirmation of the general principle that the prospective likelihood of the exercise of market power is the central inquiry required by Section 7 of the Clayton Act, and not firm size or even industry concentration in the abstract."[1093] In other words, market share and concentration, alone, are insufficient to show a Clayton Act violation. Instead, after *General Dynamics*, judicial evaluation of merger decisions require "an economically more rigorous definition of the relevant market assessment or competitive capability of the merging parties."[1094] Thus, courts will not review mergers mechanistically. Rather, careful economic analysis is necessary, and real world market power, or the lack thereof, controls; simple tests, such as per se rules, will receive scrutiny from courts, not automatic application.

772. In addition to scrutinizing mergers, antitrust laws also apply to the competitive or anticompetitive effects of joint ventures. "A joint venture is a formal combined undertaking by two firms that does not extend to a full merger of those firms. A joint venture may take the form of a partnership, a corporation, tenancy in common, or a contractual arrangement of pool resources for a common goal."[1095] Although not a full merger, parties to a joint venture that may be actual or potential competitors are tested under antitrust rules applicable to mergers, as already discussed.[1096] However, because joint ventures are less formal than mergers, judicial review tends to be less strict by using rule of reason analyses rather than broad application of *per se* rules,[1097] although per se rules have been applied to joint marketing and purchasing arrangements.[1098]

773. Antitrust evaluation of mergers can be complicated as well as expensive. Firms that wish to merge must satisfy the FTC/DOJ merger standards, a complex set of guidelines that delineate how and when mergers may proceed and how to measure market power. In order to make the process more efficient, reviews are conducted before mergers are complete, both so that firms can see whether their mergers may create a problem (and thus whether they may want to abandon the venture) and also to avoid the intertwining of the entities when an anticompetitive problem is clear (the process of entities combining before mergers are approved is known as "gun-jumping).[1099] A firm with substantial capitalization that desires to merge, acquire another firm, or form a joint venture must notify proper enforcement

1092. *Ibid.* at 509–10.
1093. Edwards, *supra* note 1083, at § 103.01[1][f].
1094. Gellhorn et al., *supra* note 1026 at 47.
1095. Edwards, *supra* note 1083, at § 103.04[1].
1096. United States v. Penn-Olin Chem. Corp., 378 U.S. 158 (1964).
1097. Texaco Inc. v. Dagher, 547 U.S. 1 (2006).
1098. *See, e.g.*, Edwards, *supra* note 1083, at § 103.04[2].
1099. Hart-Scott-Rodino Antitrust Improvements Act of 1976, Pub. L. No. 94-435; 15 U.S.C. § 18a.

agencies (including the FTC and the DOJ) prior to the consummation of the arrangement. Failure to satisfy the pre-merger requirements can result in a rescission of a completed transaction, as well as penalties.

774. The pre-merger review process with DOJ and FTC is time-consuming and information-intensive. The agencies commonly demand the production of voluminous amounts of material from the merging entities and depose numerous high-level corporate officials as well as workaday employees, to determine what the merger's effect will be. Once the agencies determine whether a proposed merger presents anticompetitive problems, negotiation ensues. The government's objective is to have the deal reshaped—such as by compelling the entities to sell off assets the combination of which might create market power—in a way that eliminates antitrust problems. The merging entities' objective is to not restructure the deal so substantially that the economies of scale and combination they seek to achieve are eroded. If the government asks for too much in terms of concessions, at least in the merging entities' view, antitrust review may kill a deal.

§2. FERC Review of Electric Utility Mergers

775. Mergers involving electric utilities receive additional competition scrutiny under U.S. law from FERC. This is true even though such mergers will receive DOJ/FTC review too, as regulated entities, including utilities, do not enjoy immunity from antitrust laws.[1100] Specifically, pursuant to Section 203 of the Federal Power Act, FERC must review the disposition of certain jurisdictional facilities. Specifically, § 203 requires FERC approval for: (1) the disposition or consolidation of jurisdictional facilities with a value in excess of USD 10 million; (2) the acquisition of generating facilities valued at over USD 10 million that are used for jurisdictional wholesale power sales; and (3) the acquisition by any public utility or electric utility holding company of securities valued in excess of USD 10 million in another electric utility or holding company.[1101]

776. The statutory standard under § 203 is broad. It specifies that FERC may approve the disposition of jurisdictional facilities only if it is in the "public interest."[1102] Section 203 further provides that FERC must determine whether a proposed transaction will result in unfair cross-subsidization of the non-utility associated company, or will result in the pledge or encumbrance of utility assets for the benefit of an associated company.[1103] Importantly, as noted, FERC's review under § 203 is in addition to antitrust review by the FTC and/or DOJ. FERC does not have jurisdiction to enforce the antitrust laws.[1104]

1100. Otter Tail Power Co. v. United States, 410 U.S. 366 (1973).
1101. 16 U.S.C. § 824b.
1102. *Ibid.*
1103. 16 U.S.C. § 824b(a); *see also* Edwards, *supra* note 1083, at § 103.01[7].
1104. California v. FPC, 369 U.S. 482 (1962).

777. FERC's policy on electricity mergers, adopted in 1996 and referred to as the Merger Policy Statement, draws heavily on the DOJ/FTC merger guidelines.[1105] As delineated in its Merger Policy Statement, FERC considers three key factors when analyzing proposed mergers: the effect on competition, the effect on rates, and the effect on regulation. The Energy Policy Act of 2005 added the further requirement that FERC must determine whether the proposed transaction will result in prohibited cross-subsidization. In 2012, FERC reaffirmed its merger policy.[1106] Then, in January 2015, FERC proposed revisions to the Merger Policy Statement concerning a utility's commitments to protect consumers from adverse rate effects resulting from mergers.[1107]

778. On the first question, a merger's effect on competition, FERC utilizes a five-step procedure for analyzing mergers. FERC assesses: (1) whether the merger will "significantly increase concentration and result in a concentrated market;" (2) whether the merger raises concerns about potential adverse competitive effects; (3) whether entry by new competitors would be timely, likely, and sufficient either to deter or counteract competition; (4) whether any efficiency gains can be achieved by other means; and (5) whether either party would be likely to fail, causing its assets to exit the market. In short, FERC attempts to ensure that proposed mergers will not harm competition.[1108]

779. On the second question, the merger's potential impact on rates, FERC requires that, if warranted, applicants propose rate protection for customers. Such protections can be provided by agreements, for example, between merger applicants and their wholesale customers. FERC has stated that there is no single mechanism for such rate protection and that negotiated settlements can serve as a mechanism for arriving at suitable rates without a formal hearing. If the parties fail to negotiate, then FERC can impose appropriate protections. In light of ongoing negotiations, FERC may conditionally approve a merger and, if rate problems remain, later revisit whether ratepayers are sufficiently insulated from harm.[1109]

780. On the third question, the merger's effect on regulation, FERC initially was most concerned about mergers transferring regulatory authority to the Securities and Exchange Commission, which, under the Public Utility Holding Company Act

1105. Order No. 592, Inquiry Concerning the Commission's Merger Policy Under the Federal Power Act: Policy Statement, F.E.R.C. Stats. & Regs. ¶ 31,044 (1996) [hereinafter, *FERC Merger Policy Statement*], *reconsideration denied*, Order No. 592-A, 79 F.E.R.C. ¶ 61,321 (1997); *see also* FPA Section 203 Supplemental Policy Statement, F.E.R.C. Stats. & Regs. ¶ 31,253 (2007) (Supplemental Policy Statement); *see* Horizontal Merger Guidelines, 57 Fed. Reg. 41,552 (Apr. 2, 1992). The Guidelines were revised in 2010 to explain their application in more detail and to make them more transparent rather than alter their methodology. *See Horizontal Merger Guidelines*, *supra* note 41.

1106. Order Reaffirming Commission Policy and Terminating Proceedings, 138 F.E.R.C. ¶ 61,109 (2012).

1107. Policy Statement on Hold Harmless Commitments, 150 F.E.R.C. ¶ 61,031 (2015).

1108. *FERC Merger Policy Statement*, *supra* note 1104.

1109. *See Ibid.*

(PUHCA, now largely repealed by the Energy Policy Act of 2005), oversaw holding companies. For mergers that would create such a shift, FERC mandates that applicants agree to abide by FERC policies, or it will hold a hearing on the merger.[1110] FERC's Merger Policy Statement also notes that where states have regulatory authority over mergers, it will rely on the state agencies to protect state interests.[1111]

781. FERC merger review, then, at its core attempts to ensure that the transaction is in the public interest. For example, FERC will not approve mergers under its jurisdiction unless the parties file tariffs offering transmission services on a nondiscriminatory common carrier basis.[1112] Similarly, FERC will condition mergers on applicants divesting certain resources, such as generation assets, in order to mitigate market power, or on applicants imposing rates on affiliates to prevent rate inflation or to guard against other forms of cross-subsidization.[1113] As with its analysis of market-based rate applications, FERC thus scrutinizes proposed transactions to determine that the merged entity will not be able to exercise horizontal or vertical market power, and that prices in affiliate transactions are at cost or at fair market value.

§3. Market Share Analyses and Consent Decrees

782. Although FERC, like DOJ and the FTC, examines market concentration in assessing proposed mergers, its application of this concept differs from its sister agencies. "FERC's policy is based on relatively simple rules regarding market concentration levels that are easy to administer, whereas the DOJ's policy is based on a more open-ended examination of a wide range of evidence that is more costly to administer."[1114]

783. DOJ/FTC, as well as FERC, assess a merger's possible competitive impact in part by assessing its likely effect on market concentration—that is, whether the merger will too centralize ownership of key assets (such as generation facilities) in a small set of companies rather than preserving a more diffuse market with many different competitors. To measure market concentration, the agencies use the Herfindal-Hirschman Index (HHI), which calculates the relative market share of different companies in the defined relevant product and geographic markets. The theory is simple: A more concentrated market should be less competitive because entities have a greater ability (a higher market share) to influence prices

1110. Because mergers are generally time-sensitive, and because FERC hearings can be extremely time-consuming, this is not an idle threat.

1111. *See FERC Merger Policy Statement*, *supra* note 1104.

1112. *See, e.g.*, El Paso Elec. Co., 68 F.E.R.C. ¶ 61,181 (1994).

1113. *See, e.g.*, Cross-Subsidization Restrictions on Affiliate Transactions, 122 F.E.R.C. ¶ 61,155 (2008); 18 C.F.R. § 35.43.

1114. Mark J. Niepfer, *Explaining the Divide Between DOJ and FERC Electric Power Merger Policy*, 32 Energy L.J. 505, 514 (2012).

by manipulating supply. Of course, any merger by definition increases market concentration because it combines the assets of two or more companies, so the question cannot be whether any given merger will increase market concentration. Instead, the question is whether it will increase market concentration too much.

784. Under the DOJ/FTC approach, as amended in 2010,[1115] certain levels determine whether a market is "unconcentrated" (HHI < 1000), "moderately concentrated" (HHI = 1000 to 1800), or "highly concentrated" (HHI > 1800). Although DOJ/FTC engage in a careful, economics-based, fact-specific inquiry in assessing mergers, for moderately concentrated or concentrated markets these guidelines specify that an HHI change created by a merger of more than 100 points potentially raises significant competitive concerns, and HHI changes of more than 200 points in a concentrated market presumably increases market power.[1116] If, however, HHI changes precipitated by the merger fall below these levels, then there is a presumptive "safe harbor" and DOJ/FTC review is less strict.[1117]

785. According to FERC rules, if the post-merger HHI is below 1000, then the merger is unlikely to have adverse competitive effects.[1118] If the analysis yields a moderately concentrated post-merger market, with HHI ranges from 1000 to 1800, then the merger raises significant competitive concerns.[1119] Finally, a highly concentrated post-merger market, with HHIs exceeding 1800, is deemed to raise significant anticompetitive concerns.[1120] FERC refers to this analysis as market "screens." That is, the HHI is not deemed the end all of inquiry, but rather is a starting point. It establishes convenient presumptions, but other evidence can be used by parties opposing a merger to demonstrate likely anticompetitive effects. If an applicant satisfies the analytic screens, they typically avoid a hearing. If, however, they fail the screens, a hearing, which can be quite time-consuming, can occur.

786. The difference in the agencies' respective policies can be seen in the review of two merger proposals: Exelon-PSEG in 2005 and in Exelon-Constellation in 2012. In the proposed PSEG-Exelon merger, DOJ determined that electricity prices would increase in two geographic markets, with concomitant consumer impacts. In order to respond to that problem, DOJ entered into a consent decree that ordered divestiture of a certain amount of generating capacity. Likewise, in the Exelon-Constellation merger, DOJ also determined that electricity prices would increase in two relevant markets and again ordered divestiture of a certain amount of generating capacity. Although concentration levels played a role in DOJ's

1115. Following DOJ/FTC's issuance of modified merger guidelines in 2010, FERC considered—but ultimately rejected—modifying its rules as well. Order Reaffirming Commission Policy and Terminating Proceeding, 138 F.E.R.C. ¶ 61,109 (2012).

1116. *Horizontal Merger Guidelines*, *supra* note 41.

1117. *Ibid.*

1118. *FERC Merger Policy Statement*, *supra* note 1104.

1119. *Ibid.*

1120. *Ibid*; *see also* Edwards, *supra* note 1083, at § 103.03[1][b].

analysis, the Department also examined the cost characteristics and the post-merger effects on the merged firm's ability to exercise market power and raise prices.[1121]

787. By contrast, FERC approved both mergers, determining that neither, once mitigation measures were taken into account, would harm competition.[1122] "The FERC compared concentration levels to the [Guidelines'] safe Harbor thresholds and found that, although there would have been some violations of the thresholds and certain time periods, the parties' proposed remedies were sufficient to clear the concentration thresholds and permit the merger to take place without an administrative hearing."[1123] Indeed, in both cases, the parties agreed to divest a certain amount of generation capacity in order to assuage FERC's concerns of post-merger market power. However, importantly, the level of divestiture approved by FERC differed from that required by DOJ.[1124] Further, FERC noted that if, post-merger, the utility withheld output from the market, such withholding could qualify as market manipulation and the merger might be revoked.[1125]

788. Understandably, the time and expense involved with mergers can be substantial. One way of reducing those costs is the entry of "consent decrees"—negotiated settlements between the regulators and the merging entities. Consent decrees provide certainty to both sides. If regulators find that a proposed merger may result in market power, the decree provides them with a legally binding mechanism for limiting that market power. Likewise, for the regulated entities, they can proceed with the merger knowing how antitrust regulation may change the ultimate structure of the proposed deal. And, once parties gain a sense of how antitrust laws are enforced over time, including through prior consent decrees, they can assess the potential anticompetitive effects of a merger beforehand, including what mitigation may be required, in order to determine whether it is appropriate to move forward with a possible transaction.

789. Indeed, consent decrees are common in energy mergers. Because their terms are negotiated between the government and the merging entities, their terms also can and do vary from one transaction to another. In a merger among oil

1121. *See generally* Niepfer, *supra* note 1113, at 516–17. *See also* United States of America v. Exelon Corp. and Pub. Servs. Enter. Grp. Inc., 71 Fed. Reg. 49,477, 49,487 (Aug. 23, 2006); United States v. Exelon Corp., 76 Fed. Reg. 81,528 (Dec. 28, 2011); John R. Morris, *The Likely Effect of the Proposed Exelon-PSEG Merger on Wholesale Electricity Prices*, 21 ELEC. J. 45 (Jan./Feb. 2008).

1122. *See* Order Authorizing Merger Under Section 203 of the Federal Power Act: Exelon Corp. and Public Service Enterprise Corp., 112 F.E.R.C. ¶ 61,011 (2005); *see also* Order Conditionally Authorizing Merger and Disposition of Jurisdictional Facilities: Exelon Corp. and Constellation Energy Grp., Inc., 138 F.E.R.C. ¶ 61,167 (2011).

1123. *See* Niepfer, *supra* note 1113, at 518.

1124. For instance, in that merger, the DOJ required that Exelon and Constellation Energy divest three power plants totaling 2,600 MW. FERC then decided that it would accept the companies proposed mitigation of divestment of 2,648 MW. Mary Powers & Valarie Jackson, *DOJ Requires Exelon, Constellation to Divest Three Maryland Plants*, Platts McGraw Hill Financial, Dec. 21, 2011; Order Conditionally Authorizing Merger and Disposing of Jurisdictional Facilities, F.E.R.C. Stats. & Regs. ¶ 61,167, 30 (2012).

1125. *See generally* Niepfer, *supra* note 1113, at 517–19.

companies in the southeast United States, for example, consolidation would have resulted in a combination of the two remaining pipelines in the area. In order to effectuate the merger, the parties agreed to divest one of those pipelines.[1126] Other consent decree conditions have included granting an easement to a competing utility system to provide natural gas service in the relevant area,[1127] and provisions to prevent discriminatory use of interconnection capacity to deny access or increase costs.[1128]

§4. State Review

790. In addition to FERC and DOJ/FTC review, state PUCs generally have jurisdiction to approve energy mergers, including consolidations, transfers of control, and amendments to their certificates of public convenience and necessity. Thus, neither DOJ/FTC nor FERC review of mergers preempts state regulatory review of such transactions; state review is another level of regulatory control. State regulatory review can include competition policy as well as utility regulation and can include both electric utilities and gas utilities.[1129] Because state PUCs and PSCs review mergers under state law, the applicable substantive standard can vary somewhat from one jurisdiction to another.

791. Typically, however, state regulators apply a fairly broad-ranging public interest standard that encompasses potential impacts on ratepayers, benefits that may flow from the merger, and the availability of reliable utility service post-merger. In this way, regulatory commissions fulfill their traditional function of assuring utilities' ongoing financial viability while also serving a public protection function.[1130] Where utility mergers can run into trouble, then, is in the need not only to obtain FERC and DOJ/FTC approval but also the approval of state regulators in multiple jurisdictions where the merged entity will function. Because state regulators tend to have their own citizens' interests at heart, they may be less amenable to seeing a proposed merger from a broader perspective, and there are both more hurdles the merger must clear and a higher likelihood of divergence in interests as the number of those hurdles increase. It thus has been said that the real roadblock in utility mergers often is in the states, not in FERC or DOJ/FTC.[1131] Arguably, this was the case with the failed Exelon/PSEG merger, where the New

1126. *See, e.g.,* Shell Oil and Texaco Inc., 5 Trade Reg. Rep. ¶ 24,362 (1998); *see also* Exxon Corp. et al., Analysis to Aid Public Comment, 64 Fed. Reg. 68,101 (Dec. 6, 1999) (consented to divestiture of retail businesses in the relevant market).

1127. DTE Energy Company, et al.; Analysis to Aid Public Comment, 66 Fed. Reg. 17,179 (Mar. 29, 2001).

1128. CMS Energy Corp.; Analysis to Aid Public Comment, 64 Fed. Reg. 14,725 (Mar. 26, 1999).

1129. *See* Edwards, *supra* note 1083, at § 103.01[7].

1130. Regulatory Assistance Project, *Electric Regulation in the US: A Guide* 3 (Mar. 2011).

1131. *Cf., e.g.,* Comments of The Electricity Consumers Resource Council et al., Inquiry Concerning Commission's Merger Policy Under the Federal Power Act, F.E.R.C. Docket No. RM96-6-000, at 2 (May 7, 1996) (criticizing FERC's merger standards as "too lax").

Jersey Board of Public Utilities' apparent distaste for the consolidation ultimately proved its undoing.

792. Likewise, most states have adopted antitrust laws of their own. Often, these antitrust laws use language that is similar or identical to the Sherman Act. However, the application and enforcement of the tests under those laws can vary. Such laws will generally apply to intrastate business activities that are beyond the reach of federal antitrust laws. To the extent that both state and federal antitrust laws apply to a particular transaction, those laws are concurrently applied, and federal preemption has generally not eliminated the application of state law.

793. Indeed, the U.S. Supreme Court in one of its most recent decisions involving antitrust and energy, reaffirmed this very principle. In *ONEOK, Inc. v. Learjet, Inc.*,[1132] a group of natural gas purchasers argued that interstate pipelines had violated state antitrust laws by manipulating the reporting of gas prices. In response, the pipelines argued that because the Natural Gas Act gave FERC authority to regulate the market in which the manipulation was alleged, that regulation preempted the state antitrust laws. The Court disagreed. Noting that the NGA carefully delineated the exercise of jurisdiction between the federal government and the states, and that the state claims targeted retail prices rather than federally regulated wholesale prices, the Court held that the state antitrust claims could proceed. The Court further reasoned that state antitrust laws were not directed at interstate pipelines, but instead, at all businesses in the marketplace. Consequently, today as before, the energy law practitioner concerned about antitrust matters must consult both federal and state law.[1133]

§5. PUHCA

794. Finally, Congress has adopted some laws that address competition effects in mergers for specific industries. Most prominent among these is the 1935 Public Utility Holding Company Act,[1134] which previously restricted utilities' freedom to create large umbrella corporate structures. PUHCA's target was a very specific kind of market power. It aimed to dismantle the behemoth utility holding companies that had sprawled across the nation in the first part of the twentieth century, and to "prevent the reappearance of the[se] huge electric and natural gas cartels . . . that abused their customers and went bankrupt in large numbers because of Enron-like speculation and accounting scams."[1135] To this end, although PUHCA included many specific restrictions, it put in place three key limits on utility mergers. First, it required Securities and Exchange Commission approval of public utility holding companies,

1132. ONEOK, Inc. v. Learjet, Inc., 135 S. Ct. 1591 (2015).

1133. *See generally* Bouknight, *supra* note 1049.

1134. Public Utility Holding Company Act of 1935, 15 U.S.C. §§ 79–79z-6, *repealed by* the Energy Policy Act of 2005, Pub. L. No. 109-58.

1135. *Ibid.*

their subsidiaries, and their mergers.[1136] Under PUHCA, such mergers had to have a geographical nexus, though the SEC began interpreting that requirement more loosely in the latter years of the statute. Second, PUHCA required "utility parent companies to incorporate in the same state where the utility operates, so that the state can regulate them, or to be regulated by the [SEC] if they operate in several states."[1137] And third, PUHCA precluded non-utilities, such as investment banks, coal operations, or oil companies, from owning utilities.[1138]

795. Eventually, repeated calls to repeal PUHCA gained purchase. Since the early 1980s, the SEC had argued that PUHCA had become unnecessary, having accomplished its chief goals, and that it had become an obstacle to mergers that otherwise would create efficiencies and benefit the market.[1139] Electric utilities also pushed for PUHCA's demise, asserting that the law limited industry development and that they needed to diversify their asset portfolios in order to limit risk.[1140] Congress finally heard these calls and, in the Energy Policy Act of 2005, largely repealed PUHCA. The full impact of PUHCA's repeal has yet to be seen. However, in the law's absence, "numerous obstacles to consolidation of the electric and gas industry" have been removed, "non-utility-related enterprises are now free to invest in public utilities without first having to divest unrelated holdings," and geographically separated utilities may now merge, such as "an electric company on the East Coast acquiring a combination gas distribution and electric company on the West Coast."[1141] Some observers have thus lauded Congress' decision, while others have bemoaned it. Certainly, however, the decision fits in well with the overall move to competitive markets in utilities, and the general baseline preference in favor of competition in the United States.

1136. James W. Moeller, *Requiem for the Public Utility Holding Company Act of 1935: The "Old" Federalism and State Regulation of Inter-State Holding Companies*, 17 Energy L.J. 343, 343–44 (1996).

1137. *Public Utility Holding Company Act (PUHCA)*, Public Citizen (last visited Apr. 23, 2015), https://www.citizen.org/cmep/energy_enviro_nuclear/electricity/deregulation/puhca/.

1138. *Ibid.*

1139. *See* Div. of Inv. Mgmt., Sec. & Exch. Comm'n, The Regulation of Public Utility Holding Companies viii-ix (1995); Amy Abel, Cong. Research Serv., Order Code RS20015, Electricity Restructuring Background: Public Utility Holding Company Act of 1935 (PUHCA) 4 (1999), *available at* http://www.ncseonline.org/nle/crsreports/energy/eng-47.cfm; *see also* Richard D. Cudahy & William D. Henderson, *From Insull to Enron: Corporate (Re)Regulation After the Rise and Fall of Two Energy Icons*, 26 Energy L.J. 35, 103 (2005); Markian M.W. Melnyk & William S. Lamb, *PUHCA's Gone: What Is Next for Holding Companies?*, 27 Energy L.J. 1, 14–15 (2006); Nidhi Thakar, *The Urge to Merge: A Look at the Repeal of the Public Utility Holding Company Act of 1935*, 12 Lewis & Clark L. Rev. 903, 913–15 (2008).

1140. Thakar, *supra* note 1138, at 913–15; *see* Am. Pub. Power Ass'n, Losing Hold at 70, Pub. Power (Sept./Oct. 2005), http://www.appanet.org/newsletters/ppmagazinedetailarchive.cfm?ItemNumber=13834.

1141. Thakar, *supra* note 1138, at 913–15.

Selected Bibliography

Aagaard, Todd S., *Environmental Law As A Legal Field: An Inquiry in Legal Taxonomy*, 95 CORNELL L. REV. 221 (2010)

ALLEY, WILLIAM M. & ROSEMARY ALLEY, TOO HOT TO TOUCH: THE PROBLEM OF HIGH-LEVEL NUCLEAR WASTE (2013)

ANDERSON, OWEN L. & EUGENE O. KUNTZ, KUNTZ LAW OF OIL AND GAS (1998)

Barr, Christopher J., *Growing Pains: FERC's Responses to Challenges to the Development of Oil Pipeline Infrastructure*, 28 ENERGY L.J. 43 (2007)

BLUE RIBBON COMM'N ON AMERICA'S NUCLEAR FUTURE, REPORT TO THE SECRETARY (2012)

BOSSELMAN, FRED, ET AL., ENERGY, ECONOMICS AND THE ENVIRONMENT: CASES AND MATERIALS (3d ed. 2010)

Carbon Pollution Emission Guidelines for Existing Stationary Sources: Electric Utility Generating Units, 79 Fed. Reg. 34,830 (June 18, 2014)

Craig, Robin Kundis, *Legal Remedies for Deep Marine Oil Spills and Long-Term Ecological Resilience: A Match Made in Hell*, 2011 B.Y.U. L. REV. 1863 (2011)

COPELAND, CLAUDIA, MOUNTAINTOP MINING: BACKGROUND ON CURRENT CONTROVERSIES, CONGRESSIONAL RESEARCH SERVICE (2015)

Cudahy, Richard D. & William D. Henderson, *From Insull to Enron: Corporate (Re)Regulation After the Rise and Fall of Two Energy Icons*, 26 ENERGY L.J. 35 (2005)

DAVIES, LINCOLN L., ALEXANDRA B. KLASS, HARI OSOFSKY, JOSEPH P. TOMAIN & ELIZABETH J. WILSON, ENERGY LAW AND POLICY (2014)

Davies, Lincoln L., *Alternative Energy and the Energy-Environment Disconnect*, 46 IDAHO L. REV. 473 (2010)

Davies, Lincoln L., *Power Forward: The Argument for A National RPS*, 42 CONN. L. REV. 1339 (2010)

Davis, Wendy B., *Out of the Black Hole: Reclaiming the Crown of King Coal*, 51 AM. U. L. REV. 905 (2002)

Dempsey, Paul Stephen, *The Rise and Fall of the Interstate Commerce Commission: The Tortuous Path from Regulation to Deregulation of America's Infrastructure*, 95 MARQ. L. REV. 1151 (2012)

DISTRIB. ENERGY FINANCIAL GROUP LLC, ABACCUS: AN ASSESSMENT OF RESTRUCTURED ELECTRICITY MARKETS (2011)

ENERGY LAW AND TRANSACTIONS (David J. Muchow & William A. Mogel eds., 2009)

FLYNN, AARON M., CONGRESSIONAL RESEARCH SERVICE, SELECTED LEGAL AND POLICY ISSUES RELATED TO COALBED METHANE DEVELOPMENT (2004)

Graffey, Elisabeth & Steven Kihm, *Does Disruptive Competition Mean a Death Spiral for Electric Utilities?*, 35 ENERGY L.J. 1 (2014)

Gresham, Timothy W. & Monroe Jamison, *Do Waivers of Support and Damage Authorize Full Extraction Mining?*, 92 W. VA. L. REV. 911 (1990)
HEMINGWAY, RICHARD W., THE LAW OF OIL AND GAS (1992)
HICKEY, JAMES E., JR., ET AL., ENERGY LAW AND POLICY FOR THE 21ST CENTURY (2000)
HIRSH, RICHARD F., POWER LOSS: THE ORIGINS OF DEREGULATION AND RESTRUCTURING IN THE AMERICAN ELECTRIC UTILITY SYSTEM (1999)
INT'L ENERGY AGENCY, WORLD ENERGY OUTLOOK (2012)
Kelliher, Joseph T. & Maria Farinella, *The Changing Landscape of Federal Energy Law*, 61 ADMIN. L. REV. 611 (2009)
Klass, Alexandra B. & Danielle Meinhardt, *Transporting Oil and Gas: U.S. Infrastructure Challenges*, 100 IOWA L. REV. 947 (2015)
KRAMER, BRUCE M. & PATRICK H. MARTIN, THE LAW OF POOLING AND UNITIZATION (3d ed. 2009)
Kramer, Bruce M., *The Legal Framework of Analyzing Multiple Surface Use Issues*, 44 ROCKY MTN. MIN. L. FND. J. 273 (2007)
KRUPNICK, ALAN J. ET AL., THE NATURAL GAS REVOLUTION: CRITICAL QUESTIONS FOR A SUSTAINABLE ENERGY FUTURE (Mar. 2014)
LOWE, JOHN S., ET AL., OIL AND GAS LAW (6th ed. 2014)
MACAVOY, PAUL W., THE NATURAL GAS MARKET: SIXTY YEARS OF REGULATION AND DEREGULATION (2000)
Manaster, Kenneth A., *An Introductory Analysis of Energy Law and Policy*, 22 SANTA CLARA L. REV. 1151 (1982)
MARTIN, PATRICK H. & BRUCE M. KRAMER, WILLIAMS & MEYERS, OIL AND GAS LAW (5th ed. 2013)
MCGREW, JAMES H., FEDERAL ENERGY REGULATORY COMMISSION (2d ed. 2009)
Melnyk, Markian M.W. & William S. Lamb, *PUHCA's Gone: What Is Next for Holding Companies?*, 27 ENERGY L.J. 1 (2006)
Moeller, James W., *Requiem for the Public Utility Holding Company Act of 1935: The "Old" Federalism and State Regulation of Inter-State Holding Companies*, 17 ENERGY L.J. 343 (1996)
Mormann, Felix, *Beyond Tax Credits: Smarter Tax Policy for A Cleaner, More Democratic Energy Future*, 31 YALE J. ON REG. 303 (2014)
Mormann, Felix, *Enhancing the Investor Appeal of Renewable Energy*, 42 ENVTL. L. 681 (2012)
MUNSON, RICHARD, FROM EDISON TO ENRON: THE BUSINESS OF POWER AND WHAT IT MEANS FOR THE FUTURE OF ELECTRICITY (2005)
Niepfer, Mark J., *Explaining the Divide Between DOJ and FERC Electric Power Merger Policy*, 32 ENERGY L.J. 505 (2012)
Order No. 436, Regulation of Natural Gas Pipelines After Partial Wellhead Decontrol, 50 Fed. Reg. 42,408 (1985)
Order No. 592, Inquiry Concerning the Commission's Merger Policy Under the Federal Power Act: Policy Statement, F.E.R.C. Stats. & Regs. ¶ 31,044 (1996)
Order No. 636, Pipeline Service Obligations and Revisions to Regulations Governing Self-Implementing Transportation; and Regulation of Natural Gas Pipelines After Partial Wellhead Decontrol, 57 Fed. Reg. 13,267 (1992)

Order No. 697, Market-Based Rates for Wholesale Sales Electric Energy, Capacity and Ancillary Services by Public Utilities, 72 Fed. Reg. 39,000 (2007)

Order No. 888, Promoting Wholesale Competition Through Open Access Non-discriminatory Transmission Services by Public Utilities; Recovery of Stranded Costs by Public Utilities and Transmitting Utilities, 61 Fed. Reg. 21,540 (1996)

Order No. 1000, Transmission Planning and Cost Allocation by Transmission Owning and Operating Public Utilities, 18 C.F.R. Part 35 (2011)

Order No. 2000, Regional Transmission Organizations, 65 Fed. Reg. 809 (1999)

Osofsky, Hari M., *Multidimensional Governance and the Bp Deepwater Horizon Oil Spill*, 63 FLA. L. REV. 1077 (2011)

Osofsky, Hari M. & Hannah J. Wiseman, *Dynamic Energy Federalism*, 72 MD. L. REV. 773 (2013)

PARFOMAK, PAUL W., CONGRESSIONAL RESEARCH SERVICE, KEEPING AMERICA'S PIPELINES SAFE AND SECURE: KEY ISSUES FOR CONGRESS (2013)

Pierce, Richard J., Jr., *Natural Gas: A Long Bridge to a Promising Destination*, 32 UTAH ENVTL. L. REV. 245 (2012)

Posner, Richard A., *Natural Monopoly and Its Regulation*, 21 STAN. L. REV. 548 (1969)

Podesta, John D. & Timothy E. Wirth, *Natural Gas: A Bridge Fuel for the 21st Century*, CENTER FOR AM. PROGRESS (2009), *available at* https://cdn.americanprogress.org/wp-content/uploads/issues/2009/08/pdf/naturalgasmemo.pdf

RAMSEUR, JONATHAN L., CONGRESSIONAL RESEARCH SERVICE, OIL SPILLS IN U.S. COASTAL WATERS: BACKGROUND, GOVERNANCE, AND ISSUES FOR CONGRESS (2010)

Raskin, David B., *The Regulatory Challenge of Distributed Generation*, 4 HARV. BUS. L. REV. ONLINE 38 (2013)

Reitze, Arnold W., Jr., *Controlling Greenhouse Gases From Highway Vehicles*, 31 UTAH ENVTL. L. REV. 309 (2011)

RICHARDSON, NATHAN ET AL., RESOURCES FOR THE FUTURE, THE STATE OF STATE SHALE GAS REGULATION (June 2013)

Rule, Troy A., *Property Rights and Modern Energy*, 20 GEO. MASON L. REV. 803 (2013)

SHERLOCK, MOLLY F., CONGRESSIONAL RESEARCH SERVICE, ENERGY TAX POLICY: HISTORICAL PERSPECTIVES ON AND CURRENT TAX STATUS OF ENERGY TAX EXPENDITURES (2011)

Spence, David B., *Federalism, Regulatory Lags, and the Political Economy of Energy Production*, 161 U. PA. L. REV. 431 (2013)

Spence, David B. & Robert Prentice, *The Transformation of American Energy Markets and the Problem of Market Power*, 53 B.C. L. REV. 131 (2012)

TOMAIN, JOSEPH P., ENDING DIRTY ENERGY POLICY: PRELUDE TO CLIMATE CHANGE (2011)

Tomain, Joseph P., *The Dominant Model of United States Energy Policy*, 61 U. COLO. L. REV. 355 (1990)

Tomain, Joseph P., *Traditionally-Structured Electric Utilities in a Distributed Generation World*, 38 NOVA L. REV. 473 (2014)

U.S. Dept. of Justice & Fed. Trade Comm'n, *Horizontal Merger Guidelines* (Aug. 19, 2010), *available at* http://www.justice.gov/atr/public/guidelines/hmg-2010.html

U.S. ENERGY INFO. ADMIN., ANNUAL ENERGY OUTLOOK 2014 WITH PROJECTIONS TO 2040 MT-21 (Apr. 2014), *available at* http://www.eia.gov/forecasts/aeo/pdf/0383(2014).pdf

U.S. ENERGY INFO. ADMIN., MONTHLY ENERGY REVIEW (Jan. 2015), *available at* http://www.eia.gov/totalenergy/data/monthly/archive/00351501.pdf

U.S. Energy Info. Admin., *International Energy Statistics* (Feb. 2015), http://www.eia.gov/cfapps/ipdbproject/iedindex3.cfm?tid=5&pid=57&aid=1&cid=regions&syid=2010&eyid=2014&unit=TBPD

U.S. Energy Info. Admin., *Rankings: Crude Oil Production, June 2014 (Thousand Barrels)*, http://www.eia.gov/state/rankings/?sid=US&CFID=13259961&CFTOKEN=4927be69fa02b9e6-2AAD71C6-25B3-1C83-5406E2FB4529D2CD&jsessionid=8430506e1ef71f55934343373e5f4a434327#/series/46 (last visited Oct. 7, 2014)

U.S. Energy Info. Admin., *What Is the Role of Coal in the United States?,* (June 2, 2014), *available at* http://www.eia.gov/energy_in_brief/article/role_coal_us.cfm

U.S. ENVIRONMENTAL PROTECTION AGENCY, STUDY OF THE POTENTIAL IMPACTS OF HYDRAULIC FRACTURING ON DRINKING WATER RESOURCES: PROGRESS REPORT (Dec. 2012), *available at* http://www2.epa.gov/sites/production/files/documents/hf-report20121214.pdf

U.S. GOV'T ACCOUNTABILITY OFFICE, GAO-13-221, PIPELINE PERMITTING: INTERSTATE AND INTRASTATE NATURAL GAS PERMITTING PROCESSES INCLUDE MULTIPLE STEPS, AND TIME FRAMES VARY (Feb. 2013), *available at* www.gao.gov/assets/660/652225.pdf

Wildermuth, Amy J., *Is Environmental Law A Barrier to Emerging Alternative Energy Sources?,* 46 IDAHO L. REV. 509 (2010)

Wildermuth, Amy J., *The Next Step: The Integration of Energy Law and Environmental Law*, 31 UTAH ENVTL. L. REV. 369 (2011)

Wiseman, Hannah J., *Risk and Response in Fracturing Policy*, 84 U. COLO. L. REV. 729 (2013)

YERGIN, DANIEL, THE PRIZE: THE EPIC QUEST FOR OIL, MONEY AND POWER (2008).

Index

The numbers here refer to paragraph numbers.

www.ingramcontent.com/pod-product-compliance
Lightning Source LLC
Chambersburg PA
CBHW071431230926
51846CB00001B/187
* 9 7 8 9 0 4 1 1 6 6 4 3 2 *